Forrest Robert Jones

Machine Design

Part II.

Forrest Robert Jones

Machine Design
Part II.

ISBN/EAN: 9783337165536

Printed in Europe, USA, Canada, Australia, Japan

Cover: Foto ©berggeist007 / pixelio.de

More available books at **www.hansebooks.com**

PART II.

FORM, STRENGTH, AND PROPORTIONS OF PARTS.

BY

FORREST R. JONES,

Professor of Machine Design in the University of Wisconsin; Member of the American Society of Mechanical Engineers; Member of the Society for the Promotion of Engineering Education; Associate of the American Institute of Electrical Engineers.

FIRST EDITION.

FIRST THOUSAND.

NEW YORK:
JOHN WILEY & SONS.
LONDON: CHAPMAN & HALL, LIMITED.
1899.

PREFACE.

THE matter presented on the following pages is confined to such subjects as the designer must deal with daily. Equations and formulas are put into such a form as to afford a ready means of application to problems under consideration. Numerical examples and data from practice illustrating principles are introduced wherever it seems that a clear understanding can be brought about in this way. The data thus presented have been gathered from numerous sources during the last fifteen years. That coming from modern practice is always given the preference, however, except where the older matter is undoubtedly the most valuable. Whenever possible, the results of practice or experiments as presented by some engineer or experimenter which fairly well represent the general experience along their lines are given in preference to abstract statements. This is believed to be the most satisfactory method, since it affords a means of studying all the facts incidental to the particular case, in addition to furnishing information fully as well as abstract statements.

The publications of engineering societies and engineering journals have been freely consulted. Much kindness has been shown by the representatives of engineering and manufacturing concerns, who have invariably been kind in answering inquiries, and many of them have furnished valuable data. Whenever possible, credit for such information is given in connection with the information itself, but much of it was of such a nature as to make it impossible to separate it from general statements.

The author desires to express his thanks to all those who have so kindly treated his petitions for information. Of those not mentioned in the body of the book Mr. Edw. L. Bateman of Fraser & Chalmers and Mr. Peter Conner of the Gisholt Machine Co. have been especially kind, the former in giving information, and the latter in giving both information and his time in making sketches and drawings of many of the figures.

The printer's proof of most of the matter closely related to the mechanics of engineering has been kindly read by Mr. E. R. Maurer, Professor of Applied Mechanics in the University of Wisconsin, who pointed out where several improvements might be made and ambiguities eliminated. By following his suggestions it is believed that the work has been materially improved.

FORREST R. JONES.

MADISON, WIS., January, 1899.

CONTENTS.

CHAPTER I.

BEARINGS AND LUBRICATION 1
1. Introductory. 2. Planer and lathe ways or V's. 3. Bearings for the ram or tool-carrier of a shaper. 4. Planer saddles, lathe tool-rests, and similar parts. 5. Cross-head guides. 6. Relative length of sliding bearings for rectilinear motion. 7. Journals, boxes, and journal-bearings. 8. Cylindrical journal-bearing. 9. Self-aligning bearings. 10. Lubrication of journal-bearings. 11. Oil- and grease-lubrication for journal-bearings. Oil-pad lubrication. Nature of rubbing surfaces. 12. Special devices for journal lubrication with oil and grease. 13. Materials for journal-bearings. 14. Frictional resistance of journal-bearings. Coefficient of journal friction. 15. Effect of changing the proportions of journal-bearings. 16. Proportions of journal bearings, and examples from practice. Problem in the design of journal bearings. 17. Step-bearings. 18. Forms of step-bearings. Forced lubrication of step-bearings. 19. Conical pivot-bearings. 20. The tractrix, curve of constant tangent, or Schiele's anti-friction curve. 21. Collar-bearings. 22. Cylindrical roller-bearings. 23. Conical roller-bearings. 24. Special forms of bearings.

CHAPTER II.

SPUR AND FRICTION GEARS 75
25. Strength of spur-gear teeth, equations and diagrams. Numerical examples. 26. Methods of strengthening gears. Buttress and shrouded teeth. 27. Short gear teeth, proportions and working loads. 28. Mortise gearing. 29. Rawhide, indurated vegetable fibre, etc., for gears. 30. Factor of safety for tooth gears. 31. Efficiency of spur gears. 32. Strength of bevel-gear teeth. Numerical examples. 33. Efficiency of bevel gearing. 34. Friction gears. 35. Cylindrical friction gears. Efficiency of friction gears. 36. Grooved friction gears. 37. Friction bevel gears. 38. Crown friction gears. 39. Double-cone, variable-speed friction gears.

v

CHAPTER III.

BELTS AND ROPES FOR POWER TRANSMISSION 112

40. Flat belts. 41. Equations for power transmission by flat belts. Numerical example. 42. Coefficient of friction and slip of leather belting. 43. Working strength of leather belting. 44. Velocity of leather belting. 45. Wear of leather belts. 46. Weight and thickness of leather belts. 47. Rawhide, semi-rawhide, and rawhide with tanned leather face, belts. 48. Cotton belts. 49. Rubber belting. 50. Leather-link belts. 51. Effect of relative positions of pulleys. 52. Special system of flat-belt driving. 53. Efficiency of flat belting. 54. Ropes for power transmission. 55. Systems of rope driving. 56. Equations for ropes transmitting power. Numerical example. 57. Grooves for non-metallic ropes. 58. Coefficient of friction of non-metallic ropes. 59. Working strength of non-metallic ropes. 60. Velocity of ropes for power transmission. 61. Wear and lubrication of non-metallic ropes. 62. Weight of hemp and cotton ropes. 63. Diameter of ropes for power transmission. 64. Effect of the relative position of pulleys upon ropes used for power transmission. 65. Efficiency of rope belting. 66. Wire rope. Results of practice. Pulley grooves for wire rope.

CHAPTER IV.

SCREWS FOR POWER TRANSMISSION 150

67. General discussion. 68. Relation between the turning moment and axial force in a square-thread screw. 69. Efficiency of a square-thread screw and collar. 70. Coefficient of friction for square-thread screws. 71. Problem in screw design. 72. Maximum stress in a screw. Example of application of formulas. 73. Angular-thread screws.

CHAPTER V.

SCREW GEARING 166

74. Common forms. 75. Worm and worm-wheel. 76. Equations for turning force and efficiency of worm and worm-wheel. 77. Tests of worm-wheel. 78. Screw gears. 79. Strength of screw-gear teeth. 80. Equations for turning force and efficiency of screw gears. 81. Coefficient of friction of screw gears.

CHAPTER VI.

SCREW FASTENINGS 181

82. Forms of screw-threads and proportions of bolt-heads and nuts.

CONTENTS. vii

PAGE

Set-screws. 83. Locking devices for nuts and screws. 84. Strength of screw-bolts. 85. Endurance of screw-bolts.

CHAPTER VII.

MACHINE KEYS, PINS, FORCED AND SHRINKAGE FITS. 201
86. Machine keys and pins. 87. Roller keys. 88. Eccentric keys or fastenings. 89. Shrinkage and forced fits. 90. Tension in and pressure against a thin ring fitted by shrinking or forcing. Numerical example. 91. Shrinkage and forced fits for thick rings and heavy parts. 92. Allowance for shrinkage and forced fits.

CHAPTER VIII.

AXLES, SHAFTING, AND POSITIVE SHAFT-COUPLINGS 216
93. Notation for equations. 94. Torsional strength of round shafts. Numerical examples of solid and hollow shafts. 95. Twist of a shaft under torsional stress. Numerical example. 96. Bending strength of round shafting. 97. Lateral deflection of shafting on account of its own weight. 98. Shaft subjected to both torsion and bending. General case. 99. Solid round shaft subjected to more than one force. Numerical example. 100. Solid round crank-shafts and other shafts acted on by a single rotative force. Numerical example. 101. Hollow round shafts. 102. Hollow round shaft acted on by more than one force. 103. Hollow round shaft acted on by a single turning force. 104. Experimental and determined value of the breaking tensile stress of round shafting subjected to bending and torsion. 105. Practically determined formulas for round shafting. 106. Shafts of symmetrical sections other than round. 107. Rigid shaft-couplings. 108. Flexible shaft-couplings. 109. Positive clutch couplings.

CHAPTER IX.

FRICTION COUPLINGS AND BRAKES 235
110. General statements. 111. Cone friction couplings. 112. Multiple-ring friction coupling. 113. Material and coefficient of friction for friction couplings. 114. Strap brake. 115. Prony brake.

CHAPTER X.

FLY-WHEELS AND PULLEYS 243
116. Applications of fly-wheels, and equations for moment of inertia and kinetic energy. 117. Determination of moment of inertia and kinetic energy of a webbed wheel. Numerical example. 118. Problem. To design a fly-wheel for a given moment of inertia, and ac-

cording to a given form. 119. Problem. To design a fly-whee' which will furnish a given amount of energy for a given variation of speed. 120. Moment of inertia of a fly-wheel with arms. 121. Stresses in fly-wheels with arms. 122. Numerical example of stress in, and enlargement of, a rotating ring due to centrifugal action. 123. Sectional-rim fly-wheels and pulleys. 124. Bursting tests of small cast-iron fly-wheels by centrifugal action. 125. Hollow cast-iron arms with wrought-iron or steel tension rods, for fly-wheels and pulleys. 126. Tangent arms for fly-wheels and pulleys. 127. Built-up plate fly-wheels. 128. Wire-wound fly-wheel. 129. Other special forms of pulleys. 130. Designs and proportions of fly-wheels and pulleys taken from practice.

CHAPTER XI.

CYLINDERS, TUBING, PIPES AND PIPE-COUPLINGS 273
131. General. 132. Tension in a thin circular cylinder due to internal pressure. Numerical example. 133. Cylinder with thick walls. 134. Bursting tests of cylinders and pipes. 135. Special forms of pipes. 136. Pipe-couplings and flanges. 137. Expansion couplings for pipes.

CHAPTER XII.

RIVETED JOINTS 286
138. Methods of making and forms of riveted joints. 139. Single-riveted joint. 140. Rivets. 141. Pitch of rivets. 142. Efficiency of riveted joints. 143. Effects of shearing, punching, and drilling plates. 144. Faulty construction and grooving of riveted joints. 145. Examples of riveted joints taken from practice.

CHAPTER XIII.

FRAMES OF PUNCHING, SHEARING, AND RIVETING MACHINES 315
146. Punching and shearing machines. 147. Stresses in a section perpendicular to the motion of the punch. 148 Numerical solution for a section perpendicular to the motion of the punch. 149. Section parallel to the motion of the punch. 150. Angular section of a punch-frame. 151. General form of a punch-frame. 152. Direct-acting hydraulic riveter.

CHAPTER XIV.

SELECTION OF MATERIALS. 331
 153. General discussion and tables of the properties of materials most common to engineering.

APPENDIX.

 A. Development of equations for an angular-thread screw. B. Graphical determination of the moment of inertia of a plane area. Approximate method. 337

FORM, STRENGTH, AND PROPORTIONS OF MACHINE PARTS.

CHAPTER I.

BEARINGS AND LUBRICATION.

1. Introductory.—In machinery the name **bearing**, or **bearing surface**, is applied to a part which presses or bears against another and moves over it at the same time.

If the surfaces are dry, or as left by the tool used to finish them, they have a strong tendency to abrade, or "cut," each other when the materials commonly used for machine members are rubbed together. To prevent this cutting, a lubricating substance, such as oil, grease, or graphite, is introduced between them.

The necessity of thorough lubrication is often of such vital importance that the bearing must be designed with especial regard to securing a continuous application of the lubricant.

BEARINGS FOR RECTILINEAR MOTION.

2. Planer and lathe "ways" or "V's."—The table or platen of a metal-working planer reciprocates on bearings which must give it an accurate rectilinear motion in order that the cutting tool may form a plane surface, or a curved surface whose elements are right lines. The V form of bearing, such as is shown in Fig. 1 at A and B, is most commonly used. In this style of bearing the weight of the table, together with the work upon it, is relied upon to hold the table in place. The cutting tool frequently exerts a side pressure tending to force the table off its bearings. The ability of

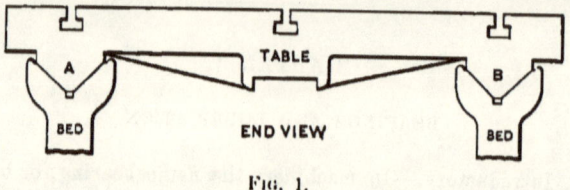

Fig. 1.

the table to resist this pressure depends on the angle θ, Fig. 2, between the sides of the V. It is plain that the smaller this angle the

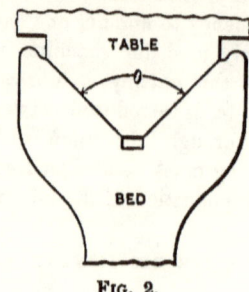

Fig. 2.

greater will be the resistance of the table to being displaced. The smaller the angle, however, the greater the total pressure between

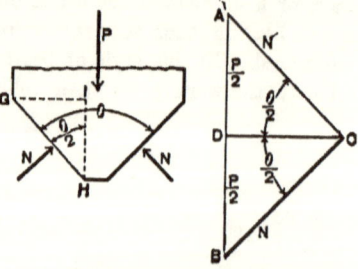

Fig. 3.

the bearing surfaces. This can be seen by the aid of Fig. 3, which represents the way on one side of the table when it is removed from the bed. The load which this V must support may be called P. The pressure against each side of the V is normal to the bearing surfaces, friction being neglected. These normal pressures are represented by N and N in the figure. The amount of N and N is found by resolving P into two forces normal to the bearing surfaces. This is done by taking AB parallel to P, and of such a length as will represent the magnitude of P according to any convenient scale of pounds per inch, or other units, and then drawing AC and BC parallel to N and N. Then $AC = BC = N$ according to the scale selected for AB. In the figure it can be seen that

$$N = \frac{P}{2} \div \sin \frac{\theta}{2} = \frac{P}{2} \csc \frac{\theta}{2}.$$

The equation shows that reducing the angle between the sides of the V increases the total pressure upon the bearing surfaces. The angle should therefore be kept as large as possible without getting it so great that the table may be thrown from the bed by the side pressure of the tool.

The V form of bearing is self-adjusting for wear, since it naturally settles down as the surfaces wear away.

For ordinary service the V's for a small planer should be made with a smaller angle θ than for a large machine. This is due to the fact that, as the size of the planer increases, the weight of the table increases more rapidly than the side pressure of the tool.

Varying the angle θ does not change the pressure per unit area on the bearing surfaces as long as the load P and the horizontal projection of the surfaces remain unchanged, for the bearing surface is increased in the same proportion as the total pressure. The power necessary to move the table increases as θ is decreased, however, and the wear on the ways is increased, so that the table will settle down more rapidly. There is also more liability to abrasion with the smaller angle, since localization of the pressure, due to unevenness of the bearing surfaces, causes a greater pressure on the high parts.

That the pressure per unit area on the bearing surface is unchanged as long as the horizontal projection of the surface and the load remain the same, whatever the value of θ, can be shown in the following manner: In Fig. 3 the horizontal width of the bearing surface GH is GI; therefore

$$GH = GI \csc \frac{\theta}{2}.$$

The total pressure upon GH is N, whose value is given in the preceding equation. The pressure per unit area, as found by dividing the total pressure N normal to the bearing surface by the total area GH, is

$$\frac{N}{GH} = \frac{P}{2} \csc \frac{\theta}{2} \div GI \csc \frac{\theta}{2} = \frac{P \div 2}{GI}.$$

This shows that the pressure per unit area is always equal to the quotient obtained by dividing the load supported upon the bearing surface by the horizontal projection of the surface, and is therefore not affected by the angle θ. The other half of the load P is supported by the opposite side of the V.

The angle θ is generally about 90° for planers having an opening from 16 inches to 24 inches square between the housings, table, and rail. A bearing surface 1 inch wide for the 16-inch, and $1\frac{1}{5}$ inches wide for the 24-inch, planer keeps in good condition under ordinary service. Planers for special work, where the piece to be machined is comparatively light, and the side pressure of the tool heavy, should have the angle θ smaller than 90°, and the width of the bearing surfaces greater than for the larger angle more commonly used. The greater width of the bearing surfaces is necessary to keep down the pressure per unit area on the bearing. Heavy planers have V angles of 110° or more.

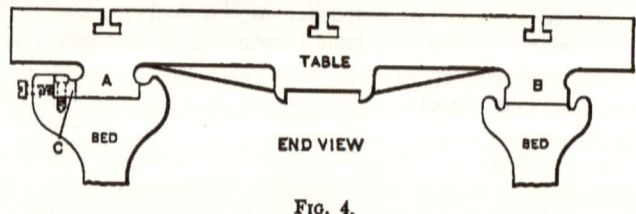

FIG. 4.

Fig. 4 shows a pair of less commonly used bearings for a planer or other reciprocating table which is partly held in place by its own weight. The cutting tool cannot displace a table having this form of ways, except when a heavy side pressure is applied against a piece of work at a considerable distance above the table. The pressure on the horizontal part of the bearings is equal to the load, instead of being greater than the load, as in the case of the V bearings. This is an advantage, since it reduces the liability to abrasion.

The strip C is adjustable horizontally to allow for wear on the vertical surfaces. If attached to the bed as shown, it can be set so as to give the table a snug running fit from end to end of the bed, even though the wear near the centre of the length of the bed is greater than at the ends, as is generally the case.

Fig. 5 shows a form of bearing used upon one side of the table

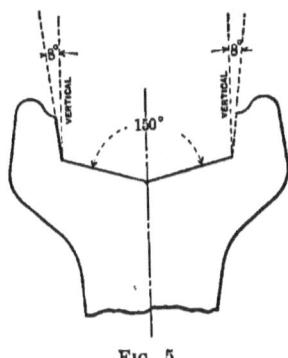

Fig. 5.

of a planer taking work 120 inches square.* The flat angle of 150° is sufficient to hold the table in position when making the return stroke, or when the tool is taking a light cut. The sides of the way, which make an angle of 8° with a vertical line, or 16° with each other, prevent displacement of the table by a heavy side pressure of the cutting tool. On account of their angularity, no adjusting device is necessary to take up side wear, for the settling of the table in the ways makes this adjustment.

V-form bearings can be continuously lubricated by placing oil-pockets, of the form shown in Fig. 6, along them at intervals. This pocket is simply a recess cast into the bed. A double-cone

---

* Used by the Wm. Sellers Co. Sketch kindly furnished by them.

roller RR is placed in the pocket so as to be partly submerged in oil. The angles of the cones are such that they will fit the table V's. The roller is supported at the ends by bearings EE which are held up by some means, as springs or counterweights, so that the roller will press lightly against the V as the table passes over the roller. The slight frictional resistance between the surfaces turns the roller, so that the oil adhering to it is carried up to the V.

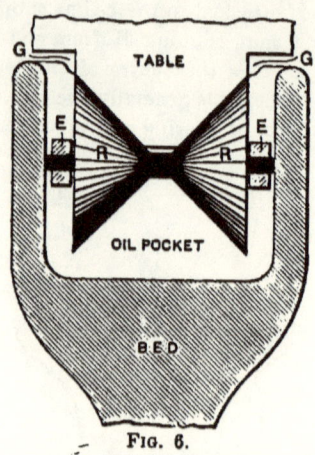

Fig. 6.

Suitable means should be provided to prevent the roller from being raised too high when the table passes from over it. After the oil is thus placed on the V, some of it is carried between the bearing surfaces by the motion of the table. The pressure between the surfaces generally squeezes the oil out, most of it going to the groove at the bottom of the V, and then flowing back to the oil-pocket; a smaller portion is forced out at the top into the grooves GG, and either flows along these grooves back to the pockets or runs down the inclined surfaces of the bearings when the table passes from over them. The groove at the bottom of the V not only serves for an oil-channel, but affords a space into which particles of foreign substances may be scraped from the exposed bearing surfaces of the bed by the end of the table V as it passes over them.

Flat bearings can be lubricated in the same manner. In this case a pair of flanged rollers, as RR, Fig. 7, mounted on the same spindle and forced lightly toward each other with a spring, so as to bear against the vertical part of the bearing, give good service. The faces of the flanges on the inner sides at F and F should be

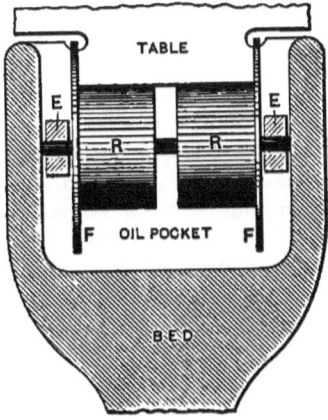

FIG. 7.

corrugated or indented, or the flanges perforated with small holes, to facilitate the carrying of the oil up to the ways.

The ways or V's forming the bearings between the carriage and bed of an ordinary engine-lathe commonly have the form of an inverted V, as shown in Fig. 8. The pressure of the cutting tool

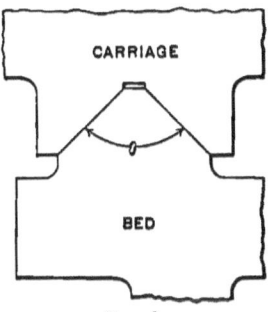

FIG. 8.

against the work often has a tendency to throw the carriage from the bed. Since the carriage is comparatively light, clamps are necessary to hold it in place. The V's can therefore have such an angle as will allow the carriage to rest on them without pressure against the clamps when the tool is taking a light cut, but the clamps may come into service when a heavy cut is taken.

The angle θ is commonly made 90° or less on engine-lathes for ordinary service. The width of the bearing surface on each side of the V is about ⅜ of an inch to 1¼ inches for a lathe which will swing a piece 24 inches in diameter.

Any device for lubricating the ways of lathes, other than putting oil on by hand, is seldom or never provided. The very frequent cut and grooved appearance of the V's of lathes seems to indicate that some simple and inexpensive method of applying oil, such as an oil-saturated porous pad fitting into a recess in the carriage, might be advantageously used to prevent excessive cutting. A great part of the cutting is undoubtedly caused by particles of the turnings getting between the bearing surfaces. The only preventative for this is to provide some protection for keeping the V's free from such foreign matter.

3. **The bearings for the ram or tool-carrier of a shaper for metal working** must guide the ram positively, so that it cannot be thrown out of position by a heavy pressure against the cutting tool in any direction. Fig. 9 is an end view of a common form of bear-

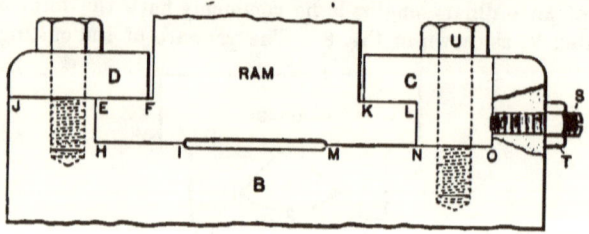

FIG. 9.

ing for this purpose. The ram reciprocates in the part B, which may be either the frame of the machine or a carrier for the ram, according to the design of the shaper.

In order to take up wear on the faces EF and HI, the clamp D is removed from the machine and a light cut taken off the surface EJ, and the clamp is then replaced. Another method is to plane

B a little lower at JE than EF and place "shims" or "liners" of thin sheet metal or paper between D and B on the surface JE; as wear occurs some of the shims are removed and the clamp drawn down tight again. It is essential that the clamp be firmly held against the part B. Wear along the surface KL and MN is taken up in the same manner by cutting away the surface NO, or by removing liners which had been placed between the clamp and part B. Wear along EH and LN is taken up by the set-screw S, which is used to force the clamp C against the ram. T is a lock-nut for holding the set-screw in place, the cap-screws U being loosened for this purpose.

Another form of shaper-ram bearings that has been much used, but which seems to be giving way to that just described, is shown in Fig. 10. The bearing surfaces are EF and HI on one side, and

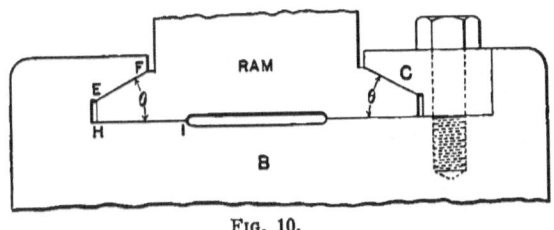

Fig. 10.

the similar surfaces on the other. The objectionable feature of this bearing is that, on account of the wedge-like form of the bearing surfaces, which make the angle θ with each other, the pressure between them is apt to be so great locally that cutting will occur. Otherwise this bearing presents the excellent feature that it can be adjusted for wear in all directions by simply lowering the clamp C slightly, or by forcing it towards the ram with a row of set-screws placed as the one at S in Fig. 9.

The angle θ, Fig. 10, is commonly from 30° to 45° in practice.

The width of the bearing surface EF is about 1 inch on a shaper having $\theta = 30°$ and a maximum stroke of 18 inches.

Another method of taking up the side wear in a ram having bearing surfaces at right angles is shown in Fig. 11, where the bearing strip E is held against the ram by a line of set-screws S, each provided with a lock-nut. Some method of preventing end motion of the strip must be supplied. This may be done by having the end of one or more of the screws made cylindrical, so as to fit into a small hole in the strip. Screws with conical points, fitting into conical countersinks in the strip, will hold it in place, but the

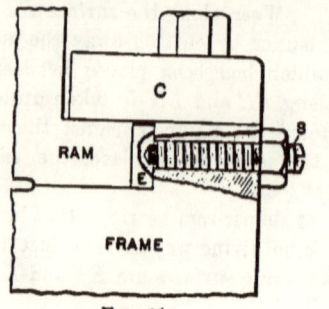

Fig. 11.

strip has a tendency to slip off the screws, thus causing unnecessary pressure between the rubbing surfaces.

On account of the thinness of the strip, the pressure is somewhat localized at the set-screws. The consequent wearing away of the strip more rapidly at these places than between them, makes it necessary to spring the strip by setting up the screws, in order to hold the ram rigidly when taking a heavy cut. This causes a constant pressure of the strip against the ram, even when on the return stroke. The consequent wear and loss of power are disadvantages.

Another method of adjusting the bearing strip, which almost entirely prevents its springing so as to press constantly against the ram, is to tap the screws into it and allow them to pass through smooth holes in the frame; liners can then be placed back of the strip, and the screws tightened to hold it back firmly against them.

4. **For planer-saddles, lathe tool-rests, and similar parts,** where one of the sliding parts is much longer than the other, the "gib" shown at G in Fig. 12 can be conveniently used. In Fig. 12, B may be taken to represent a portion of a planer cross-rail with the upper portion of the tool-carrying saddle resting upon it. The gib is tapered on two sides, from end to end; the other sides are parallel. One of the sides forming the taper rests against the

corresponding sloping surface of the saddle, and the other against the rail. A stud S, tightly screwed into the saddle, and having two lock-nuts T and T, is used for adjusting the gib so as to secure the desired fit between the moving parts. Wear on the vertical faces is taken up by the clamp C. The dead weight of the saddle

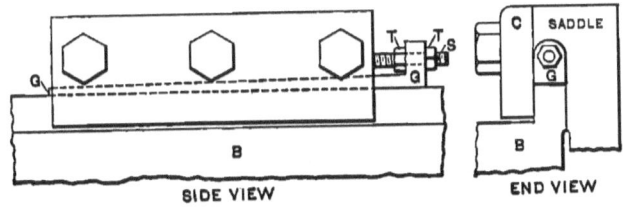

FIG. 12.

and the parts appertaining to it is carried on the horizontal surface under the gib. The taper of the gib is generally between $\frac{1}{4}$ and $\frac{3}{8}$ of an inch per foot.

The tool-slide which moves on the carriage of a lathe may be made with a gib G, as in Fig. 13. Here the gib bears against the

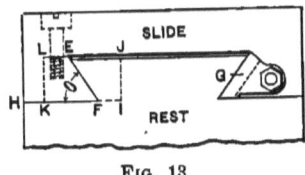

FIG. 13.

inclined surfaces so that, by adjusting it alone, wear in all directions is taken up. The taper of the gib may be the same as for rectangular bearings, viz., $\frac{1}{4}$ to $\frac{3}{8}$ of an inch per foot.

The angle θ generally lies between 45° and 60°. For a lathe which will swing 24 inches in diameter, EF may be 1 inch, and FH 1$\frac{1}{2}$ inches; for one swinging 12 inches in diameter, the corresponding dimensions are $\frac{5}{8}$ inch and 1 inch.

Another device that is often used for rectangular bearing surfaces is that of Fig. 14, where the right-angled piece G, resembling a box-square, is adjustable with both a vertical and a horizontal row of set-screws, thus compensating for wear in all directions.

The thin strip E, Fig. 11, can be used on angular bearings such as shown in Fig. 13. If the set-screws are kept at right angles to the outer vertical side of the slide, their points should be either

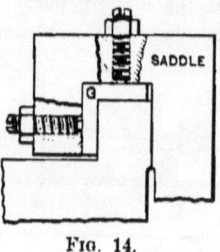

Fig. 14.

conical, so as to have line contact with the side of the strip, or else should be fit into a hole drilled into the strip. The object in both these cases is to get as much contact surface between the points of the screws and the strip as possible, in order to prevent crushing the metal at the place of contact.

Another method of adjustment is shown by the dotted lines at the left side of Fig. 13. This is by using an angular strip $EFKL$ which bears against the surface LK of the slide, and is drawn upward by screws passing through smooth holes in the slide and into the threaded holes of the strip. Or, leaving the slide solid, as indicated by the full lines, the same device can be used on the rest, the strip being $EFIJ$ in this case. The screws must be put in from the opposite direction for this construction; they are not shown in the figure. The gib should be placed, when possible, where it does not receive the pressure of the cutting tool.

5. Cross-head guides for engines, pumps, and similar machines having a crank and connecting-rod, are frequently made of the form shown in Fig. 15, which is an end view of the guides BB and piston-rod A. The bearing surfaces are at CDE and FGH. They must take the pressure due to the angularity of the connecting-rod with the piston-rod, this pressure being in either direction along a line DG, not drawn. Since there is very little or no side pressure at right angles to DG, the angles θ can be made large. Adjustment for wear can be made either by moving apart the bearing surface of the cross-head or bringing those of the frame B together.

To do this, the cross-head must either be made of several pieces, or the guides made of separate pieces and attached to the frame.

Fig. 16 shows another form of cross-head guides. The bearing surfaces here are parts of cylindrical surfaces, which are represented as arcs of circles in the figure. The centres of these arcs are at O and O' on a line through B and B', the radial distances OB and $O'B'$ being less than half the distance BB' between the bearing sur-

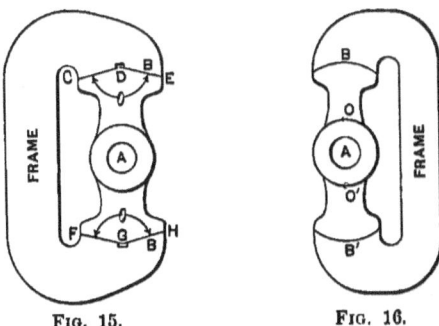

Fig. 15.   Fig. 16.

faces. It can be readily seen that this construction prevents the cross-head from rotating about A.

In contrast with this construction for preventing the turning of the cross-head in the guides, some prominent engine-builders bore the guides concentric with the cylinder, in order to allow the cross-head to adjust itself to the crank-pin and connecting-rod.

Flat bearing surfaces normal to the line BB', Fig. 16, are frequently used for cross-head bearings. In order to prevent lateral motion the cross-head generally extends over the edges of the guides and has a bearing against their sides.

6. The relative length of sliding bearings for rectilinear motion, and their positions at the ends of the stroke, should be taken into consideration when designing bearings which must perform considerable service during the life of the machine of which they form a part. The guides and cross-head bearings of an engine or pump may be taken as an example. It will first be assumed that, at the end of the stroke, the cross-head, which is much shorter than the guides, does not reach the ends of the parallel faces of the guides. The result is that shoulders will be formed just where the end of the cross-head stops, because the material will be worn away up to that point, but not beyond it. This may be a very serious fault. If the cross-head shoes are lengthened, or the guides shortened, so that the end of the shoe will just reach the end of the guide at the end of the stroke, no shoulder will be formed, but the wear will not be so great at the ends of the guides as back some distance, for only a very small portion of the length of the shoe rubs over this part of the guide. A cross-head running between parallel guides in this manner will therefore wear loose at the middle of the stroke while still having a close fit at the ends. By allowing the cross-head to pass over the ends of the guides the wear throughout the length of the latter becomes more uniform, but still greater at the middle than at the ends.

Uniform wear from end to end of the guides can be secured by making the cross-head of the same length as the stroke, and running it on guides of its own length. By this means half the length of the cross-head will project beyond the guides at each end of the stroke, so that the length of bearing surface passing over each end of the guides, for a movement from one end of the stroke to the other, is one half the length of the cross-head; the whole length of the cross-head passes over the middle of the guides for the same motion. The tendency is ordinarily for the wear to be more rapid at the centre than the ends. This cannot occur, however, with the long cross-head, since it either covers the entire length of the guides or, when not at the centre of the stroke, presses more heavily against the guides towards their ends than at the middle. The great length of the cross-head is apt to be more objectionable, however, in most cases, than irregular wear of the guides.

BEARINGS FOR ROTARY MOTION.

Journals, Boxes, and Journal-bearings.

7. Definitions.—When two parts of a machine rotate with regard to each other, as a wheel and its axle, the part of the one which is enclosed by, and whose surface rubs against, the other, is called the **journal**; the part which encloses the journal is called the **box**, or, less specifically, the **"bearing."** The name "bearing" or journal-bearing is very commonly applied to the whole assembly of parts embodying both the journal and its box.

8. The cylindrical journal-bearing is the most common form. When the pressure upon it is always in nearly the same direction, the box may be made to extend only partly around the journal. The boxes used on the axles of railway cars furnish an example of this kind. These boxes encompass about one-third of the circumference of the journal. This is sufficient to hold the axle in place, and furnishes almost as much efficient bearing surface as if the box covered half of the circumference of the journal.

A difficulty that sometimes assumes serious proportions is liable to be met when a box, extending half-way around the journal, is used, especially if the box is well fitted to the journal before going into operation. The nature of this trouble can be described with the aid of Fig. 17, in which B is the box and A is the supporting

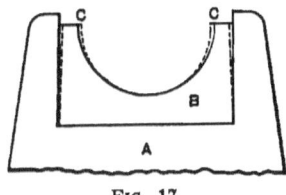

FIG. 17.

frame. Before operation the box is of the form indicated by the full lines, the semicircle CC just fitting the journal. When the journal is rotated in the box, the latter is apt to have a tendency to take the form shown by the broken lines, the points CC trying to approach each other. This action causes the sides of the box to press unduly hard against the journal. The result is increased frictional resistance and wear between the rubbing parts.

One remedy for this is to make a very loose fit along the surfaces near C and C. Another and more certain one is to provide some means for holding the box back against the frame so that the parts CC cannot move away from it; this may be done with bolts passing through the frame and into the box, or by dovetailing the sides of the frame and box together.

When the pressure acting on a bearing is reversed, as indicated by the arrow-heads in Fig. 18, the box must generally completely surround the journal. The adjustment for wear under such conditions should be in the direction of the forces. For the vertical pressures in Fig. 18 the separation should be on a horizontal plane through the centre of the journal. The short lines at D and D show the separation on this plane. The small shoulders at the ends of these lines are to hold the cap C in the proper position. Adjust-

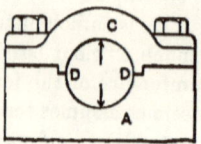

Fig. 18.

ment is made by bringing the cap down nearer the frame as wear occurs. Liners placed between the cap and frame may be removed for this purpose, or some of the cap or frame cut away. If the pressure is horizontal and reversed, the separation should be made on a vertical plane.

In many cases the wear is both vertical and horizontal. The horizontal steam-engine with its heavy fly-wheel furnishes an example of this. The downward pressure caused by the weight of the fly-wheel, as indicated by the vertical arrow in Fig. 19, wears away

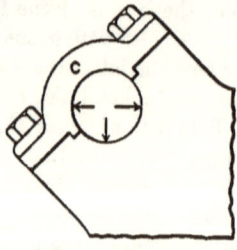

Fig. 19.

the bottom of the box. The pressure, indicated by the horizontal arrows, is due to the steam-pressure on the piston; this causes wear on the sides of the bearing. When close fitting of the running parts is not of great importance, a box divided at an angle with both lines of pressure, as shown in the figure, can be adjusted with a fair degree of satisfaction. For more accurate adjustment the box must be divided into three or more parts.

Fig. 20 shows a box divided into four parts, and adjustable

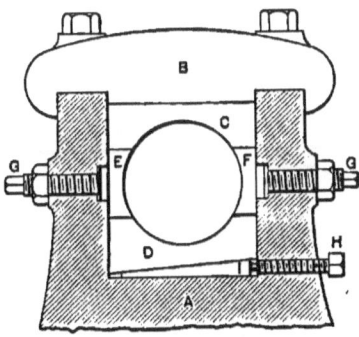

FIG. 20.

both vertically and horizontally. The four parts of the box, C, D, E, and F, are enclosed by the frame A and the cap B. The wedge I, at the bottom of the box, is used for the vertical adjustment, the top C being held down by the cap B. The side screws G and G are for adjusting the side pieces E and F horizontally; these screws are made with enlarged ends, for bearing against the sections of the box, in order to prevent crushing of the metal where they come in contact. With a box of this form the axis of the journal can always be kept in the same position, whatever the direction of the wear. This is an important consideration in some classes of machinery.

A unique and very compact adjustable journal-box is shown in Fig. 21.* It consists of an outer shell or casing A enclosing an adjustable lining B, made in two parts. Both the shell and the lining have the same eccentricity, so that when the lining is placed in the shell with the thickest part of one next the thinnest part of the other, the outer surface of the shell and inner surface of the lining are concentric. The complete box can therefore be placed in a hole bored in the frame of a machine, and turned about its own axis without displacing the axis of a journal fitting into it. The

* Used by the Straight Line Engine Co., Syracuse, N. Y.

18 FORM, STRENGTH, AND PROPORTIONS OF PARTS.

lining does not completely encircle the journal, but the circumference is completed by a wedge D at the thickest part, together with

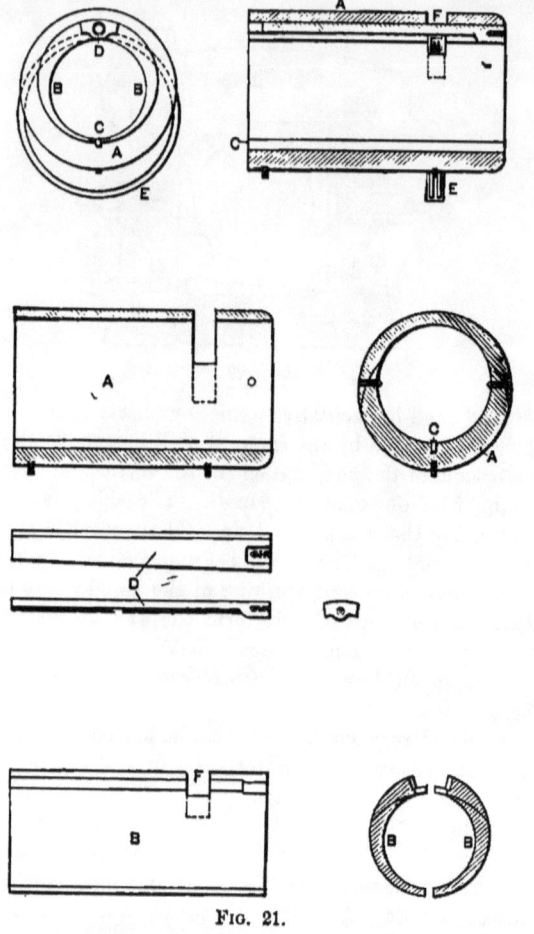

FIG. 21.

thin strips or liners C placed at the thinnest part, when the bearing is new. As wear occurs these strips are removed from the thin side and placed alongside the wedge, thus reducing the diameter of the bore of the box, and at the same time moving the centre of the bore slightly nearer the wedge side of the bearing. By placing the thin side of the bearing where the greatest wear occurs, each

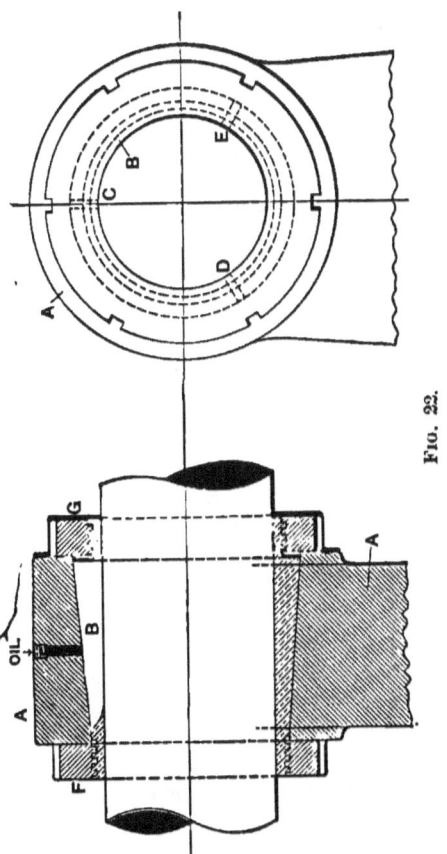

Fig. 22.

adjustment of the lining tends to bring the journal back to its original position concentric with the outer surface of the shell.

When but a slight adjustment for wear on the journal and box is required, and compactness is desired, the one-piece box of Fig. 22 is serviceable. The box B is made of a single piece, cylindrical in the bore, and coned on the central part of its outer surface to fit the conical hole in the frame A of the machine; it is cut through at C, D, and E, from the large end to near the small one; the ends are threaded for nuts F and G. When new, these nuts are adjusted so that the inner surface takes the circular form and size to which it was bored when the box was solid. In order to take up wear on the rubbing surfaces, G is loosened and F tightened so as to draw the box into the frame, thus springing the sides together so as to diminish the diameter of the bore; G is then tightened to hold the box firmly in place.

Fig. 23 represents what is probably the simplest form of adjust-

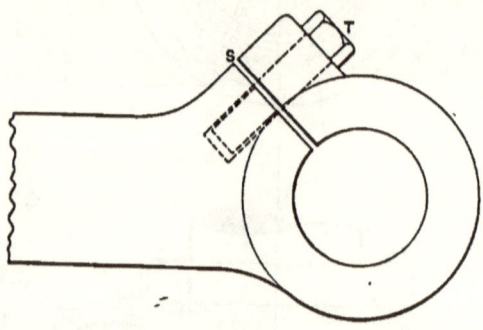

FIG. 23.

able box that can be devised. The slot S, cut along one side, allows it to be adjusted by the cap-screw T. This design is also frequently used where it is desirable to have a pin pass easily through the bore when putting the parts together, but held tightly in place during the operation of the machine.

In order to prevent excessive end motion of a journal in its box, collars are frequently placed at the end of the journal, forming a part of it, as in Fig. 24. It frequently occurs that if the collars are fitted so as to allow free running, but no end shake, of the parts, there will be excessive pressure and binding between the collars and the box when the machine is put into operation. This is probably due to the fact that the heat generated by the rubbing of the surfaces over each other raises the temperature of the thin shell more rapidly than that of the journal, consequently expanding it more rapidly, and causing it to press against the collars as stated. The higher coefficient of expansion of the material of the box may also have a slight tendency to increase the binding when a material

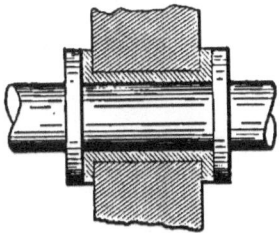

Fig. 24.

having this property is used. The remedy is plainly to make a loose fit when constructing the machine.

A bearing having a journal with a single collar is shown in Fig. 25. This is a form that is used on the spindle of a lathe where it rests on the head-stock at the end next the face-plate or live-centre. A smaller adjustable collar is placed at the other end to keep the spindle from moving endwise, toward the right in this case.

9. Self-aligning bearings are desirable in many classes of machinery. Fig. 26 represents a common and compact form. The outer surface of the box or sleeve B is partly made up of a portion of the surface of a sphere. The supporting part A has a concave spherical surface to conform to that of the sleeve. This device allows the axis of the sleeve to adjust itself in line with the journal, and maintain such adjustment, even if the shaft of which the journal forms a part is temporarily sprung, or permanently bent so as to wabble. Movement of the support A has no effect on this alignment of the box. Some means of preventing the box from rotating in the support must be provided. This can be done by a pin C fitting tightly in the sleeve and loosely in the support.

It is not necessary to have the spherical surface extend completely around the sleeve; two diametrically opposite segments of the surface can be used, as shown in Fig. 27, where B is the sleeve having the spherical segments under the spherically concave points of the two screws C and C, which are supported by the frame A.

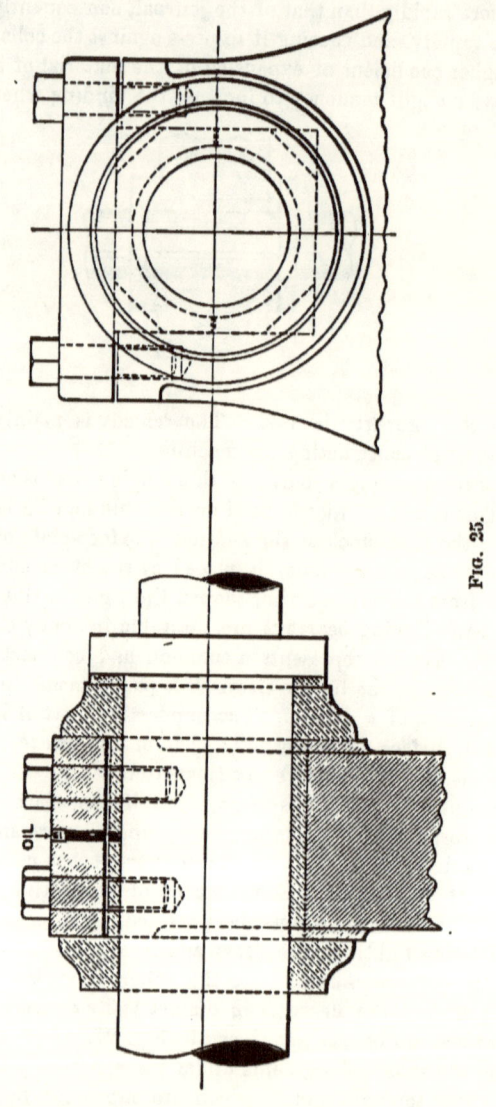

Fig. 25.

The lips D and D, extending around the segments, prevent the sleeve from rotating, by striking against the end of the screws. (See also Fig. 29.)

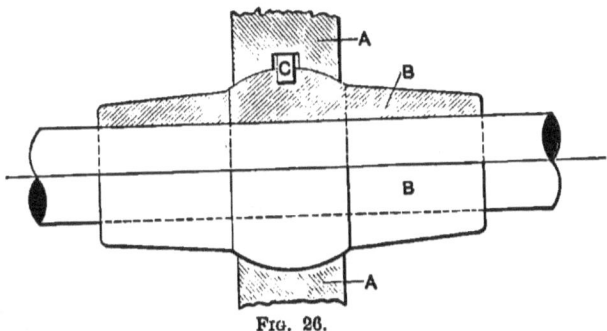

Fig. 26.

A self-aligning bearing can readily be made without the use of a spherical surface, but it is necessarily not so compact. Fig. 28 shows such a bearing, which is also adjustable both vertically and horizontally. The sleeve B is supported on the points of two

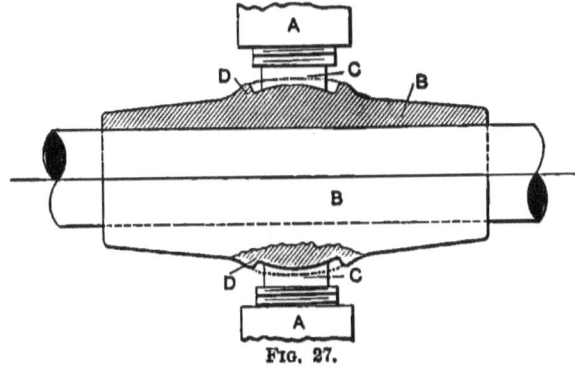

Fig. 27.

coaxial screws which pass through threaded holes in the extremities of the arms of the yoke C; the lower part of the yoke passes through a smooth hole in the supporting frame A, and has lock-nuts D and D for adjusting and locking it in place, still leaving it free to turn in the frame.

10. Lubrication of journal-bearings.—Either of the three forms of lubricants—liquid, pasty, and solid—may be applied to a journal bearing. The liquid lubricants consist chiefly of oils, the pasty of

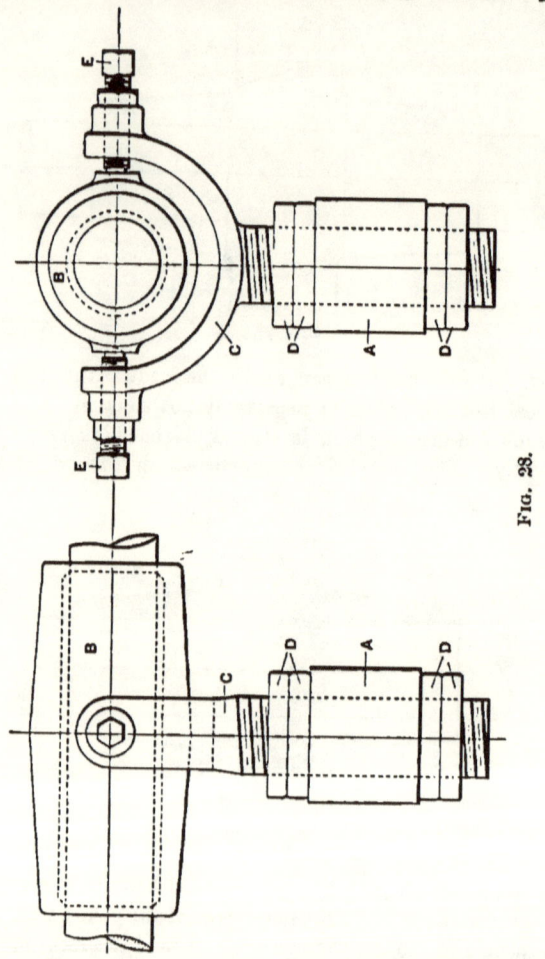

Fig. 28.

grease, and the solid of graphite. The oils are by far the most commonly used.

11. Oil- and grease-lubrication for journal-bearings.—Both experimental investigation * and practice show that when an oil-lubricated journal is run under a uniform load, constantly applied in the same direction, a film of oil adheres to the journal and is carried between the rubbing surfaces, unless the pressure between these surfaces is so great as to squeeze or rub it off. As long as the film of oil is maintained between the rubbing surfaces, the lubrication is effective; but when the oil is so completely squeezed out, or burned by the heat due to friction, as to break the film and allow the metallic surfaces to come into contact, abrasion and consequent seizure or destruction of the surfaces follow. Seizure can occur only when the box completely or nearly surrounds the journal, and is sufficiently rigid to grip and hold it when the metals come into contact. Probably the seizure is often due to sudden heating and expansion of the material forming and lying just beneath the rubbing surfaces, when the lubricant has ceased to separate them.

The film of oil must withstand the greatest pressure that occurs at any part of the bearing. If a box presses vertically downward on its journal, the maximum pressure will usually be at or near the top and centre of its length. From this point the pressure gradually decreases towards the sides and ends.

If a hole is drilled in the middle of the box at the top, so as to communicate with the atmosphere, the oil will be forced out of it as the journal turns. If a pressure-gauge is attached to the opening, it will show the pressure acting on the oil at that place,† which is probably about the same as would exist if the hole had not been made. Any attempt to introduce oil at this point in the usual manner, as with a drip-cup, would clearly be unsuccessful. Even if a groove were cut along the top of the box for some distance, but not reaching the ends, it would only collect the oil and force it up through the opening. The oil must therefore be applied at some other place. If an oil-cup which gradually feeds the oil on the bearing is used, it will give the most satisfactory results when the opening through the box is on the side where the surface of the

---

* Tower's experiments: Proc. Inst. of Mech. Engrs., 1883, p. 632; also other experiments.
† Tower's experiments.

journal is approaching the top or point of highest pressure of the rubbing surfaces. This applies especially to boxes which only partly surround the journal. If the box completely encircles it, satisfactory lubrication can be obtained by applying the oil at any point where it will pass through the opening to the rubbing surfaces, provided suitable oil-grooves for distributing the oil over the surfaces are cut into either the box or journal, or both. Such grooves are generally cut into the box.

When a "half-box," as one which covers only half or less of the circumference of the journal is called, rests on top of the journal, it can be lubricated very successfully by letting the lower part of the journal run in a reservoir of oil. This is called **oil-bath** lubrication. The oil adhering to the journal is carried up between the rubbing surfaces and thoroughly distributed over them. If the half-box is placed under the journal, the oil can be most successfully applied at the top.

Oil-pad lubrication is an excellent substitute for bath lubrication when the latter cannot be advantageously applied on account of waste of oil or inability to provide a suitable reservoir for retaining it. The oil-pad may be made of a piece of porous woven material which is saturated with or dips into oil and presses against the journal. An excellent, and at the same time comparatively inexpensive, material which is used instead of the woven pad, is the cotton "waste" so commonly used for cleaning machinery. It is almost wholly, possibly wholly, used in this country for lubricating car-axle journals. The waste and oil are placed in a closed box beneath the journal, filling the space so that there is good contact between the waste and journal. The capillary action of the waste carries the oil to the journal in sufficient quantity to lubricate it.

Mr. E. Charbal states that woollen wicking was found better and more economical than cotton for railway service, the delivery of oil being from 50% to 100% better, and the renewals only 68% of those for cotton.[*]

While experimenting with a pad-lubricated journal 4 inches in diameter and 8 inches long, running against a "half-box" at 266

---

[*] Min. of Proc. Inst. Civ. Eng., 1894–95, Part II., p. 412.

revolutions per minute under a total load of 15132 pounds = 756.6 pounds per square inch of projected area, Mr. John Dewrance found a vacuum of 28.4 inches of mercury, corresponding to 13.9 pounds per square inch between the rubbing surfaces, at a point where the rotating part had passed the position of maximum pressure.*

The nature of the rubbing surfaces of a bearing is of importance. In general it may be stated that the smoother they are the more satisfactory service will they give. It has been found that by burnishing the surface of a journal with a roller burnisher pressed hard against it while rotating in a lathe, so that the burnisher rolled on the bearing surface, the frictional resistance to the rotation of the journal when placed in its bearing was lowered and more satisfactory operation secured. It is the practice of the C., M. & St. P. Ry. to roll the journals of their car-axles while turning the wheel-fit in the lathe. It is found decidedly beneficial.

In contrast with the above, it is not an uncommon occurrence to find that a journal, turned and finished in the ordinary manner by polishing or grinding, which runs hot despite the best attention and care in both its manufacture and operation, can be put into satisfactory working order by rubbing it with emery-cloth so as to make the scratches of the emery parallel to its length; a file has been used satisfactorily for the same purpose by draw-filing the journal so as to make the scratches of the file parallel to the axis of the journal, as with the emery-cloth.†

There seems little reason to suppose otherwise than that the smoothest, truest surface that can be obtained is the best for a bearing, both in point of low frictional resistance and great durability. But if the surfaces fit very truly together over a considerable area which has no oil-grooves or other indentations, the oil may not be carried between them in sufficient quantities for satisfactory lubrication. It is in such a case that the scratches of the emery-cloth or file have a beneficial effect. The scratches act as little reservoirs or pockets which carry the oil in between the surfaces and distribute it over them. An increased number of oil-grooves in the bearing, properly arranged, or a few in the journal, so as to cut the

---

* Minutes of Proc. of Inst. Civ. Eng., vol. cxxv., p. 362; *Engineering*, London, Jan. 1, 1897, p. 20; *The Engineer*, London, Jan. 8, 1897, p. 42.
† Trans. Amer. Soc. Mech. Engrs., vol. vi., pp. 849-857.

rubbing surfaces into small areas, would doubtless have the same effect as the numerous scratches. Oil-grooves, sufficiently large not to fill with dirt and gummy oil, should be provided plentifully in a bearing having considerable area and doing heavy service. It might seem that this should be applied to railway-car boxes. There is a reason why it could hardly be successfully done, however, without objectionably great expense. The reason is as follows: Owing to the nature of the service they perform, it is scarcely possible to examine the rubbing surfaces of the bearings from the time the boxes are put in place until they are worn out. The allowable wear is so great that grooves of a practicable depth would be worn out of the box long before it had completed its life. And even if this were not true, the probability is that the grooves would be filled with dirt in a short time, and thus become useless; this is also true of grooves cut in the journal, which is a less desirable place.

An intermittent or a reversed force, causing pressure on a bearing, offers an opportunity for the lubricating oil to get between all parts of the rubbing surfaces. On account of this beneficial action greater bearing-pressure can be used than for a constant pressure in one direction. This is markedly shown in practice in the main bearing and crank-pin of the ordinary form of double-acting steam-engine.*

For very heavy, continuous bearing-pressures and slow speeds grease is better than oil for lubricating journal-bearings. The greater viscosity, or "body," of the grease prevents it from being squeezed out from between the rubbing surfaces as quickly as the thinner oil, and slow speeds do not carry the oil in between the rubbing surfaces with sufficient rapidity to overcome the squeezing-out action.

In journal-bearings, it may be taken as a rule that the faster the speed and lower the bearing-pressure, the thinner or less viscous the lubricant that will give the best service. It is frequently found that a change from a thick to a thin lubricant, or *vice versa*, will overcome difficulties of heating and abrasion of a journal-bearing.

A case has been cited where a journal which ran hot in its box when abundantly supplied with oil, cooled down and worked satisfactorily when the supply of oil was diminished.

---

* Trans. Amer. Soc. Mech. Engrs., vol. vi., p. 856.

12. Special devices for journal-lubrication with oil and grease.
—On account of the importance of securing a constant supply of oil on a bearing when working hard, some of the typical appliances for this purpose are worthy of description. One of the simplest and most commonly used for a bearing whose journal rotates, is a light ring, of a greater diameter than the journal, which is hung over, and rests on, the top of the journal, a part of the top of the box being cut away for this purpose. The lower part of the ring dips into a chamber of oil below the journal. Since the weight of the ring rests on the top of the journal, the rotation of the latter causes the ring to turn also, so that the oil adhering to it is carried up and deposited on the journal. Such a ring-oiling device is shown in Fig. 29.* By providing a large and deep oil-reservoir, connected

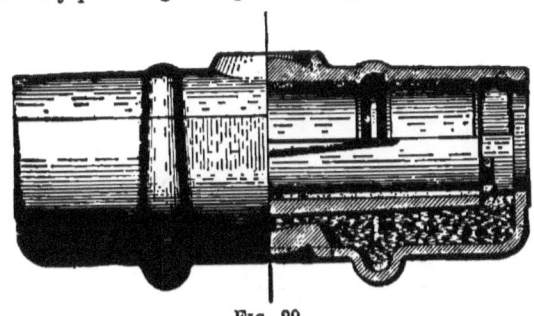

Fig. 29.

by suitable channels to both ends of the box, so that the oil will be led back to the bottom, the particles of foreign matter gathered up by the oil while passing over the journal have an opportunity to settle out, leaving the oil clean for its next application.

A short, endless chain is frequently used instead of the oil-ring. It has the advantage that, in certain forms of bearings having the box cast in one piece with the reservoir, it can be introduced through a smaller opening than the ring, or even without any external opening through the side of the box. This fact makes it especially applicable to line-shaft boxes and similar light bearings which are not attached to a heavy frame. The point of superiority sometimes advanced for the chain, that it will carry a more copious

---

* Bearing manufactured by Rice Machinery Co.

supply of oil to the bearing, is hardly worthy of consideration, for the ring will furnish an ample supply.

A recent and apparently excellent substitute for the woven pad or cotton waste, mentioned in the preceding section, is a block of hard wood, concaved on one end to fit the journal, and having broad but very thin slits from this surface to the opposite end, which dips into oil. The concave end is held against the journal by a spring, and the capillary action of the slits carries the oil up to the surface of the journal.

Wicking, similar to that of candles or lamps, is used for lubricating by placing it so that the lower end is in oil and the upper touches, or hangs over, the journal; capillary action carries the oil up to the journal.

When the box of a bearing rotates around the journal, as when a pulley turns on a shaft, oil can be readily applied as shown in Fig. 30, provided the shaft does not extend far beyond the bearing on

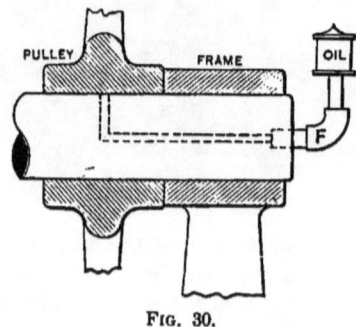

FIG. 30.

one side, at least. A hole is drilled into the end of the shaft as far as the centre of the pulley, and then another from the surface of the shaft to meet it. The oil-cup is attached by a piece of pipe which screws into the end of the shaft. This arrangement answers when the shaft is held rigidly in the supporting frame, so that it never revolves. If the shaft turns at times, a running pipe-coupling can be placed at F and the oil-cup supported by the frame.

A comparatively short box on a long shaft, when the conditions of operation are such that one rotates continuously and the other at intervals, as a pulley on a line- or counter-shaft, presents the most difficult of all problems in oil-lubrication. Numerous appliances have been devised, but none seems to have come into general use. When the parts are not readily accessible, and are comparatively light, as line- and counter-shafts and their pulleys in most shops and factories, a bearing whose box is made of a material which is itself a lubricant, is the most satisfactory (see § 13). A grease-cup having a piston which is pressed against the contained grease by a spring, so as to gradually force the grease between the rubbing surfaces, is often used when the parts are easily accessible, or heavy; it can be used in almost every case, but the attention required makes it frequently undesirable. If the hub of a pulley can be enlarged in diameter without inconvenience, a recess can be made in it, with an opening to the journal, and an absorbent pad fitted into it so as to press against the journal; by saturating the pad with oil at intervals lubrication sufficient for light loads can be secured.

Grease-lubrication is often used for a bearing whose box does not rotate. One of the most common methods of applying it is by means of a grease-cup having an opening or oil-hole leading straight to the shaft, and a copper or brass pin which partly fills the hole. The pin stands vertically, with its lower end resting on the journal, being held in contact with it by its own weight. The grease in the cup completely surrounds the pin. As the journal becomes warm by running, the heat is carried through the pin to the grease, warming and melting it, so that it flows down to the journal. The slight heating of the pin by its own rubbing against the journal is hardly appreciable. While this device can be quite safely relied upon to keep the bearing from getting dry and cutting, it does not afford economy of power, since the journal must have enough frictional resistance to generate heat for melting the grease. The end of the pin is apt to wear a groove in the journal, which may be objectionable.

A crank-pin which revolves about a crank-shaft in the ordinary manner, can be lubricated the most successfully by a device of the general nature of that shown in Fig. 31. A hole is drilled into the end of the pin, and another, to intersect it, from the rubbing surface at about the middle of its length; the latter hole should start from the part most distant from the crank-shaft. A piece of tubing is screwed rigidly into the end of the pin, with its free end extending towards the centre of the crank-shaft, where a hollow ball is attached, the centre of the ball being coincident with the centre line of the crank-shaft. A hole *B* in the side of the ball, drilled concentric with the crank-shaft, so as to remain stationary when the parts rotate, permits the introduction of oil by any suitable

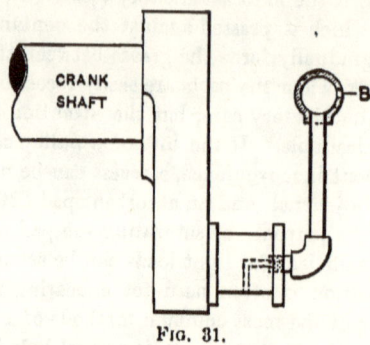

FIG. 31.

means, as an oil-cup with a drip-tube extending through the hole into the ball. After the oil is introduced it is carried out to the rubbing parts by centrifugal force.

The pin of a double crank or pair of disks can be continuously oiled by means of a slight modification of the device just described. Fig. 32 illustrates the method. A groove *C* is turned in the crank or disk, or a grooved collar attached to it. An oil-hole leads from the bottom of the groove to the surface of the crank-pin at *E*. Oil that is put into the groove will be carried to the crank-pin bearing when the parts are rotating. The drip from the inner end of the crank-shaft bearing *B* may be thus used for the crank-pin, or oil can be fed into the groove by a tube leading from an oil-cup down to the under side of the crank-shaft.

Reciprocating journal-bearings, such as that of a cross-head pin of a horizontal steam-engine or pump, are most commonly lubricated by a method represented by the following device: A piece of wicking is either suspended from an attachment to the frame of the machine, or stretched over a curved surface, and a supply of oil is fed to the wicking so as to keep it completely saturated and almost ready to allow a drop to fall from it. A metallic scraper or "wiper" is fastened to the parts to be lubricated, which, as it is brought under the wicking at each stroke of the moving parts, scrapes or picks some of the oil from it; the oil is led from the wiper to the rubbing surfaces through suitable oil-ways.

There are numerous modifications of this device, a notable one consisting chiefly of a thin strip of metal attached to an oil-cup so

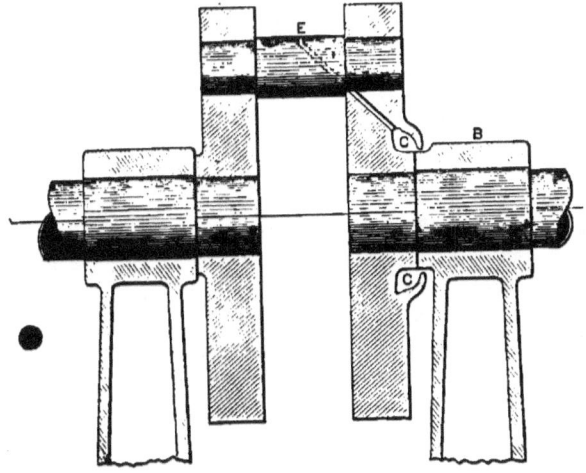

FIG. 32.

that the oil fed from the cup will gather at the lower end, which is pointed. Another thin strip of metal, bent back upon itself so as to almost come together near the ends, is fastened to the reciprocating parts so that the ends will pass on each side of the point to pick off the drop of oil and then let it flow down to the rubbing parts. These oiling devices are generally placed so as to act at one end of the stroke, thus preventing the throwing of the oil which would occur in high-speed machinery if it were picked off near the middle of the stroke.

A less common device for oiling a reciprocating journal consists of a pair of telescoping tubes, the end of one being fastened to the centre of the reciprocating pin, and one end of the other to a standard on the frame of the machine. The end of the standard to which the tube is attached is above the line of travel of the part to be lubricated. The fastenings at the ends of the tubes are made to permit rotary motion; this, together with the telescopic action, furnishes a means of providing a continuous passageway from an oil-cup on the standard to the journal. Suitable holes in the latter complete the path for the oil to the rubbing surfaces.

When the parts to be lubricated reciprocate vertically, an open cup can be attached to one of them, and the oil dropped directly into it from a feeding-cup on the frame.

As a precaution against injury to an oil-lubricated bearing, should it run dry from neglect or failure of the oiling device, sufficient grease to lubricate it for a reasonable period is placed in a suitable recess formed in the box. As long as the lubrication is sufficient to keep the bearing cool, the grease remains; but as soon as the temperature rises the grease melts and flows to the journal.

Oil-grooves should be made with regard to the effective distribution of the oil over the rubbing surfaces. If a horizontal journal always rotates in the same direction in its box, it is not the best practice to cut the grooves longitudinally, for the oil is then aided but slightly, if at all, in its flow along the grooves by the action of the revolving surface. By cutting the grooves as right- and left-hand threads of very rapid pitch, both starting from the oil-hole in the direction that the journal turns, the oil will be drawn along the grooves by the journal, thus securing both a more rapid circulation and effective lubrication. This applies especially to ring- and chain-oiled bearings.

13. Materials for journal-bearings.—As long as a suitable lubricant is continuously introduced between the rubbing surfaces of a journal and its box, almost any of the metals common to engineering practice can be used for these surfaces. Some will stand much higher bearing-pressures than others, but this depends generally upon the strength of the material. The serious injury that is almost certain to overtake many of the metals when copious lubrication fails, even for an instant, precludes them from use except for slow speeds or light bearing-pressures. Others, while running without injury under both speeds and pressures comparatively high, offer so much frictional resistance to the rubbing motion, that the power lost in overcoming it is so great as to be objectionable in some classes of machinery. A bearing that is called upon to perform heavy service continuously should, and often must, be made of a material, or materials, which will continue to run even though the lubrication is defective, and whose frictional resistance to the rubbing of the parts over each other is low. This last quality is generally, but not always, a necessity for continuous, heavy service; it is always desirable, however. If the frictional resistance is high, the material must either have good heat-conducting capacity, so that it will carry away the frictional heat, or else a special cooling device, such as a hollow box with water circulating through it, may become necessary. A crude method of cooling, which is much used in rolling-mills on the bearings of the rolls, is to let a stream of water flow over the bearing. This would clearly be unsatisfactory where flowing water is objectionable.

For heavy service, mild steel and wrought iron both run satisfactorily on brass or bronze. The journal is generally necessarily made of steel or iron on account of the required strength to endure such service. Brass and bronze exert a comparatively low frictional resistance to the rubbing of the steel or iron over them, and are both softer than iron and steel. This latter is an important property for the material of the box, for, should the lubrication fail, the iron or steel journal will continue to rub over it for a considerable time without serious injury to the bearing, this being a not infrequent occurrence with the journals and half-boxes in railway practice; or, if the box completely encircles the journal, it may seize and stop it without serious injury to the rubbing parts, since the softer metal will not cut into the surface of the steel or iron, and so long as the latter remains comparatively smooth, it will not cut and groove the softer metal to a serious extent. This cannot always be taken as true for the harder grades of the alloys just mentioned.

Babbitt metal is an excellent alloy for a steel or iron journal to run upon. On account of its softness and weakness, it can be used only as a lining to a supporting box or shell, when the service is heavy. Its frictional resistance is probably the lowest of any material that can be conveniently and economically used for journal-boxes. It will not injure the journal if lubrication fails, and it will not seize upon it. In case of excessive heating, it will melt and run out of the box. This property is sometimes used to advantage in high-speed machinery where sudden stoppage by seizure would be disastrous.

Cast-iron boxes for steel and iron journals are quite extensively used, especially for line-shafting. The pressure between the rubbing parts must be kept low, however, and even then the parts are in great danger of serious injury if the lubricant is not constantly applied. A continuous-oiling device, such as a ring or chain dipping in a reservoir of oil, becomes a necessity when the speed and pressure are at all high. If abrasion once begins, it is almost certain to injure both surfaces considerably, the cast iron generally suffering the most. The appearance of a bearing so injured seems to indicate that a particle of the cast iron is ground off and imbedded in the journal so as to slightly groove the box, and then the loose pieces of metal continue the grinding action in connection with the grooving, which roughens the surface of the journal still more and causes the destruction to go on, cutting away the box until the machinery is stopped. The brittle and crumbly nature of cast iron makes it exceedingly liable to great injury when abrasion once begins. The more close-grained the cast iron, the less liable is it to cutting. In order to keep the pressure per unit area as low as possible, cast-iron self-aligning boxes are made very long—as long as four times the diameter of the bore, or even longer in proportion.

The comparatively low cost of cast iron, and the ease with which it can be cast into form and machined, make it a desirable material from the point of first cost; hence the reason for its somewhat extensive use for journal-boxes, as before mentioned.

Chilled cast-iron journal-boxes are used on some classes of machinery not requiring accurate fitting of the rubbing parts. The bore of the box is cast against a chill which gives it a fairly accurate form. On account of the extreme hardness of the chilled surface, it is almost impossible to finish it with tools, but it may be ground smooth. In its almost exclusive application to the less accurate classes of machinery, the chilled cast-iron boxes are used without finishing the bore other than by cleaning it carefully. The hardness of the surface eliminates the liability to the abrasion and continned cutting common to iron which has been cast in the ordinary sand mould, and against sand or clay cores.

Cast-iron journals are used on the rolls of rolling-mill machinery. Each roll is cast as a single piece, body and ends. The journals or necks are cast in sand, so they have the ordinary crystalline structure and softness of sand castings. These journals run successfully on brass, bronze, or Babbitt-lined boxes. The speeds and total pressures are frequently quite high, but the journals are of large diameter, so that the bearing-pressure per unit area is comparatively low.

In some classes of metal-working machine-tools, where great rigidity is desired, a cast-iron spindle is used, whose journals run on boxes of the same material. When so used, the speed of rubbing is generally low, and the bearing-pressure per unit area always so. If well lubricated at first, the rubbing surfaces soon take on a glazed, glass-like surface, which, once thoroughly fixed, will stand much neglect in the way of lack of proper lubrication. The same is true, possibly even to a greater extent, when the cast-iron journal runs on Babbitt metal, or on cast iron and Babbitt metal in alternate strips.

Hardened steel, or case-hardened iron or mild steel, journals and boxes, ground to form, are used frequently where it is especially desirable to maintain accuracy in the running parts. These hard parts run well together when the rubbing surfaces are smoothly finished. They will also run excellently on any of the materials that can be used on steel, wrought iron, or cast iron. Their cost is much greater than that of the unhardened materials.

A **self-lubricating material** for journal-boxes has been invented in recent years. It is made of graphite held together by wood fibres. The process of making is to mix together the pulverized graphite and wood pulp in a bath of water, run the mixture into moulds having the form of the piece wanted, and then subject it to heavy pressure, forcing the water out through minute openings in the moulds for this purpose. The compressed "fibre-graphite" is then treated with oil and baked to give it desirable qualities. The material thus manufactured is strong enough to work under pressures ordinarily required for line- and counter-shaft boxes, pulleys, crank-shafts of small engines, etc. No lubricant is required or should be used, for the graphite is itself a solid lubricant. The application of oil or grease to a fibre-graphite box, not only does no good, but is harmful. The cause of their harmful action is as follows: The surface of a journal is always somewhat rough. When it is first revolved in a fibre-graphite box, this roughness abrades the box to some extent, and the small particles of graphite thus loosened collect in the depressions on the journal, finally filling them flush with the higher parts of the metallic surface, thus forming the best possible rubbing surface. If oil is now applied to the bearing, it loosens these small particles from the journal and floats them away, leaving the surface in the original condition; the continued application of oil will not allow particles of graphite to collect on the journal, and consequently the wearing away of the box continues much more rapidly than if it were left dry.

The fact that the fibre-graphite bearing requires no attention whatever when running satisfactorily, and that oil and grease are entirely absent, makes it a desirable material for places difficult of access, and when oil and grease may be harmful.

Some of the minerals are used in their natural condition for journal-boxes. Probably these are confined to the more close-grained ones almost exclusively. They are used without any lubricant for much the same reason as just mentioned for fibre-graphite. When an iron or steel journal first runs on a stone box, the latter grinds off the surface of the metal, which lodges in the interstices and irregular depressions of the stone, filling them and making a smooth surface which works satisfactorily. If oil or water is applied, the particles of metal are floated out of the concavities of the stone, and the grinding action is again started.

The action of a stone box and steel journal can be readily understood by rubbing the blade of a knife over a clean oil-stone until the stone becomes glazed and stops cutting or grinding away the steel, and then applying oil, thus causing the stone to cut freely again.

It is believed that no materials except the more precious stones are used in their natural state for bearing surfaces in this country. These are used only for light machinery, such as watches, clocks, electric meters, chronographs, etc. It is said that the spindles of large grindstones are quite commonly run on stone boxes in the cutlery-manufacturing establishments in England.

Journal-bearings that are submerged in water and cannot be conveniently enclosed so as to be lubricated with oil or grease, frequently have the boxes made of wood. Lignum-vitæ, letterwood, and camwood are the varieties chiefly adopted. The former seems to be the most largely used. It is generally fixed with the end of the grain against the journal. The water itself is a good lubricant for steel, wrought iron, or bronze, running on wood. It is therefore only necessary to provide a way for the water to get between the rubbing surfaces. This is often done by using thin strips of the wood set in a metal frame or casing so as to run parallel with the length of the journal, and extending out a short distance from the casing toward the journal, thus leaving small passageways between them for the water.

The wood has another point in its favor, which is that there is little or no electrolytic corrosion of the journal or shaft passing through it, such as is always present when the journal runs on a metallic box. This corrosion is greatest in sea-water. It affects the journal most seriously at the end of the bearing, a groove forming around it at that locality. The steel of the shaft, the bronze or brass box, together with the salt water, present all the elements of a galvanic cell, and it can only be expected that electrolytic action will occur. When the shaft and box are in direct contact, the electrical resistance through the water at their exposed surfaces is low, and consequently the electrolysis is rapid. The casing holding the strips of wood can be made of the same material as the journal, thus reducing or eliminating the corrosion. Moreover, the fact that the journal and metal of the box are not in direct contact may be beneficial in reducing corrosion, but this can be of little moment if the box is supported on metallic framing which in any way has contact with the shaft of which the journal is a part.

14. Frictional resistance of journal-bearings.—If a horizontal journal rotates in its box under a vertical load P, and the sum of all the elementary frictional forces acting tangent to its rubbing surface, to oppose rotation, is f, then $f \div P = \mu$ is called the **coefficient of journal friction** for the bearing.

The quantity f may also be defined as the force which must be applied to a rotating piece at a distance from the axis of the journal equal to its radius, and normal to the axis, in order to keep it rotating uniformly.

In an oil-lubricated bearing, the frictional resistance partakes partly of the nature of the friction between two unlubricated solid bodies rubbing over each other, and partly of that of a solid moving through a considerable volume of liquid. The value of μ therefore lies between that for solid friction of the rubbing materials of the bearing, and that of the same surfaces passing through a volume of the lubricant. Any variation of the rubbing surfaces, the kind of oil, or the quantity that is applied, will affect the value of μ.

Many experiments have been made to determine the value of μ and its method of variation with change of speed and pressure. These experiments, made by different investigators, are unquestionably nearly all, probably all, thoroughly reliable. The values obtained for μ, and the method of its variation, are so greatly different in nearly all cases that it seems impossible to arrive at any definite conclusion regarding them. The following facts seem to be fairly well established for mild steel running on brass, bronze, or white-metal alloys, operating under a steadily applied load:

1st. In an oil-lubricated journal-bearing, the coefficient of friction μ at first decreases as the velocity of rubbing increases from zero, then increases with further increase of speed. The speed at which this change in the variation of μ occurs becomes higher as the bearing-pressure is increased.

2d. In an oil-lubricated journal-bearing, the coefficient of friction μ decreases, rapidly at first, then more slowly, and finally begins to increase, as the load is increased from zero up to the limit of working.

3d. In an oil-lubricated journal-bearing, for the higher speeds, the coefficient of friction μ decreases as the temperature rises, until the lubricant either becomes so thin as not to sufficiently separate the rubbing surfaces, or is disintegrated; but for low speeds μ increases as the temperature rises.

A series of tests for the coefficients of friction of the alloys given in Table I were made by Messrs. Joseph Kuhn and Robert T. Mickle.* One journal each of wrought iron, steel, and cast iron

* "Variation of the Coefficient of Friction with Different Loads and Bearing Metals." *Engineering News*, May 18 and 25, 1893, pp. 468 and 494.

BEARINGS AND LUBRICATION. 41

TABLE I.*
COMPOSITION OF BEARING-METAL ALLOYS. PERCENTAGES.

Alloy.	Copper.	Tin.	Lead.	Zinc.	Phosphorus.	Remarks.
A	79.70	10.00	9.50	....	.80	Penn. Ry. standard phosphor-bronze.
B	74.90	9.40	9.45	5.45	.80	Same as A with 6% zinc added.
C	78.45	9.80	11.75	....	....	Baltimore & Ohio Ry. standard bearing alloy.
D	74.90	9.30	11.15	5.35	....	Same as C with 6⅜% zinc added.
E	87.50	12.50	....	....	....	Copper 7, tin 1.
H	85.71	14.29	....	....	....	Copper 6, tin 1.
I	83.34	16.66	....	....	....	Copper 5, tin 1.

* *Engineering News*, May 25, 1893, p. 496.

TABLE II.†
TESTS OF FRICTIONAL RESISTANCE OF ALLOY A.

Pressure, pounds per square inch.	Velocity of Rubbing, feet per minute.	Wrought-iron Journal.		Steel Journal.		Cast-iron Journal.	
		Temperature, Fahr. deg.	Coef. of Friction, μ.	Temperature, Fahr. deg.	Coef. of Friction, μ.	Temperature, Fahr. deg.	Coef. of Friction, μ.
35	246	76.0	.0176	75.5	.0151	77.8	.0159
	503	89.0	.0199	88.0	.0181	91.5	.0179
	1066	127.5	.0204	112.8	.0256	115.8	.0173
65	230	85.3	.0085	83.8	.0095	84.5	.0092
	483	102.1	.0112	96.9	.0076	103.0	.0103
	968	136.0	.0122	123.6	.0107	129.3	.0137
125	222	89.3	.0044	87.3	.0054	92.6	.0120
	451	100.9	.0074	99.1	.0065	100.8	.0078
	805	137.3	.0092	133.8	.0048	133.5	.0082
250	247	86.6	.0038	87.0	.0050	89.8	.0057
	500	105.4	.0025	107.9	.0042	104.8	.0040
400	234	91.5	.0018	94.0	.0035	92.1	.0038

† *Engineering News*, May 25, 1893, p. 495.

was used. The dimensions of all were the same, viz., 2 feet in circumference and 1⅜ inches long. The boxes were each 1.42 inches long and 2⅟₈ inches wide, the width being measured at right angles to the axis of the journal; the projected area was 4 square inches. The journal was moved endwise $\frac{3}{16}$ of an inch 26 times a minute. The lubricant was "Extra" lard oil, as used by the Pennsylvania Railway Company. The results of the tests are given in Tables II and III.

TABLE III.*
COMPARATIVE TESTS OF BEARING ALLOYS UNDER PRESSURE OF 250 POUNDS PER SQUARE INCH, AT A SPEED OF 500 FEET PER MINUTE.

Alloy.	Velocity, feet per minute.	Wrought-Iron Journal.			Steel Journal.			Cast-Iron Journal.		
		Temperature, Fahr. deg.	Wear, milligrams per 10000 feet travelled.	Coef. of Friction, μ.	Temperature, Fahr. deg.	Wear, milligrams per 10000 feet travelled.	Coef. of Friction, μ.	Temperature, Fahr. deg.	Wear, millimetres per 10000 feet travelled.	Coef. of Friction, μ.
A	500	105.4	33.4	.0025	107.9	41.2	.0042	104.8	22.5	.0040
B	475	117.8	51.7	.0040	125.8	47.1	.0061	121.0	35.2	.0069
C	478	110.5	29.8	.0055	113.5	29.5	.0040	122.4	54.9	.0091
D	458	113.0	43.5	.0056	117.8	68.7	.0049	127.8	35.0	.0070
E	473	113.1	49.5	.0055	113.1	31.4	.0052	116.1	9.9	.0074
H	468	111.1	7.7	.0043	112.6	25.9	.0045	116.1	16.6	.0066
I	500	115.0	25.7	.0043	117.3	32.1	.0038	120.1	13.9	.0046

* *Engineering News*, May 25, 1893, p. 495.

In Table II it can be seen that, for the wrought-iron journal, when working under 250 pounds per square inch, μ falls from .0038 to .0025 when the velocity of rubbing increases from 247 to 500 feet per minute; evidently the speed at which μ begins to increase has not been reached. The same is true for the steel and cast-iron journals for the same pressure and speeds. At a pressure of 125 pounds per square inch, the wrought-iron journal seems to have passed the speed for the change in the variation of μ from decrease to increase; the cast-iron journal seems to have nearly reached the turning-point, for μ increases from .0078 to only .0082 when the speed changes from 451 to 865 feet per minute. In nearly all the other cases the turning-point has been passed. There is, as shown by the table, a considerable increase of temperature in all the cases of increase of speed, but the rise of temperature runs fairly uniform for similar increases of speed.

It was found, in a test of commercial-power-transmission machinery, that the coefficient of friction increased rapidly after a rubbing speed of 2.9 feet per second = 174 feet per minute was reached.*

Table II also shows a continuous decrease in μ as the pressure increases for an approximately constant speed. The pressure does not reach a value high enough to cause μ to begin to increase.

Fig. 33 † shows, for a rubbing speed of 314 feet per minute and

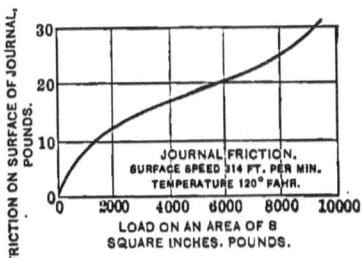

FIG. 33.

a temperature of 120° Fahrenheit, a decrease of μ up to a bearing-pressure of about 800 pounds per square inch, and a slight increase of μ for increasing pressure above 1000 pounds per square inch, of projected area of the journal.

Fig. 34 shows how μ decreases as the temperature rises, for a speed of 314 feet per minute and a load of 1000 pounds per square inch of the projected area of the journal.

A striking example of the effect of the condition of the rubbing surfaces upon the coefficient of friction μ, has been stated by Mr. A. H. Emery.‡ The materials rubbing together were hardened

---

* "Experiments upon Friction in Electric Motors and Transmission Shafting," by S. Hanappe. Minutes of Proc. Inst. Civ. Eng., vol. CXXII., p. 496.

† Fig. 33 may be found in *Digest of Physical Tests*, July, 1897, p. 175; Fig. 34 on p. 174.

‡ Trans. Amer. Soc. Mech. Engrs., vol. VI., p. 852.

steel and non-hardened steel. By changing the hardened-steel rubbing surface, which had been ground on a fine, solid emery-wheel, to the fine finish produced by a polishing-wheel, μ was changed from 30% to 3%; the pressure per square inch and the lubricant remained the same, and the same pieces of steel were used in both cases. The pressures were exceedingly high, reaching 50000 pounds per square inch in some cases.*

When no liquid lubricant is used, but the solid surfaces of the journal and box rub directly against each other, as in the fibre-

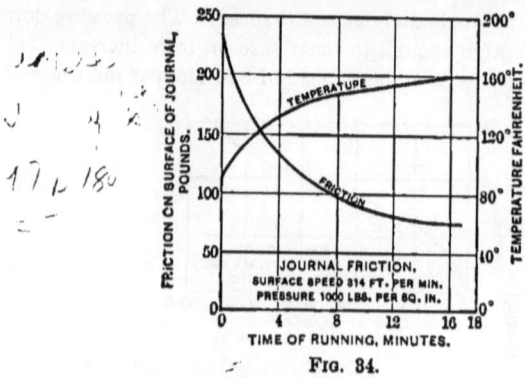

FIG. 34.

graphite bearing, it is probable that there is little, if any, change in the value of μ with change of load and speed, except such as may be caused by deformation of the box, unless there is excessive heating.

The power which is expended in overcoming the frictional resistance of a journal is transformed into an equivalent amount of heat at the rubbing surfaces. This heat must be conducted away, first by the lubricant to the rubbing surfaces, when a lubricant is used, and then through the materials of the journal and box, so that it may radiate, or be carried away by some cooling substance, as water, when special means of cooling are provided. The oil, therefore, must have, as one of the qualities of a good lubricant, considerable heat-conducting capacity when used for heavy service. It is always desirable, however, that its unctious properties shall be

---

* Mr. Emery states in a private communication that this pressure was used on the wedge-shaped jaws of a testing-machine holder.

such as to reduce the generation of heat to a minimum, thus reducing the power expended in overcoming friction.

15. Effect of changing the proportions of journal-bearings.—In the majority of cases where a bearing is to perform heavy service, the diameter is made as small as the requisite strength will permit. The length is then made such that the bearing-pressure per unit area will be low enough for satisfactory operation.

In order to examine the effect of changing the diameter of a journal, let it be assumed that a bearing, running with a solid lubricant, and designed to operate under a given load and speed, heats to excess. It therefore becomes necessary to change its size in order to secure cool running. First, assume that the diameter is increased, the length remaining unchanged. For convenience in pointing out the effect of this change it will be assumed that the diameter is doubled. The frictional resistance to rotation $= f$, acting tangent to the surface of the journal, is unchanged; for the coefficient of friction μ is the same in both bearings before excessive heating occurs, and the load is the same for both. The distance through which f acts, during one revolution of the larger journal, is twice that for the smaller. If $D =$ the diameter of the smaller, and $2D$ that of the larger, journal, the energy converted into heat by the frictional resistance during one revolution will be $\pi D f$ for the smaller, and $2\pi D f$ for the larger; that is, twice as much mechanical energy is wasted, and twice as much heat produced, in the larger as in the smaller journal. The larger bearing has twice as much area of material at the rubbing surface for carrying away the heat, as the smaller; therefore the quantity of heat to be conducted away per unit area of the material is the same for both. This shows that the liability to excessive heating is just the same in both bearings, and that, while increasing the diameter brings no improvement in the way of cooler running, it doubles the waste of energy in the bearing.

Again, it may be assumed that the length of the journal is doubled, the diameter remaining $D =$ that of the original journal. The frictional resisting force f is the same as before; the diameter is also the same. The mechanical energy transformed into heat is therefore the same as before, and equals $\pi D f$. The area of the material through which the heat is conducted away from the rubbing surfaces is twice as great in the longer as in the shorter bearing; hence the heat that must pass through a unit area of the material is but one half as great in the elongated bearing as in the original. The longer journal is consequently less liable to excessive heating; the mechanical energy lost in it is the same as in the shorter. In a general way, it may be said that doubling the length, without changing the diameter, of a bearing running with a solid lubricant, reduces its liability to excessive heating by one half, without changing the power necessary to overcome its frictional resistance.

When excessive heating of an oil-lubricated bearing cannot be prevented by the best lubrication, or by putting the rubbing surfaces in their best condition, together with proper precautions for cooling, etc., it is evident that there is need of change in the dimensions. It is assumed that the journal is to run at a given rotative speed and under a given total load. If the bearing-pressure per square inch is very high, and the velocity of rubbing comparatively low, it may be remedied by increasing the diameter; or, if the bearing-pressure is low, and the velocity of rubbing high, decreasing the diameter may prove effective. The strength of the journal must be considered, of course. Lengthening the journal will reduce the liability to heating; but if, on account of too great length, the pressure becomes localized, as at one end, the trouble may be increased. On account of the numerous causes affecting the successful running of an oil-lubricated bearing, it does not seem possible to arrive at any more definite conclusions than these. Each problem must be considered with regard to the conditions affecting it, and dealt with accordingly.

16. Proportions of journal-bearings, and examples from practice.—The most recent and valuable formulas for the proportions of the bearings of engines have been given by Professor John H. Barr of Cornell University.* The formulas, or at least the constants in them, are derived from American practice in engine-construction. The engines are divided into two classes: "low-speed" for Corliss and other long-stroke engines usually making not more than 100 to 125 revolutions per minute, and "high-speed" for those having a stroke from one to one and a half times the piston

---

* Trans. Amer. Soc. Mech. Eng., vol. XVIII., 1897, p. 756.

diameter, and a rotative speed of 200 to 300 revolutions per minute. The data for the formulas were obtained from about eighty separate engines classed as high-speed, and about eighty-five classed as low-speed. The high-speed engines ranged from 20 to 240 H.P., and the low-speed engines from 45 to 740 H.P. The practice of thirteen builders is represented in the high-speed, and of twelve builders in the low-speed, engines. The equations are given for the maximum, minimum, and mean sizes of the parts.

The notation is:

A = area of piston, square inches;
B = a constant;
C = a constant;
D = diameter of piston, inches;
H.P. = rated horse-power;
K = a constant;
L = length of stroke, inches;
M = a constant;
N = revolutions per minute;
S = steam-pressure, taken at 100 pounds per square inch above exhaust, as a standard pressure;
$a = dl$ = projected area of journal, square inches;
d = diameter of journal, inches;
l = length of journal, inches.

MAIN JOURNALS:

$$d = C\sqrt[3]{\frac{\text{H.P.}}{N}}; \quad \ldots \ldots (1)$$

$$l = Kd. \quad \ldots \ldots \ldots (2)$$

The projection of the journal equals

$$dl = MA. \quad \ldots \ldots \ldots (3)$$

The values of the constants for the main journals are:

	High-speed Centre-crank Engine. For each of two journals.			Low-speed Side-crank Engine. One journal only.		
	Mean.	Maximum.	Minimum.	Mean.	Maximum.	Minimum.
$C =$	7.3	8.5	6.5	6.8	8.0	6.0
$K =$	2.2	3.0	2.0	1.9	2.1	1.7
$M =$	.46	.70	.37	.56	.64	.46

48 FORM, STRENGTH, AND PROPORTIONS OF PARTS.

CRANK-PIN:

$$l = C\frac{\text{H.P.}}{L} + B; \quad \ldots \ldots \ldots \quad (4)$$

$$d = \frac{KA}{l} \quad \text{or} \quad dl = KA. \quad \ldots \ldots \quad (5)$$

The values of the constants for the crank-pin are:

	High-speed Engines.			Low-speed Engines.		
	Mean.	Maximum.	Minimum.	Mean.	Maximum.	Minimum.
$B =$	2.5″	2.5″	2.5″	2.0″	2.0″	2.0″
$C =$	.30	.46	.13	.6	.8	.4
$K =$	.24	.44	.17	.09	.115	.065

CROSS-HEAD PIN:

$$a = dl = CA; \quad \ldots \ldots \ldots \quad (6)$$

$$l = Kd. \quad \ldots \ldots \ldots \ldots \quad (7)$$

The values of the constants for the cross-head pin are:

	High-speed Engines.			Low-speed Engines.		
	Mean.	Maximum.	Minimum.	Mean.	Maximum.	Minimum.
$C =$	.08	.11	.06	.07	.10	.054
$K =$	1.25	2.0	1.0	1.3	1.5	1.0

Problem in the design of journal-bearings.—The application of the above formulas may be illustrated in the following problem:

It is required to determine the dimensions of the main bearings, crank-pin, and cross-head pin of a centre-crank engine having a stroke $L = 14$ inches, piston diameter $D = 12$ inches (corresponding to $A = 113.1$ square inches), and rated at 100 H.P. for a speed $N = 250$ revolutions per minute.

This comes under the high-speed class.

Main journals:

$$d = C\sqrt[3]{\frac{\text{H.P.}}{N}} = 7.3\sqrt[3]{\frac{100}{250}} = 7.3 \times .7368 = 5.38″;$$

$$l = Kd = 2.2 \times 5.38 = 11.84″.$$

This corresponds to a value of $M = .56$ in the equation $dl = MA$.

Crank-pin:

$$l = C\frac{\text{H.P.}}{L} + B = .3\frac{100}{14} + 2.5'' = 2.14 + 2.5 = 4.64'';$$

$$d = \frac{KA}{l} = \frac{.24 \times 113.1}{4.64} = 5.85''.$$

Cross-head pin:

$$dl = CA = .08 \times 113.1 = 9.048 \text{ sq. in.};$$
$$l = Kd = 1.25d.$$

Therefore;

$$d \times 1.25d = 1.25d^2 = 9.048;$$
$$d^2 = 9.048 \div 1.25 = 7.238;$$
$$d = 2.69'';$$
$$l = 1.25d = 1.25 \times 2.69 = 3.36''.$$

The nearest convenient working dimensions would naturally be used in practice.

With regard to the difference in the allowable working pressures for constant, as compared with intermittent, loads, Mr. Babcock made the following statement:[*] "I found that in crank-pins with good fitting I could allow as high as 1200 pounds maximum to the square inch; pins, perhaps 4 to 6 inches diameter, running up to 60 or 70 revolutions, would stand that continuously without getting warm. The main journal of the same engine would not stand over 300 pounds to the square inch without getting warm."

Regarding the locomotive Lady of the Lake, Druitt Halpin states[†] that, at the beginning of the stroke, the total pressure on the crank-pin $3\frac{1}{2} \times 3$ inches is 28140 pounds, which gives $28140 \div (3 \times 3\frac{1}{2}) = 2680$ pounds per square inch of projected area. The modifying effect of the reciprocating parts is not considered. In

---

[*] Trans. Amer. Soc. Mech. Engrs., vol. VI., 1885, p. 856.
[†] Minutes of Proceedings of the Inst. Mech. Engrs., 1883, p. 657.

the same engine the main bearings are 6 × 6 = 36 square inches, and carry 8 tons each, or 498 pounds per square inch.

Mr. Beauchamp Tower quotes Mr. Tomlinson as saying that 300 pounds per square inch is undesirable in locomotive-axle bearings, while 1000 pounds per square inch, and considerably more, can be used on the crank-pins.*

Mr. Henry Davy states that in pump-bearings, for speeds of rubbing up to 12 feet per minute, 600 pounds per square inch for continuous pressure in the same direction, and 1000 pounds per square inch for intermittent pressures, can be satisfactorily used.†

In a freight locomotive recently built by the Schenectady Locomotive Works for the Boston & Maine Railroad Company, the bearing-pressures, due to the weight of the parts supported, calculated from data given in *Engineering Review* of March 26, 1898, are: 215 pounds per square inch of projected area for the driving-wheel axles, and about 250 pounds per square inch for the tender-bearings. The driver-axle bearings are 8 inches diameter by 10 inches long; the tender-journals $4\frac{1}{4}$ inches diameter by 8 inches long.

Dr. C. B. Dudley states that in railway practice, bearing-pressures as high as 350 to 400 pounds per square inch are used.‡

President Joseph Tomlinson of the Institute of Mechanical Engineers is recorded as saying: "The practical limit at which he had arrived was $2\frac{1}{2}$ cwt. per square inch; and if more than this pressure were allowed to the bearings of a locomotive engine, notwithstanding their freedom to wabble from side to side, they would not run cold. . . . Whenever he had an engine which would not run cold, and in which the weights upon the bearings could not be changed, he had put a bigger axle in and thereby cured the heating directly." §

Several of the leading builders of large engines, such as are used for direct driving of electrical generators, state that the highest pressure they consider safe for crank-shaft bearing is 150 pounds per square inch of projected area of the journal. The same value

---

* Proc. Inst. Mech. Engrs., 1884, p. 30.
† *Ibid.*, 1883, p. 654.
‡ Journal of the Franklin Institute, 1892, p. 83.
§ Proc. Inst. Mech. Engrs., 1891, p. 131.

is given as the safe limit for small high-speed engines by the builders of this class of machinery.

Mr. Edwin Reynolds of the Edward P. Allis Co., is quoted as follows regarding journal-bearings:* "The square root of the speed in feet per second multiplied by the pressure per square inch of projected area, should not exceed 500." In reply to an inquiry, he is further quoted in the same place as saying: "It is true I have used the rule you mention for a limit and never go up to 500, except in vertical engines where the steam pressure in the cylinder is sufficient to lift the shaft against the cap. 350 to 375 should be the limit for horizontal engines. This method for determining size and proportions has proved satisfactory in a very large number of machines."

Step-bearings.

17. When the weight of a vertical shaft and the parts attached to it, together with whatever end thrust may come on it, are supported by a box which bears against its end, and at the same time prevents it from moving sidewise, the whole combination of the rubbing parts and others in the immediate neighborhood is called a **step-bearing.**

18. **Forms of step-bearings.**—The simplest form of step-bearing is shown in Fig. 35, where the step B is bored out to fit the end of

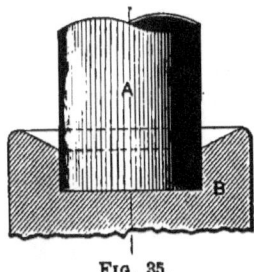

Fig. 35.

the vertical shaft A. In this case the bearing is made up of only two parts. The sloping part at the top of the box holds the oil that is used for lubrication. Radial grooves are generally cut across the

---

* *American Machinist*, Oct. 16, 1898, p. 379.

bottom of the shaft to allow the oil to gain access to all parts of the horizontal bearing surface. Bearings of this class are commonly used where either the speed of rotation is low, or the pressure is light. Pillar-cranes are ordinarily supplied with such a bearing for stepping the mast. In this application the pressure may be high, for the speed of rotation is very low.

When the shaft in Fig. 35 makes one revolution, the parts on the end farthest from the centre rub over the box through a distance equal to the circumference of the shaft, while those near the centre rub over a much smaller distance, and the geometrically central point does not have any motion over the box. On account of this inequality of rubbing there will be a corresponding unevenness of wear, so that if the parts are fitted together accurately when new, so as to make the pressure uniform over the end of the shaft, the outer portion will wear away most rapidly in service, thus causing the pressure to become heaviest at the centre, and lightest at the outer part, of the rubbing surfaces. The pressure at the centre may become so intense as to crush the material at that point. Even if this does not occur, abrasion and cutting are likely to take place. The inequality of wear and pressure may be partly obviated by removing some of the material at the centre of the rubbing parts, leaving a pair of annular rings for the rubbing surfaces. This is advisable in nearly all cases.

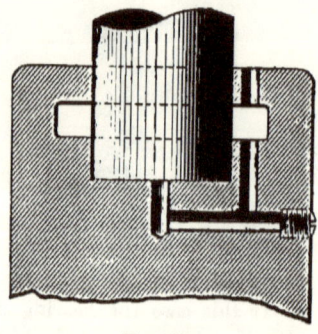

FIG. 36.

If the speed of the shaft is high, difficulty will be experienced in lubricating it on account of the centrifugal action of the rotating part throwing the oil from the centre and not allowing it to return again, unless some special provision is made for its doing so. Such provision can be readily made, however, as shown in Fig. 36, by making an oil-passage from the top of the step to the centre of the bottom of the bearing. This arrangement forms a small centrifugal pump, which draws the oil in at the bottom of the bearing through the oil-passage, throws it to the outer part of the bore, and forces it to the groove around the top of the step, so that it will again be ready to start on the same circuit. Complete and free circulation may be thus secured.

For heavier duty, either on account of increased speed or pressure, this form of bearing may be made more durable by placing a number of disk-shaped washers between the end of the shaft and the box, as shown in Fig. 37. One set of these washers is generally made

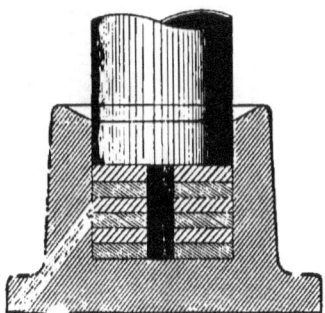

Fig. 37.

of some hard material, such as steel, and the other set of a softer material, as brass or bronze. The two kinds are then placed alternately, so that each washer rubs against a material different from itself. If the shaft is of mild steel and the box of cast iron, which is a common construction, the top washer is often fastened to the shaft, and the bottom one to the box, thus making all of the wear come upon the washers. The number of pairs of bearing surfaces over which the wear is distributed is one more than the number of free washers. The series of washers permits a slower speed of rubbing between each pair of the surfaces, and, in case abrasion should begin between any pair, the rubbing motion will cease there until the oil has an opportunity to get between them, or until repair can be made, without serious injury to the machine or the necessity of stopping it. The washers generally have a hole bored through the centre, and are grooved radially for oil. The same device for securing circulation of the oil as is shown in Fig. 36 can be used, of course. Hardened and ground tool-steel, or case-hardened and ground mild-steel, washers running on brass or bronze give most excellent service.

In machinery where the shaft and box cannot be accurately aligned, or where they may get out of line from some cause, such as settling of the supporting parts or springing of the shaft, lenticular washers with spherical faces of the form shown in Fig. 38 may be used. By making them smaller in diameter than the

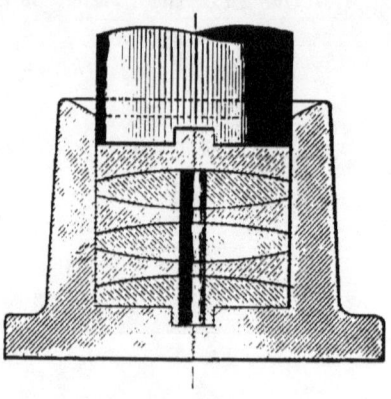

FIG. 38.

bore of the box, they will adjust themselves to a perfect bearing for any relative position of the shaft and box, within the limits for which they are designed. As with the flat washers, however, the wear will be more rapid at the parts more remote from the centre.

The step-bearings for turbine water-wheels running on vertical shafts often have a lignum-vitæ step B, Fig. 39, which supports the metallic shaft. The step is made crowning on top, and the end of the shaft cupped to fit it, the rubbing surfaces being spherical in form. Water-lubrication is easily obtained by cutting radial grooves in the rubbing surfaces, for the bearing is surrounded by water, which will flow into the grooves. If the speed is high, the water can be made to circulate freely through the grooves by boring a hole through the centre of the wooden step from the bottom

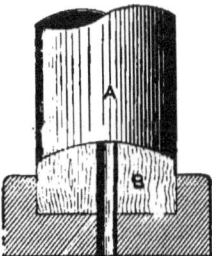

Fig. 39.

upward to the rubbing surfaces, as shown. The centrifugal action will throw the water out through the grooves to the circumference of the bearing, and at the same time draw it in through the hole in the centre of the step.

Exceedingly heavy service, both as to speed and pressure, is sometimes required of a step-bearing. It is not advisable, or even practicable in many cases, to meet this requirement by increasing the diameter of the bearing; for not only does the wear increase with the diameter, as has already been pointed out, but there is also a corresponding increase of frictional resistance with its accompanying increase of power loss in the bearing. Moreover, the liability to abrasion and cutting is also increased. Some other means must therefore be adopted for securing the desired qualities.

Forced lubrication affords what seems to be the most satisfactory solution for securing the successful operation of step-bearings for heavy duty. This can be applied most easily and economically in the water-lubricated bearing with the wooden step, Fig. 39, when used in connection with a water-wheel. In such a case it is only necessary to connect a pipe to the bottom of the hole in the bearing, and lead it to the water in the fore-bay. Practically the whole head of water is thus made available for forcing the lubricant, which is the water itself, in between the rubbing surfaces. This assumes that the bearing is below the wheel, so that the lubricating water can flow freely from the bearing to the tail-race. The oil-grooves, if used, should extend from the centre only part way to the circumference. If extended clear across the rubbing surfaces, the lubricant would be forced through them without performing its function.

To secure a higher pressure, for forcing the lubricant into the bearing, than the head of water will give, a force-pump may be attached to the pipe connected with the bearing.

Any of the step-bearings shown in the preceding figures can be lubricated with oil in a manner similar to that just described for using water, it being necessary to provide a reservoir for catching the oil which overflows from the top of the bearing, and a pump for taking up this oil and forcing it into the bearing again.

Pivot-bearings.

19. Conical pivot-bearings of the form shown in Fig. 40 are extensively used in light machinery. The ease with which they can be adjusted for wear is the chief factor in bringing them into use for light work. The wedge-like action of the point, and the unequal wear on the rubbing surfaces, prevent their use for heavy machinery to any considerable extent.

The pressure acting over the conical bearing surfaces may be assumed as acting at two diametrically opposite points for the purpose of finding the amount of the total normal pressure between them. Thus, in the figure, for the thrust P the normal pressure $N = (P \div 2) \csc (\theta \div 2)$, and the total normal pressure is

$$2N = P \csc \frac{\theta}{2}.$$

The intensity of pressure per unit area is the same as if the load were supported on a flat-ended shaft of the same area as the projection, on a plane normal to its axis, of the bearing surface of the pivot. In other words, the angularity of the pivot-point does not affect the pressure per unit area on the bearing surfaces. This is shown by dividing the total normal pressure $2N$ by the area of one of the rubbing surfaces. The area A of the conical rubbing surface of the pivot, Fig. 40, is

BEARINGS AND LUBRICATION.

$$A = 2\pi R \times \tfrac{1}{2}(VB) - 2\pi r \times \tfrac{1}{2}(VC)$$

$$= \pi R \times R \csc \frac{\theta}{2} - \pi r \times r \csc \frac{\theta}{2}$$

$$= \pi(R^2 - r^2) \csc \frac{\theta}{2}$$

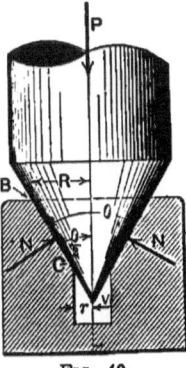

FIG. 40.

Hence for the pressure p per unit area

$$p = \frac{2N}{A} = \frac{P \csc \dfrac{\theta}{2}}{\pi(R^2 - r^2) \csc \dfrac{\theta}{2}} = \frac{P}{\pi(R^2 - r^2)}.$$

The angle θ does not enter the last member of this equation; hence p is not affected by the angle of the cone.

The wear will be more rapid on the parts having the greatest radial distances from the axis, and consequently the greatest rubbing action. The result is that after use the pressure between the rubbing surfaces will be greater near the point of the cone than near the base. For this reason a portion of the box is generally cut away at the centre, as indicated in the figure.

It is common practice, in light machinery, to use a conical bearing at each end of a shaft or spindle. It will then withstand side pressure as well as thrust.

58   FORM, STRENGTH, AND PROPORTIONS OF PARTS.

20. The "tractrix" or "curve of constant tangent" is the only theoretically correct outline for a step- or pivot-bearing, since it is the only one that will wear uniformly over the rubbing surfaces, and thus maintain a uniform pressure per unit area between the surfaces. It is also called **"Schiele's anti-friction curve,"** after its discoverer. The nature of the curve can best be described by the method of drawing it. This can be done on a piece of smooth horizontal paper in the following manner: In Fig. 41 the line AB is taken for the directrix of the curve. A beam-

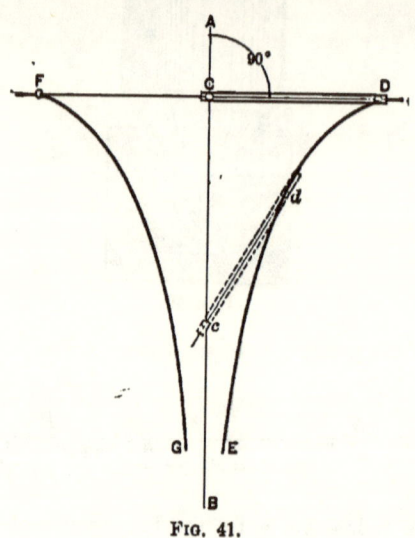

FIG. 41.

compass is placed so that one point, C, is on the directrix, and the other, D, lies on a line drawn normal to the directrix at C. C is then moved along AB toward B so that D trails freely after it, care being taken to hold the beam-compass by the point C so that D is not thrown out of the path it will naturally follow when there is nothing to press it aside. If D is a smooth, round pencil-point, it will trace the tractrix DE. A wedge-pointed pencil may also be used at D by placing it so that its edge takes the direction CD. When using this kind of a pencil-point, it is best to draw another curve, FG, on the opposite side of the directrix, starting with C at the same point as before. They can then be tested for accuracy by folding the paper along AB, so that they will coincide if correctly drawn.

From the method of drawing the tractrix, it is evident that, when the compass is in any position cd, the line joining the points c and d must be tangent to the curve at d; the distance cd, being that between the points of the beam-compass, remains constant; therefore, on any line drawn tangent to the tractrix, the distance from the point of tangency to its intersection with the directrix is a constant. Hence the name "curve of constant tangent."

The tractrix can be continued to an indefinite length, but in practice only a portion of it is used. This is generally taken from D toward E as far as desired. Fig. 42 represents the end of a shaft

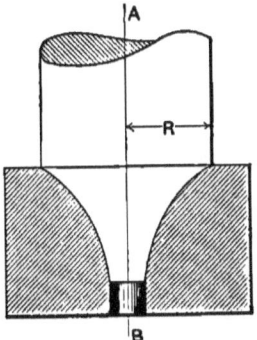

FIG. 42.

turned down to this form of profile, and resting upon the supporting part of the bearing. The directrix coincides with the centre-line of the shaft, and the largest radius is equal to the distance between the compass-points that were used for describing the curve.

The proof that the tractrix will wear away uniformly over the entire rubbing surface is as follows: In Fig. 43 a narrow band or ring of the bearing shown in Fig. 42 is represented. The width of the band, measured along the rubbing surface, is s; this width is assumed to be so small that the curved surface may be treated as if it were conical in form, but it is necessarily magnified in the draw-

60 FORM, STRENGTH, AND PROPORTIONS OF PARTS.

ing so that the parts may be seen. The mean radius of the ring is r, and its horizontal thickness is t. The line R has the direction of the curve at its mean radius, and the length of R, between the point of tangency and the directrix AB, is the same as the greatest radius of the bearing, as shown in Fig. 42.

It has been shown in the preceding section that the pressure on a conical pivot-bearing is not affected in its intensity per unit area by the angle of the cone; therefore, in a correctly fitted tractrix-

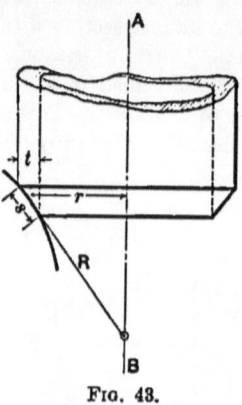

Fig. 43.

bearing, the pressure per unit area is the same over the entire rubbing surfaces. The wear upon each unit area of one surface may therefore be taken as proportional to the distance through which it rubs against the other while making one revolution. For convenience it may be assumed that

$v =$ volume of material worn from a unit area of one rubbing surface when the surfaces rub over each other through a unit distance under any given pressure.

The area of the rubbing surface of the ring is $2\pi rs$; the distance through which a point on the ring rubs during one revolution is $2\pi r$; therefore the volume of the material worn from the ring during one revolution of the shaft equals

$$V = 2\pi rs \times 2\pi r \times v.$$

BEARINGS AND LUBRICATION.

In order to see more clearly how the shaft will settle vertically on its support when this amount of material is removed from the small band which has been selected, it may be assumed that this band is the end of a thin tube, of a thickness t, cut entirely free from the rest of the bearing. The sectional area of this tube, on a plane perpendicular to its axis AB, is $2\pi rt$. In the figure it can be seen that, in the similar right triangles, one having the hypothenuse s and side t, and the other the hypothenuse R and side r, $t:s = r:R$, or $t = \dfrac{sr}{R}$; therefore the area of the cross-section of the tube equals

$$A = 2\pi rt = 2\pi r \frac{sr}{R}.$$

The amount K by which the tube is shortened during a revolution is the quotient found by dividing the quantity of material removed by the area of the cross-section of the tube, which gives

$$K = \frac{V}{A} = \frac{2\pi rs \times 2\pi r \times v}{2\pi r \dfrac{sr}{R}} = 2\pi v R.$$

This equation shows that the shortening of the tube is independent of its radius r, and is represented by the continued product of the constant 2π, and the quantities v and R, which are also constants for a given bearing, load, and solid lubricant. The shortening of this tube is therefore the same as that of any other that may be cut from the bearing, which shows that the complete bearing wears away so that a uniform pressure is maintained between the rubbing surfaces, and the bearing will retain its original form.

The equation of the tractrix, which may be used for obtaining points through which to draw the curve, can be developed as follows: In Fig. 44, AB is the directrix of the curve, p any point on the tractrix, and $R = CD$ the tangent at p; the coordinates of p are x and y. Taking the differential portion ds of the curve at p, and the corresponding dy and dx, it can be seen that, in the similar right triangles, one having the hypothenuse ds and sides

dx and dy, the other the hypothenuse $R = CD$ and sides x and $\sqrt{R^2 - x^2}$,

$$\frac{dy}{dx} = \frac{\sqrt{R^2 - x^2}}{x}.$$

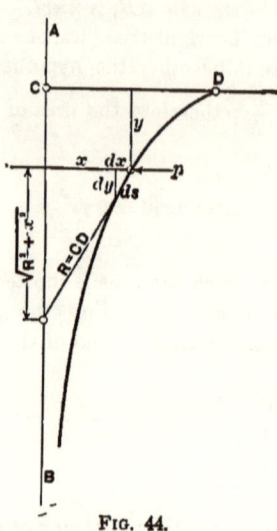

FIG. 44.

Integrating this expression, and taking into account only the positive signs, gives

$$y = R \log_\epsilon \frac{R + \sqrt{R^2 - x^2}}{x} - \sqrt{R^2 - x^2}$$

The natural system of logarithms must be used in this equation. This is indicated by the subscript ϵ. ($\epsilon = 2.7182818$.) The natural logarithm of a number may be found either in a table of natural logarithms, or by multiplying the common logarithm of the number (i.e., the logarithm to the base 10, which is written $\log_{10}$) by $\log_\epsilon 10 = 2.3025851$.

In the equation it is convenient to put $R = 1$ when solving for corresponding values of x and y. Any system of units or scale of drawing can be adopted for plotting the curve so as to give it the required size. By putting $R = 1$ the equation becomes

$$y = \text{natural log } \frac{1 + \sqrt{1 - x^2}}{x} - \sqrt{1 - x^2},$$

or

$$y = 2.3025851 \text{ common log } \frac{1 + \sqrt{1 - x^2}}{x} - \sqrt{1 - x^2}.$$

For a point having $x = .6$, the solution for y is, with natural logarithms,

$$y = \text{nat. log } \frac{1 + \sqrt{1 - (.6)^2}}{.6} - \sqrt{1 - .36}$$
$$= \text{nat. log } \frac{1 + \sqrt{.64}}{.6} - .8$$
$$= \text{nat. log } 3 - .8$$
$$= 1.0986 - .8 = .2986.$$

The solution by common logarithms differs from this only in taking the logarithm of 3. Thus, by common logarithms,

$$y = 2.3025851 \text{ com. log } 3 - .8$$
$$= 2.3025851 \times .4771213 - .8$$
$$= 1.0986 - .8 = .2986.$$

Other values of the coordinates, calculated as above, are given in Table IV.

TABLE IV.

COORDINATES OF TRACTRIX.

x	y	x	y
1.0	0	.4	.650
.9	.031	.3	.920
.8	.093	.2	1.313
.7	.182	.1	1.997
.6	.298	0	∞
.5	.451		

The tractrix-bearing can be lubricated, when running at high speed, in the same manner as the step-bearing, Fig. 36. It is possible that, when oil-lubrication is used, the wear is not uniform on account of the variation of the coefficient of friction with speed of rubbing at the different radial distances.

This form of bearing is but little used, the only apparent reason being that it is difficult to make.

Collar-bearings.

21. Collars extending out from the surface of a shaft are generally provided for taking the end thrust of a shaft that cannot be supported by a step-bearing. Fig. 45 shows a shaft with such collars resting in a box of suitable form to receive it.

If, instead of a number of collars, a single one were used to resist the thrust, the area of its side pressing against the box would have to be equal to or greater than the combined areas on one side

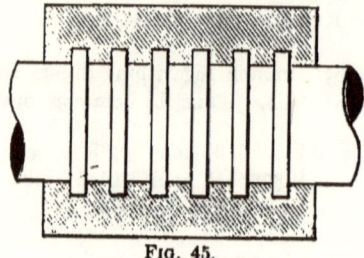

Fig. 45.

of all the collars shown, in order to keep the pressure between the rubbing surfaces the same per unit area. This would necessitate making the collar of a comparatively large diameter. This increase of diameter would bring into action the uneven wearing so marked in flat step-bearings, and also increase the frictional resistance to turning, in about the same proportion that the mean diameter of the rubbing surface of the single collar is greater than that of each collar when several are used. This is one reason, probably the most important one in most cases, for using the multiple-collar thrust-bearing instead of a single collar; another reason is that a greater economy of space can be secured with the smaller diameter of the multiple-collar bearing.

When the thrust is always in the same direction, as indicated by the arrow in Fig. 46, the bearing may be shortened, while still

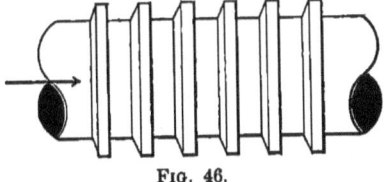

FIG. 46.

retaining equal strength and wearing surface, by making the collars of the form shown in this figure, where the flat faces are of the same area as in the preceding figure, but are brought nearer together.

In order to obtain the best service from a collar-bearing working under heavy pressure and at high rubbing speed, each collar should have its own individual bearing-ring, separately adjustable. By this means, which is the common practice for large machinery, each ring can be adjusted for equal bearing-pressure, even though the wear on the different ones may be unequal; and in case heating should occur at any collar, its ring can be adjusted to partly relieve the pressure between them until they run cool again, or it may even be removed for repair while the machinery is operating, the load being carried by the remaining collars in the meantime. Elaborate collar-bearings are used on the propeller-shafts of screw-propelled vessels.

When used for heavy service, collar-bearings are lubricated by an oil-bath, or by pipes leading to holes in the bearing-rings which open on the rubbing surface of the ring; four pipes and openings are often used on a single ring, placed at equal distances apart around the ring. Suitable oil-grooves are cut in the rubbing surface, starting from the oil-opening and running zigzig a short distance from it in a direction corresponding to that in which the collar rubs over it.

Table V represents the practice in thrust-bearings on merchant and naval vessels built in the yards of the Newport News Shipbuilding & Dry Dock Co. The horseshoe bearing has each part, which comes into contact with the collars of the shaft, made in the shape of the letter U, or a horseshoe. This form allows the bearing parts to be removed and replaced readily.

TABLE V.*

COLLAR THRUST-BEARINGS.

Steel collars running against white metal of bearing-rings; oil-bath lubrication.

	Gunboats.		Battle-ships.	Merchant Ships.				
Horse-power of one engine............	875	800	5000	600	250	3500	3800	400
Revolutions per minute	300	283	120	115	110	115	80	110
Diam. of shaft, inches	$5\frac{7}{8}$	$5\frac{7}{8}$	14	8	6	$12\frac{3}{4}$	$15\frac{1}{4}$	7
Inner diam. of rubbing surface of collar, inches........	6	6	$14\frac{1}{4}$	$8\frac{1}{4}$	$6\frac{1}{4}$	$13\frac{1}{4}$	16	$7\frac{1}{4}$
Outer diam. of rubbing surface of collar, inches........	$8\frac{7}{8}$	$8\frac{7}{8}$	$21\frac{1}{4}$	$12\frac{1}{4}$	$9\frac{1}{4}$	19	24	11
Number of collars...	9	9	11	6	6	15	11	6
Estimated total thrust of one engine exerted on one collar-bearing, pounds ...	13750	13100	79800	12300	6390	55900	71200	9600
Method of cooling...	Water circulation	Water circulation	Water circulation			Water circulation	Water circulation	
Type of bearing.....	Collar	Collar	Horse-shoe	Collar	Collar	Collar	Collar	Collar

* Data kindly furnished by Mr. C. B. Orcutt, President of the Newport News Shipbuilding & Dry Dock Co., builders of the battleships Kearsarge, Kentucky, Illinois, etc.

The practice of the Marine Iron Works of Chicago, for their smaller work, is shown in Table VI and Figs. 47, 48, and 49. Fig. 47 is a steel shaft with seven collars integral with the shaft. The box has a Babbitt lining cast around the collars so as to fit them. An oil-trough at the top of the bearing, from which oil-holes lead to the top of each collar, furnishes a means of lubricating. The data relative to this bearing are given in Table VI. This bearing is designed for a 50-H.P. engine at 300 revolutions.

Table VI.*

COLLAR THRUST-BEARINGS.

Type of Bearing	Fig. 47.	Fig. 48.	Fig. 49.
Horse-power of engine	50	25	75
Revolutions per minute	300	400	275
Diam. of shaft, inches	2⅜	2¼	4
Inner diam. of rubbing surface of collar, inches	2⅜	2¼	4
Outer diam. of rubbing surface of collar, inches	4¾	4⅝	9
Number of collars	7	1	1
Estimated total thrust on collar-bearing, pounds	1630	1040	2160
Material of collar	Steel	Cast iron	Steel
Material of bearing-rings	Babbitt metal	Bronze washer	Bronze washer
Method of lubricating	Oil-trough above journal†	Grease-cups. Solid oil in cups placed on oil-holes	Grease-cups ‡

* Data kindly furnished by the Marine Iron Works, Chicago, Ill., except the estimated total thrust on collar-bearing, which was calculated by the writer.
† Oil-holes lead from trough to edges of collars.
‡ Chamber in bottom part of box can be filled to make an oil-bath if desired.

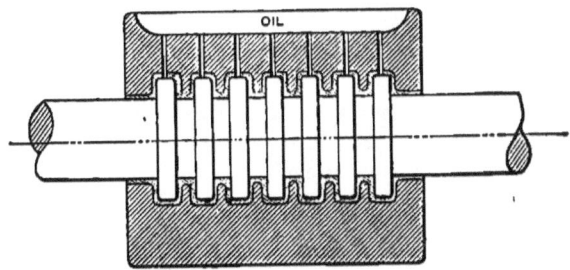

Fig. 47.

68　FORM, STRENGTH, AND PROPORTIONS OF PARTS.

Fig. 48 is a thrust-bearing for a 25-H.P. engine making 400 revolutions per minute. It consists of two cast-iron collars fastened to the shaft on each side of a cast-iron box-bearing. Loose collars of bronze are interposed between each end of the box and the collar adjacent to it. It is lubricated by means of grease-cups attached

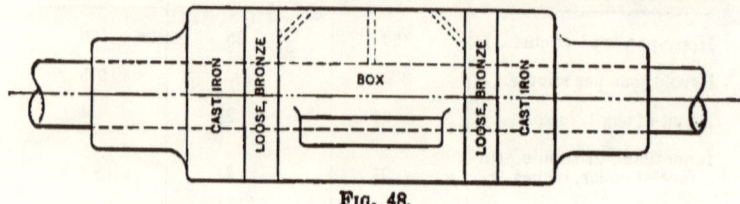

Fig. 48.

to oil-holes leading to the journals and the faces of the collars. The remaining data are given in Table VI.

Fig. 49 is for a 75-H.P. engine making 275 revolutions per minute. It has a steel collar, fastened to the shaft, which bears

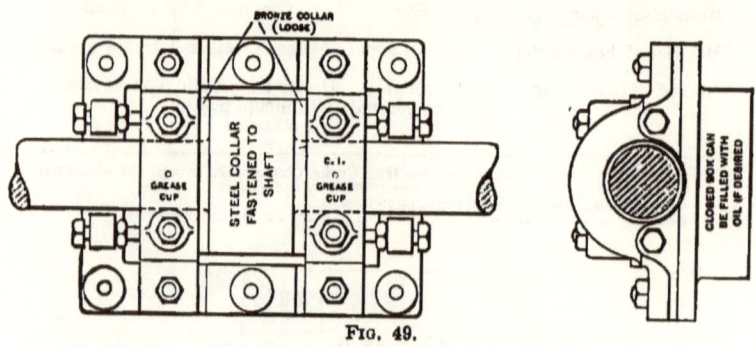

Fig. 49.

against loose bronze collars, which in turn bear against cast-iron boxes. Lubrication is secured by grease-cups, and, if desired, by filling the reservoir at the bottom of the box with oil, thus obtaining bath lubrication. Other data are given in Table VI.

Roller-bearings.

22. Cylindrical roller-bearings for working conditions similar to those which an ordinary journal-bearing is designed to meet are used to some extent. The shaft or journal is surrounded by several cylindrical rollers, which roll inside of an accurately bored casing. The diameter of the casing is slightly greater than that of the journal plus twice that of the rollers, so that the latter will roll freely when not on the side of the journal where the load forces it against them.

The journal, rollers, and casings must all be accurately cylindrical, and have their axes parallel, in order to work correctly. It is not possible to keep the rollers parallel to the journal without some additional device for that purpose. What appears the simplest, and a very effective, method is to use a slotted bushing or "cage" somewhat longer than the rollers, and bored to have a free-running fit on the shaft; the slots are cut longitudinally, and each is wide enough to allow a roller to drop into and turn freely in it. The cage rotates less rapidly than the journal. From ten to twenty rollers are commonly used.

The rubbing of the rollers against the cage causes some frictional resistance. Numerous devices have been applied to prevent such resistance, one being to use no cage, as just described, but to groove or score the ends of the rollers, and place balls between them, the balls themselves running in a retaining-ring at each end of the bearing.

If, from wear or any other cause, a roller gets out of line with the journal and casing, it will, when under pressure, have a tendency to move endwise in the cage, assuming that a cage is used, on account of its inclination to the axis of the journal and the natural attempt to move at right angles to its own axis. Such motion is prevented by the end of the roller striking the end of the slot in the cage; this produces undesirable wear and frictional loss at the end of the roller. A bearing working under heavy pressure may be rapidly destroyed when once it has worn enough to let the rollers get out of alignment. The destruction is generally most rapid where the end of the roller presses against a rigid cage. It would seem that this end wear might be largely reduced, or even wholly prevented in many cases, by having a light spring at each end of the roller, which would allow it to move endwise freely for a short distance when on the working side of the bearing, and then bring it back to mid-position endwise while on the loose side. With balls running in the grooved ends of the rollers, the end thrust does not cause destructive wear, since the thrust is taken by the ball pressing against the side of the groove.

There is another trouble that is liable to occur if a roller gets out of alignment, especially if it is long and the material is brittle. As soon as the roller gets out of position the line of its contact with the journal is curved instead of straight. The roller must therefore bend and become liable to fracture by the bending. A broken roller is almost certain to cause rapid destruction of the bearing. Several short rollers, lying end to end in the same slot of the cage, are sometimes used for long bearings.

Practice shows that cast-iron casings are not suitable for roller-bearings, on account of their rapid destruction after they have worn appreciably; steel bushings are found to be satisfactory.

A special roller, Fig. 50, has been designed to obviate the

FIG. 50.

danger arising from the fracture of a roller, either from getting out of line or unevenness of the surfaces of the bodies in contact. It is made by winding a steel ribbon about a mandrel, in the same manner that a strip of paper may be wound about a round pencil, so that the edges of successive convolutions just clear each other.*
For service where shocks and jars are frequent and heavy, as in street-railway service, the casing is lined with a bushing made in the same manner. The cage for a hollow-roll bearing can be made of wires or rods passing through the rolls and attached to rings around the journal at each end of the rolls.

The coefficient of friction of a properly adjusted roller-bearing is exceedingly low, especially at slow speeds. It decreases as the diameter of the rollers increases, and is about proportional to the load. It very commonly happens, however, that this coefficient keeps increasing with the age of service of the bearing, reaching many times its original value before the bearing is worn out. The

---

* Made by the Hyatt Roller Bearing Co.

chief cause of this increase is the uneven wearing of the working parts, possibly the most serious phase of this particular trouble being where the journal or casing wears to a slightly conical form on account of lack of uniformity in the quality of the material.

The load which a roller-bearing will carry is greater for high speeds than can be put on the common journal-bearing with sliding surfaces, practically the same for medium speeds, and less for very low speeds.

Table VII gives the results of tests made by Wm. Sellers & Co. on several Hyatt roller-bearings.

TABLE VII.

TESTS OF HYATT ROLLER-BEARINGS.*

Coefficient of friction μ given in the body of table.

Dimensions of Bearings.	Total Load, pounds	Revolutions per Minute.				
		5	25	48	128	214
Diam. of journal 1⅝". Length of bearing 3". Bearing bored .009" smaller than sum of two liners, two rollers, and shaft.	1000 2000 3000	μ	μ	μ .16964 .08974 .06739	μ	μ ....
Diam. of journal 1⅝". Length of bearing 3". Bearing bored .004" larger than sum of two liners, two rollers, and shaft.	1000 2000 3000	.02958 .01874 .01249	.01578 .00986 .00789	.00789 .01080 .01020	.00785 .00789 .00789	.00789 .00592 .00526
Diam. of journal 8". Length of journal 12". Bore of bearing .023" larger than sum of two liners, two rollers, and shaft.	10000 20000 30000 40000 50000	.02399 .01923 .01856 .01805 .01708	....	.03826 .02386 .01950 .01795 .01692	.02540 .01560 .01280 .01240	....
Hyatt commercial; 2" shafting-box	500 1000 1500 2000	.03156 .02762 .01841 .01282	.03156 .01973 .01710 .01775	....	.02367 .01578 .01578 .01578	.01972 .01381 .01315 .01282

* *Age of Steel*, April 10, 1897, p. 17; *American Machinist*, June 24, 1897, p. 20.

Two street-railway cars, one fitted with Hyatt roller-bearings and the other with common journal-bearings, were tested at Providence, R. I. The roller-bearings showed a saving of 13% of the total power used to drive the car with common bearings.*

23. Conical roller-bearings are used to resist the end thrust of a shaft or other part, performing the same function as the more common form of step-bearing already described. Fig. 51 shows the

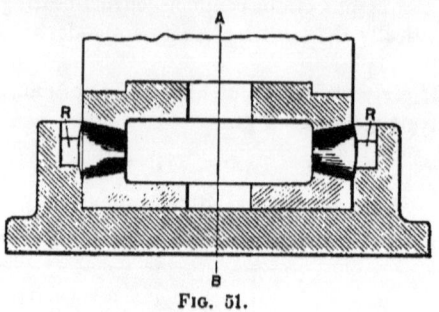

Fig. 51.

typical form of such a bearing. The rollers are truncated cones of such a taper and so placed that their vertices, as well as those of the conical surfaces on which they roll, all coincide on the axis AB of the shaft. It is not necessary that the rollers shall have their axes at right angles to that of the shaft, but they may be inclined at any angle within practical limits. It is often convenient to make either the step or the end of the shaft flat on the surface where the cones roll, and the other bearing surface coned to suit the rollers.

On account of the taper form of the rollers, there is a tendency to force them out radially from between the shaft and step. The retaining-ring R, which encircles them, is to prevent such displacement. It should just touch the bases of the cones when they are in position, and have an easy running fit in the step. The ring will then travel about the centre of the bearing at about the same speed as the rollers, which is half that of the rotating shaft when the step and end of the shaft are coned to the same angle.

If no other device than the retaining-ring is used to hold the rollers in place, there is a tendency for them to twist around so that their axes will not intersect that of the shaft, even when the rollers are of such a diameter that the proper number will just lie tangent

---

* *American Machinist*, Oct. 21, 1897, p. 794.

to each other when in place; and when wear has occurred they will twist slightly, the twist generally being in the direction that will cause a heavier pressure against the retaining-ring. In small bearings it is better to use fewer rollers and hold them in position by a cage, which may be a thin round plate, perforated with as many holes as there are rollers, the holes being of such a form and so placed that the rollers will just fit into them and be held in the proper position when the cage is in the bearing.

For large bearings or heavy pressures, and especially when the rollers are very short in comparison with their diameter, a "spider," consisting of a solid body pivoted at the centre of the bearing, and having arms extending out through holes bored in the rollers concentric with their axes, is used. The flanged ends of the arms prevent the rollers from moving radially outward. The arms should be rigid enough, or be connected at their outer ends in such a manner as, to hold the rollers accurately in place.

The rollers should be small in comparison with the diameter of the bearing in order to keep their tendency to move out radially as small as possible. It would hardly be advisable to make the apex angle, embraced between two diametrically opposite elements of the conical surface, greater than 15° in any case; 10° or less is a more suitable angle. It should not be forgotten, however, that the rolling resistance of the cones increases as the diameter decreases.

A case is cited where a step-bearing with conical rollers working under a pressure of 104.5 pounds per square inch had a coefficient of friction $\mu = .0025$ when running on steel.*

Special Forms of Bearings.

24. The bearing shown in Fig. 52 is often used on light machinery, such as circular saws, wood-planing machines, etc. It is something of a combination between a journal- and thrust-bearing, and serves in a measure the purpose of both.

The collars are integral parts of the shaft. The box is divided lengthwise, and lined with a soft bearing metal, which is cast in while the journal is in place.

A bearing of a similar nature is made by cutting grooves into

---

* *Cassier's Magazine*, May, 1897, p. 66. Several forms of roller-bearings are illustrated and described in this article.

the shaft to serve the same purpose as the collars. It answers very well when only light service is required, but the shaft is weakened so as to be unfit for heavy service.

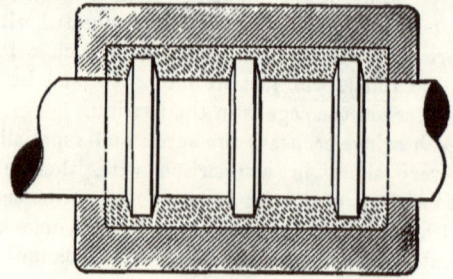

FIG. 52.

Fig. 53 is a taper bearing provided with lock-nuts a and b for adjusting it and taking end thrust. The box surrounding it is

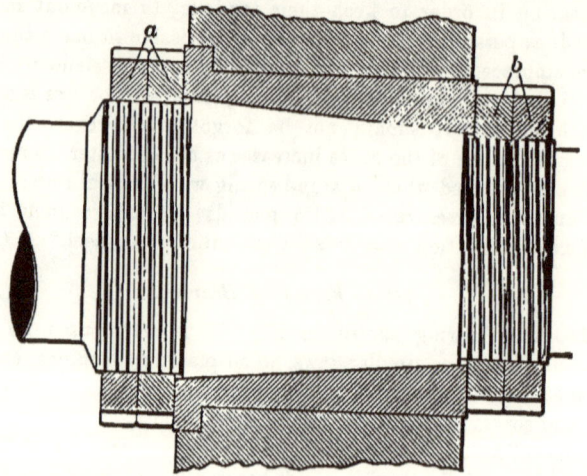

FIG. 53.

solid. The taper is commonly about 1½ inches in diameter per foot of length. This form of bearing is chiefly used on light machinery running under light journal-pressure.

CHAPTER II.

SPUR- AND FRICTION-GEARS.

SPUR-GEARS.

25. Strength of spur-gear teeth.—When a rotating spur-gear transmits power to its mate, there must be pressure between the pairs of teeth that are in contact. If the teeth were correctly formed and accurately spaced, the pressure necessary to turn the driven gear would be nearly uniformly distributed between the pairs of teeth in contact. The elasticity of the teeth might theoretically affect the uniformity of the distribution, but not to an extent to be worthy of consideration in practice.

On account of the inaccuracy of spacing that always exists to some extent, and the loss of the correct form of the teeth by wear, it is customary to assume, for purposes of designing, that the entire force that is required to rotate the driven gear is applied by a single

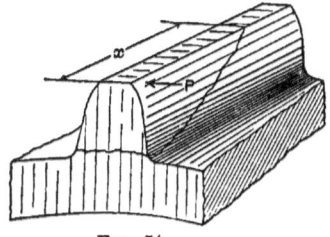

FIG. 54.

tooth of the driver to its mate on the driven. If there is a probability that the teeth may be thrown out of alignment by the springing of the shafts or movement of the supports, or that foreign substances may come between the teeth, the pressure may be localized at the top of a tooth at one end, as at P in Fig. 54, and break it off as shown by the irregular line of fracture. To allow for such a condition, it is customary to assume that, however wide the face of the gear, the strength of the tooth is no greater than if the gear-face had only a width x measured across the face of the gear. The value of x is different for teeth of the same pitch and height, but of different profile, and cannot be accurately determined in any case. For the more common forms of gears having teeth whose height is roughly .7 of the circular pitch, it is generally safe to take x as 1.5 to 2 times the circular pitch, according to whether the teeth are narrow or broad at the base. The wider the face of the gear, the less rapidly will it wear; and if accurately made and rigidly supported, the strength will be about proportional to its width for widths as great as six or more times the circular pitch.

When a pair of teeth first come together, there is a shock caused by the blow as they strike each other. The intensity of the shock increases with the speed at which the teeth travel; hence gears running at high speed will not stand as much pressure between their teeth as those running slower. There may also be shocks caused by loads suddenly applied to the driven gear, as in rock- and ore-crushers, and rolling-mill machinery. The gear should be proportioned to resist such shocks.

The pressure between a pair of teeth would be normal to their surfaces along the line of contact, if there were no frictional resistance acting to prevent the combined motion of rolling and slipping of the one over the other. This frictional resistance always does exist, however, to an extent depending on the material of the teeth, the finish of their working surfaces, and whether they are lubricated or dry, clean or covered with dirt and grit.

Fig. 55 shows the profiles of a pair of teeth A and B just as they come into contact when A is driving B, the rotation about the centres of the gears being in the direction indicated by the arrows. The point of contact is at C, and the common normal to the tooth curves passes through the pitch-point P. The pressure between the teeth would be in the direction CP if there were no friction between them. The friction causes the line of pressure to be inclined to CP by some angle PCH, whose value depends on the coefficient of friction of the tooth surfaces. The lever-arm, about the centre of the driven gear, of the force acting along CH, is shorter than that of a force acting along CP; hence the force that acts along

CH to turn the driven gear against a given resistance must be proportionally greater than that which would act along *CP*.

E and *D* are the positions of the same pair of teeth just as they are on the point of separating. As before, the common normal to the curves at the point of contact *F* passes through *P*, and the frictional resistance causes the pressure between them to make some

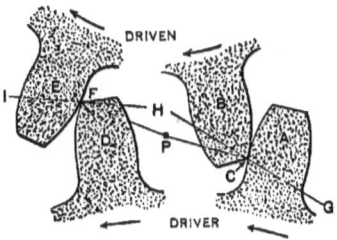

FIG. 55.

angle *PFH* with this normal. In gears that are not lubricated, or are lubricated and covered with dirt and grit, it is possible that the coefficient of friction may be large enough to increase the angle *PFH* to such an extent that the direction of pressure *FH* will become tangent to the gear-face at the top of the tooth; or, what is practically the same, normal to the radial line which passes through the centres of the tooth and gear. This radial line will be hereafter referred to as the median line of the tooth.

A tooth of the driven gear is subjected to the greatest stress when in the position *B*, just as it comes in contact with its mate, for the pressure is then applied at the greatest distance from its base; and a tooth of the driving gear is working under the greatest stress when in the position *D*, just as the pair are separating.

For the tooth *B*, on the driven gear, it can be seen that the pressure against it, acting along a line inclined to the median line of the tooth, produces both radial compression and bending stresses in the material. This is true whether friction is considered or not. Friction increases the proportion of compressive to bending stress.

The tooth *D*, on the driver, also has both compressive and bending stresses caused by the pressure against it when friction is not considered. Friction reduces the ratio of compressive to bending stress, and if the friction becomes great enough to bring the line of pressure normal to the median line of the tooth, the compressive stress is eliminated, and the action is a purely bending one.

If such a material as cast iron, which is much weaker in tension than compression, is used for gears having equal diameters and the same form of teeth, D will be more apt to break than B; if the material is one that is equally strong in both tension and compression, B would probably fail first.

The force P that will break a gear-tooth when exerted normal to its median plane at the top of the tooth, and uniformly distributed across the face of the gear, as in Fig. 56, may be found by the

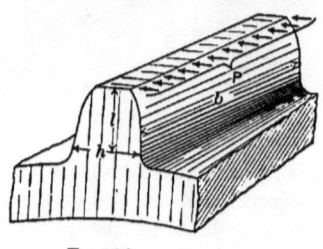

FIG. 56.

formula for a cantilever. The quantities entering into this formula when applied to a gear-tooth are:

$P =$ force applied tangent to top of tooth;
$S =$ maximum fibre-stress per unit area in the material;
$v =$ shearing stress per unit area in the material;
$b =$ breadth of gear-face;
$h =$ thickness of tooth at breaking section;
$l =$ distance from top of tooth to breaking section.

The formula is

$$P = \frac{Sbh^2}{6l}.$$

On account of the filleting which is commonly used at the bottom of the tooth, the distance l of the breaking section from the top of the tooth is less than the total height of the tooth. The location of the breaking section can be found for any tooth, whether the profiles on both sides of the median line are similar or not, by taking sections normal to the median plane and at different distances from the top of the tooth, introducing the l and h of each in the formula, until the one that gives a minimum value of P is found. When the profile of the tooth is symmetrical about the median line, the value of P can be obtained more directly by drawing a parabola whose vertex is at the centre of the top of the tooth, and whose sides are tangent to the tooth curves. The value of P for any section parallel to the direction of P is the same as for all other sections, this being the property of a cantilever of parabolic profile and uniform breadth.

In any system of interchangeable gears, such as the cycloidal with constant size of generating circle for a given pitch, or the involute with constant angle of obliquity, the thickness of the teeth at the base is less for gears of small diameter than those of large; consequently the teeth of the smaller gears are weaker.

The curve, Fig. 57, was obtained by calculating, according to the formula for the cantilever just given, the strength of cycloidal teeth generated by a describing circle having a diameter half that of a 15-tooth gear of the pitch adopted, and a fillet at the bottom of the tooth of a radius equal to one sixth of the width of the space at the addendum circle.* This is practically the form of tooth adopted by the Brown and Sharpe Mfg. Co. for their interchangeable gears. The length of the tooth from top to bottom is 0.6866 times the circular pitch, and the thickness on the pitch-circle equals half the pitch. It can be seen by the diagram that the teeth on a 120-tooth gear are twice as strong as those on one having 12 teeth. By multiplying or dividing the reading on the scale of "pounds pressure at tooth-point" by a suitable constant, the curve can be made to apply to any pitch, width of gear-face, and fibre-stress.

With regard to the effect of the position of the line of contact between a pair of engaging gears, it is shown by construction that, in a pair of perfectly made and aligned gears having 13 teeth each, at the instant contact changes from one to two pairs of teeth, or

---

* "Diagrams for the Relative Strength of Gear Teeth," by Forrest R. Jones. Trans. Amer. Soc. Mech. Engrs., vol. XVIII., p. 766.

vice versa, which is the time when the greatest stress is brought upon the material of the tooth, the pressure is applied at such a distance outside of the pitch-line of the tooth subject to the greater bending moment, as to allow an increase of pressure of more than 70% of that which could be applied at the top of the tooth. With

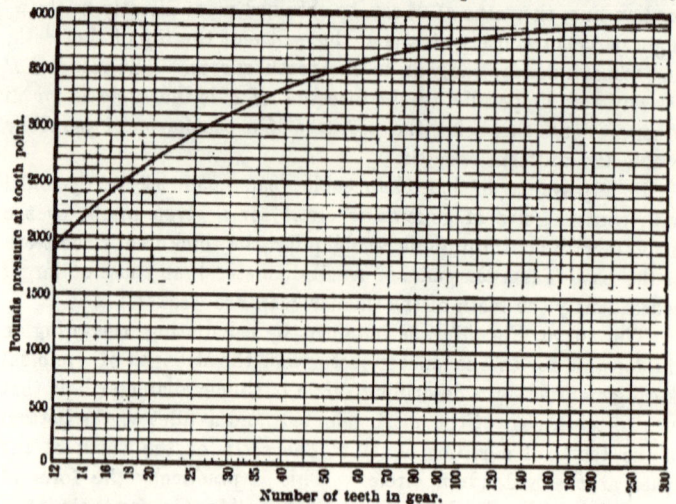

Diagram showing the pressure at the point of a single tooth in a gear of ½ diametral (=6.28 inches circular) pitch and 1 inch face, which will produce a fibre stress of 6000 pounds per square inch in the material of the gear.

FIG. 57.

a greater number of teeth in either or both gears, a greater increase is allowable.

In calculating the strength of a tooth by the cantilever formula, no account is taken of the shearing stress on the section under consideration. The value of this shearing stress per unit area of the section normal to the median plane is

$$v = P \div \text{area of section.}$$

Combining this with the fibre-stress due to bending gives for the maximum tension or compression, both being the same in value,

$$\text{Maximum stress} = \frac{S}{2} + \sqrt{v^2 + \frac{S^2}{4}}.$$

SPUR- AND FRICTION-GEARS.   81

The value of the "maximum stress" for a given pressure P is but 4% greater than that of S obtained by the formula for the cantilever. This 4% excess is for a rack-tooth, which is the strongest of all, internal gears excepted; it grows smaller as the number of teeth decreases. On account of its comparatively small value, when compared with the uncertain elements which enter into any attempt to calculate the strength of gear-teeth, it may be safely neglected.

The maximum shearing stress, according to the formula for combined shear and flexure,

$$\text{Maximum shear} = \sqrt{v^2 + \frac{S^2}{4}},$$

is about 54% of the maximum tension.

The curves of Figs. 58 and 59 were laid out from Fig. 57 by making use of the fact that, when the number of teeth and width of a gear remain constant, the strength of the teeth varies as the circular pitch; and by taking account of the fact that, when the number of teeth remains constant, and the width of the gear-face bears a constant ratio to the circular pitch, the strength of the teeth varies as the square of the pitch, the curves of Figs. 60 and 61 were plotted.

In the diagrams, Figs. 58, 59, 60, and 61, each of the diagonal lines, some of which are straight and some broken, represents a different fibre-stress in the material of the tooth. The break in some of the lines has no significance, being made only as a means of shortening the diagrams to a convenient length. It should be noted, however, that a change of pounds per inch is thus made necessary on the pressure-scale, the change occurring at the pressure-line where the angle is made in the diagonals. Figs. 58 and 59 are for gears having a face 1 inch wide. They are of exactly the same nature, one for small and the other for large pitches, and might have been combined in a single diagram. Such a combination would make it difficult to read values for the smaller pitches unless the whole diagram were excessively large. The same is true of Figs. 60 and 61.

The diagrams, Figs. 58, 59, 60, and 61, are for use in finding any one of the four factors, circular pitch, pitch diameter, pressure

82   FORM, STRENGTH, AND PROPORTIONS OF PARTS.

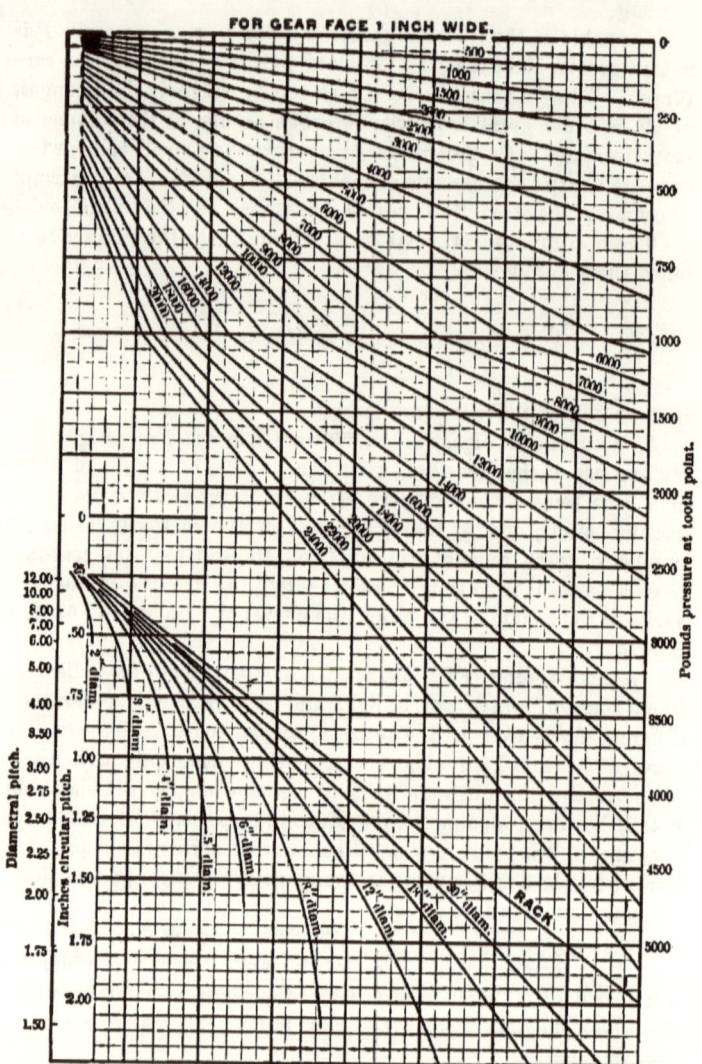

Diagram showing relation between pressure at tooth point, stress in outer fibre, and pitch of a gear tooth.   Figures in diagram above "RACK" indicate fibre stress—those below, diameter of gear.

FIG. 58.

SPUR- AND FRICTION-GEARS. 83

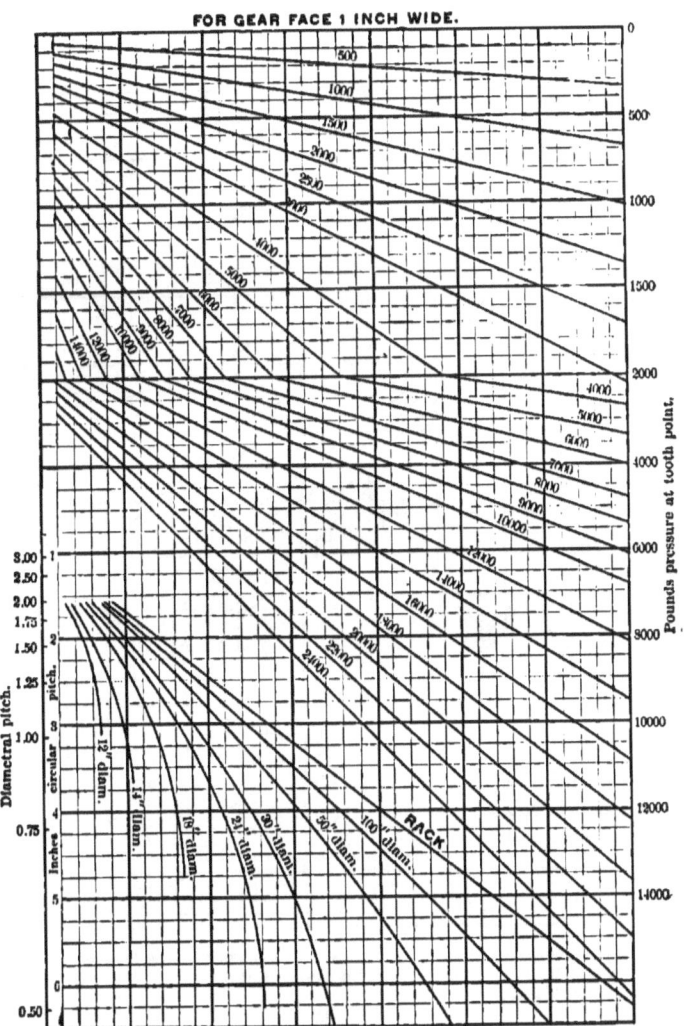

Diagram showing relation between pressure at tooth point, stress in outer fibre, and pitch of a gear tooth. Figures in diagram above "RACK" indicate fibre stress those below, diameter of gear.

FIG. 50.

84 FORM, STRENGTH, AND PROPORTIONS OF PARTS.

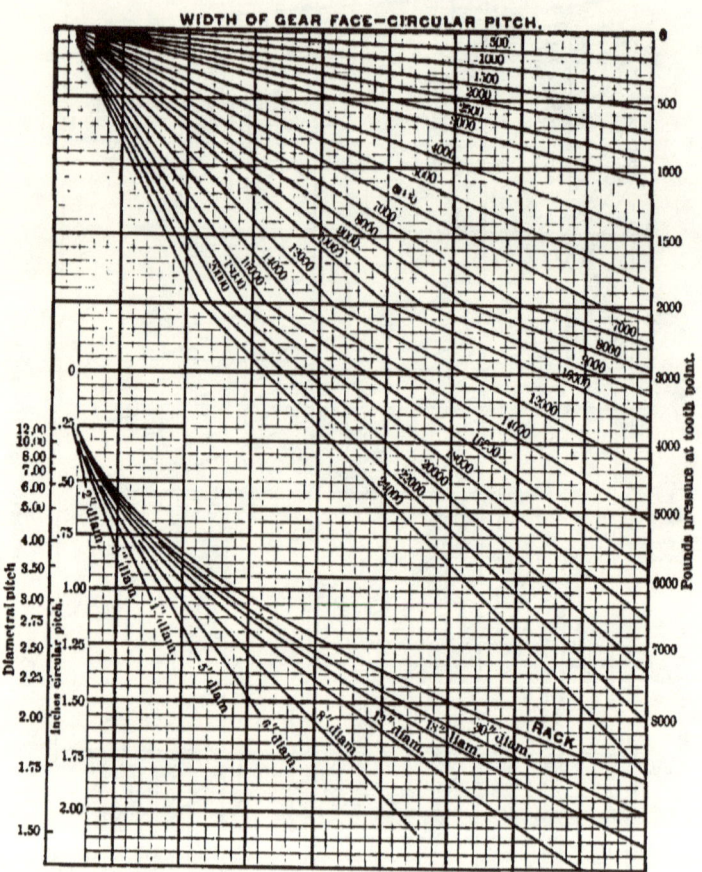

Diagram showing relation between pressure at tooth point, stress in outer fibre, and pitch of a gear tooth. Figures in diagram above "RACK" indicate fibre stress—those below, diameter of gear.

FIG. 60.

SPUR- AND FRICTION-GEARS. 85

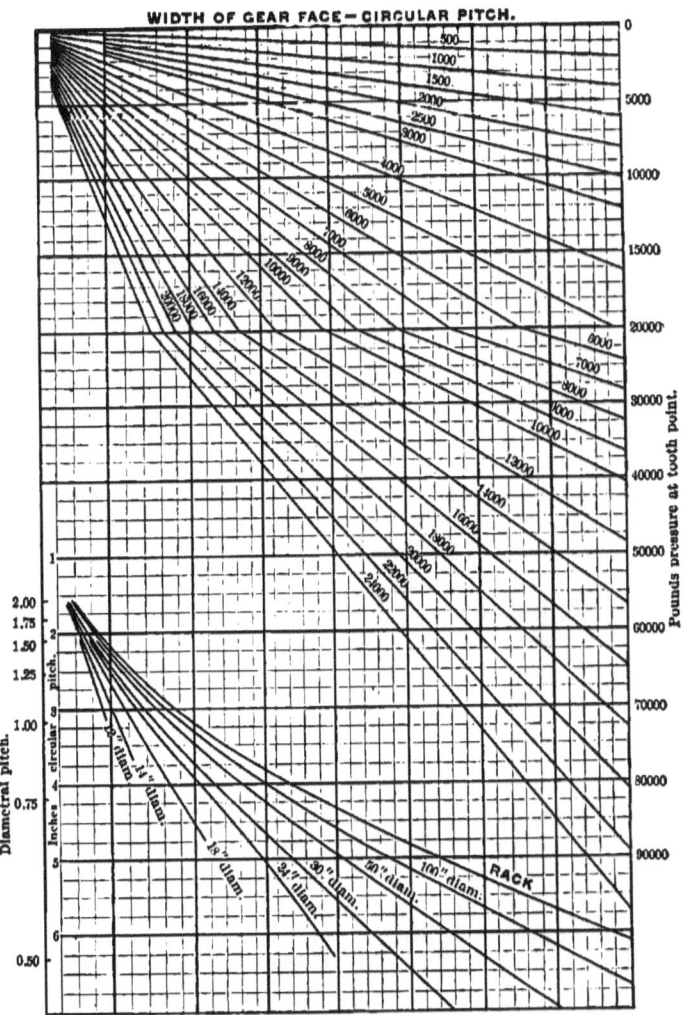

Diagram showing relation between pressure at tooth point, stress in outer fibre, and pitch of a gear tooth. Figures in diagram above "RACK" indicate fibre stress—those below, diameter of gear.

FIG. 61.

at tooth-point, or fibre-stress, when the other three are known or assumed.

Examples.—Let it be required to find the pitch of a gear 18 inches in diameter, 4 inches face, working at 4000 pounds maximum stress in the material, to withstand 2000 pounds pressure at its point. This gives a pressure of $2000 \div 4 = 500$ pounds per inch of width of face. In Fig. 58 on the right-hand scale, marked "pounds pressure at tooth-point," find the 500-pound line and follow it to its intersection with the 4000-pound fibre-stress line; from this point drop down to the curve marked "18 inches diameter," and thence to the scale on the left of the diagram, thus obtaining a reading of 2.25 diametral pitch, or about 1.4 inches circular pitch.

Again, suppose a pressure of 4550 pounds is to act on a gear of 8 inches pitch diameter and 7 inches face, the limit of fibre-stress being 6000 pounds per square inch. The pressure per inch of width in this case is $4550 \div 7 = 650$ pounds. By the same method as before, following the 650-pounds pressure-line to its intersection with the 6000-pounds fibre-stress line, and thence toward the bottom of the diagram, it is found that the vertical line does not cut the 8-inches diameter-line, but falls to the right of it, the latter terminating at 1.5 diametral pitch, this being the greatest pitch that can be used when the number of teeth is not less than twelve, which is the lower limit in the diagrams. The fact that the vertical and diameter lines do not intersect shows that no gear having twelve teeth or more can be designed to fulfil the conditions given. With a fibre-stress slightly greater than 6000 pounds per square inch, however, a gear of 1.5 diametral pitch will answer.

The use of Figs. 60 and 61 for a width of face equal to the circular pitch is the same as the preceding examples, except that the pressure for a width of face equal to the circular pitch is read on the scale of pressures at tooth-point, instead of the pressure per inch of width, as was done before. Thus, for a gear 30 inches in diameter, whose face width is to be three times the circular pitch, to work at a fibre-stress of 14000 pounds per square inch, under a pressure of 90000 pounds at the tooth-point, we have $90000 \div 3 = 30000$ pounds pressure on a width of face equal to the circular pitch. Following the 30000-pounds line from the right-hand side of Fig. 61 to its intersection with the 14000-pounds diagonal line, and thence to the 30 inches diameter curve, gives 5.75 inches circular pitch, which is something less than .5 diametral pitch.

The following formula has been developed from Fig. 57 by Prof. John H. Barr.*

The notation, slightly changed from his, is:

C = circular pitch, inches;
D = pitch diameter of gear, inches;
F = width of gear-face, inches;
N = number of teeth in the gear;
P = pressure on teeth, pounds;
S = tensile or compressive fibre-stress in the material of the gear, pounds per square inch.

$$P = CFS\left(0.106 - \frac{0.678}{N}\right). \qquad \qquad (8)$$

From this the following equations may be deduced:

For a gear-face 1 inch wide,

$$P = CS\left(0.106 - \frac{0.678}{N}\right); \qquad (9)$$

$$C = \frac{PN}{S(0.106N - 0.678)}. \qquad (10)$$

For use when N is known.

$$P = CS\left(0.106 - 0.215\frac{C}{D}\right); \qquad (11)$$

$$C = D\left[0.246 - \sqrt{.0605 - 4.65\frac{P}{SD}}\right]. \qquad (12)$$

For use when D is known.

For a width of gear-face = circular pitch,

$$P = C^2S\left(0.106 - \frac{0.678}{N}\right); \qquad (13)$$

$$C = \sqrt{\frac{PN}{S(0.106N - 0.678)}}. \qquad (14)$$

For use when N is known.

$$P = C^2S\left(0.106 - 0.215\frac{C}{D}\right) \qquad (15)$$

For use when D is known.

---

* Trans. Amer. Soc. of Mech. Engrs., vol. XVIII., 1897, p. 776.

When the width of the gear-face is equal to the circular pitch, or bears a given ratio to it, the equation involving the diameter of the gear contains both the cube and square of C; hence no equation having a general solution for obtaining C when D, P, and S are known can be obtained from the original form given above. The last equation given can be used tentatively, of course, by substituting assumed values of C in it and solving, continuing until a satisfactory value is found.

In designing gears it is customary to assume that the pressure against the teeth is the same in amount as would have to be applied tangent to the pitch-circle of the gear in order to drive it. The difference between the pitch radius and the radial distance to the point where the pressure is applied to the tooth surface is not great enough, in any ordinary system of gears, to need consideration.

When, for a gear, the pitch radius R in inches, and the greatest torsional moment M in inch-pounds to be exerted upon it, are known, the force P that must act tangent to the pitch-circle can be found by the equation

$$P = \frac{M}{R} \text{ pounds.} \quad \ldots \ldots \quad (16)$$

If the highest rate of working, expressed in horse-power (H.P.), or in inch-pounds of energy E transmitted per minute, together with the number of revolutions per minute V, and pitch radius R inches, are known, P can be found by the equation

$$P = \frac{33000 \times 12 \text{ H.P.}}{2\pi R V} \text{ pounds,} \quad \ldots \ldots \quad (17)$$

or

$$P = \frac{E}{2\pi R V} \text{ pounds.} \quad \ldots \ldots \quad (18)$$

Table VIII, representing experiments made by the Brown & Sharpe Mfg. Co. upon cut gears of their own manufacture, shows the observed breaking pressure of the teeth and the calculated fibre-stress which would exist if the pressure were all applied at the top of one tooth, normal to its median plane, and uniformly distributed across the face of the gear.

TABLE VIII.
BREAKING LOADS FOR CUT CAST-IRON GEARS, EXPERIMENTALLY DETERMINED.

Diametral Pitch.	Circular Pitch, Inches.	Width of Face, inches.	Number of Teeth.	Pitch Diam., inches.	Revolutions per Min.	Velocity at Pitch-circle, feet per min.	Observed Breaking Pressure at Pitch-circle, pounds.	Stress in Outer Fibre, pounds per sq. inch.
10	.3142	1 1/8	110	11	27	78	1060	33000
8	.3927	1 1/4	72	9	40	94	1460	29000
6	.5236	1 5/8	72	12	27	85	2220	24000
5	.6283	1 1/2	90	8	18	85	2470	20500

The following formulas, 19 to 32, based upon the assumption that the pressure is always normal to the tooth surface, have been developed from investigations made by Mr. Wilfred Lewis.* In them the force is assumed as being effective at the point where the normal to the top of the tooth curve intersects the median line of the tooth. The pressure between the teeth is resolved into two components at this point—one radial, and the other perpendicular to the median line of the tooth. The radial component is neglected, only the one normal to the radius and exerting a purely bending action on the tooth, being considered.

Formulas 19 to 32 are applicable to cycloidal gears developed by a generating circle whose diameter is half that of a 12-tooth gear of the same pitch as that under consideration, and to involute gears in which the normal to the tooth curve at the pitch point makes an angle of 75° with the line drawn from the pitch point to the centre of the gear.

The proportions of the teeth to which the following formulas apply, taking the circular pitch C as the unit of measurement, are: thickness on pitch-line $= .47C$; addendum $= .3C$; clearance $= .05(C+1)$.

Two methods of filleting the bottom of the teeth are represented. In one the fillets are as large as will just clear the tops of the teeth

---

*Proceedings Engineers' Club of Philadelphia, vol. x., 1893, p. 16 ; *American Machinist*, May 4, 1893, p. 3, and June 8, 1893, p. 7; Trans. Amer. Soc. Mech. Engrs., vol. xviii., 1897, pp. 776, 781.

of an intermeshing rack, and in the other the radius of the fillets is equal to the clearance between the top of a tooth and the bottom of a space.

For convenience of application, the formulas are given for both a gear-face one inch wide, and for a width of face equal to the circular pitch. It is assumed that the pressure is uniformly distributed across the face of the gear. The notation is the same as in the preceding equations.

The formulas deduced from Mr. Lewis's investigation of strength of gear-teeth are:

FOR A GEAR-FACE ONE INCH WIDE:

Large Fillets.

$$P = CS\left[0.124 - \frac{0.684}{N}\right]; \quad \text{For use when } N \text{ is known.} \quad (19)$$

$$C = \frac{PN}{S(0.124N - 0.684)}. \quad (20)$$

$$P = CS\left[0.124 - 0.218\frac{C}{D}\right]; \quad \text{For use when } D \text{ is known.} \quad (21)$$

$$C = D\left[0.284 + \sqrt{.0807 - 4.6\frac{P}{SD}}\right]. \quad (22)$$

Small Fillets.

$$P = CS\left[0.124 - \frac{0.888}{N}\right]; \quad \text{For use when } N \text{ is known.} \quad (23)$$

$$C = \frac{PN}{S(0.124N - 0.888)}. \quad (24)$$

$$P = CS\left[0.124 - 0.282\frac{C}{D}\right]; \quad \text{For use when } D \text{ is known.} \quad (25)$$

$$C = D\left[0.22 - \sqrt{0.049 - 3.57\frac{P}{SD}}\right]. \quad (26)$$

FOR A WIDTH OF GEAR-FACE = CIRCULAR PITCH:

Large Fillets.

$$P = C^2 S \left[0.124 - \frac{0.684}{N} \right]; \quad \quad \quad \quad \quad \quad \quad \quad (27)$$

$$C = \sqrt{\frac{PN}{S(0.124N - 0.684)}}. \quad \text{For use when } N \text{ is known.} \quad \quad (28)$$

$$P = C^2 S \left[0.124 - 0.218 \frac{C}{D} \right]. \quad \text{For use when } D \text{ is known.} \quad \quad (29)$$

Small Fillets.

$$P = C^2 S \left[0.124 - \frac{0.888}{N} \right]; \quad \quad \quad \quad \quad \quad \quad \quad (30)$$

$$C = \sqrt{\frac{PN}{S(0.124N - 0.888)}}. \quad \text{For use when } N \text{ is known.} \quad \quad (31)$$

$$P = C^2 S \left[0.124 - 0.282 \frac{C}{D} \right]. \quad \text{For use when } D \text{ is known.} \quad \quad (32)$$

Cases that require a large amount of power to be transmitted, when the diameter and speed of a gear are so limited by the conditions to be fulfilled as to require an excessive and objectionable breadth of face, can sometimes be met by using two or more gears of the required diameter, on the same shaft, with faces of such a breadth that the sum of all their breadths is equal to that which would be required for a single gear rigidly supported and perfectly aligned. When several gears are used on the same shaft in this manner, they should not be placed so that their teeth are in line, but "stepped" by placing the teeth of each successive gear in advance of (or behind) those of its neighbor by a distance equal to the circular pitch divided by the number of gears on the shaft; such an arrangement gives smoother running and less liability to breakage.

26. Methods of strengthening gears.—For cases where very strong teeth are required, and where the pressure against the teeth is usually or always in the same direction, the form of tooth shown in Fig. 62 can be used, in which A is the working side. When

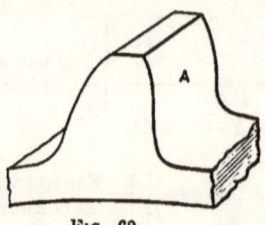

FIG. 62.

used for such a purpose as hoisting, where the machinery is reversed and driven backward to lower the hook or other device for suspending the load, the side opposite A, as well as A, should be made of some correct tooth outline; but when A is the only side that works, the opposite side may be made of any outline that will give strength and, at the same time, clear the tops of the teeth of the intermeshing gear.

For cycloidal gears this **buttressed tooth** can be obtained by using a large describing circle for the face and flank of A, and a small one for the corresponding parts of the opposite side, when it is desired to make the latter of a form suitable for backward driving. Involute gears of this form require a large base-circle for the working side A, and a small base-circle for the opposite side.

The strength of such a tooth, as compared with that of the ordinary form having the same pitch, height, and breadth, can be determined quite approximately by developing a tooth of each form, and comparing their thicknesses at or near the bottom; when a large fillet is used at the bottom of the tooth, the weakest plane lies a short distance above the bottom. The strength of the teeth is approximately proportional to the squares of their thickness at the sections lying the same distance from the top. Having once obtained the ratio of strengths, it can be used for all teeth of the same form, as long as the pitch and breadth of gear face of those compared are equal to each other.

"**Shrouding**" is another method of adding strength to teeth. It consists of adding an annular ring or disk to one or both ends of a gear. This shroud may extend either partly or entirely to the tops of the teeth. It forms an integral part of the gear-casting or forging. Since the shroud forms a rigid support for the ends of the teeth, shearing of the metal between the ends of the tooth and the shroud must occur at this point in order to allow the tooth to break near the bottom as a cantilever. A full shroud at each end of a gear evidently increases its strength more than when part of the tooth stands above the shroud. The thickness of a shroud, measured parallel to the axis of the gear, is generally at least as great as that of a tooth at the pitch-line.

The strengthening effect of a shroud depends upon the breadth of the gear, a narrow one being more strengthened than a wide one of the same pitch and diameter; for, while in both cases the shearing resistance added by the shrouds is the same, it forms a greater ratio to the cantilever resistance as the gear grows narrower.

When a pinion engages with a large gear, the former wears more rapidly on account of having its teeth come into mesh more frequently; and, if they are of the same material, the pinion consequently grows weaker more rapidly than the spur-gear, in addition to having been the weaker of the two at first, on account of the form of its teeth; hence the pinion is the one to be shrouded if this device is used at all. Two meshing spur-gears of the same size and material should both have equal shrouds extending about half-way to the top of the teeth, provided it is necessary to strengthen them.

27. Short gear-teeth are much stronger than those having the proportions commonly used, in which the height of the tooth is roughly 0.7 of the circular pitch. The shorter teeth also run more quietly. Modern practice is adopting them to a considerable extent when the gears are not intended to be interchangeable. A tooth height of about 0.4 of the circular pitch is commonly used. Such teeth are especially satisfactory for cast gears used without machine-finishing on the working surface of the teeth.

Mr. C. W. Hunt gives the following proportions and, in Table IX, working loads for the involute gears adopted by the company bearing his name.* They are used for coal-hoisting engines and similar machinery, which generally do not have solid foundations. The teeth are cast to form and used without machine-finishing.

---

* Trans. Amer. Soc. of Mech. Engrs., vol. XVIII., 1897, p. 787.

94   FORM, STRENGTH, AND PROPORTIONS OF PARTS.

The notation is changed to agree with that of the formulas given above.

TABLE IX.
WORKING LOADS FOR SHORT-TOOTH UNCUT CAST-IRON GEARS.

C Circular Pitch, inches.	P'' Working Load, pounds.	P''' Maximum Load, pounds.	C Circular Pitch, inches.	P' Working Load, pounds.	P'' Maximum Load, pounds.
1	1320	1650	2¼	6700	8300
1¼	2300	2600	2½	8300	10500
1½	3000	3700	2¾	10000	12500
1¾	4100	5000	3	12000	14800
2	5300	6600			

The addendum and dedendum are each equal to 0.2 of the circular pitch C; the clearance equals $.05(C+1)$ inches; the width of the gear-face is $2C+1$ inches.

Non-metallic Spur-gears.

28. Mortise-gearing is made by keying or pinning wooden teeth into cast-iron rims designed to receive them. Fig. 63 shows end and

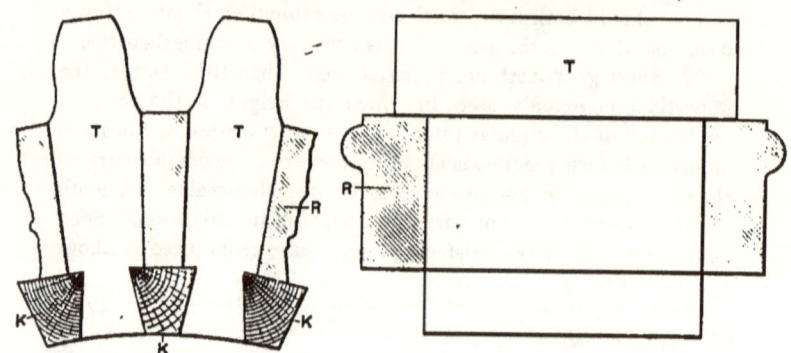

FIG. 63.

side views of a mortise-gear, the rim R being in section, so as to show full end and side views of the wooden teeth T and the wooden keys K which hold the teeth in the tapered openings through the rims. Fig. 64 is another form, the tooth-shank here being smaller than the bottom of the tooth, thus making a shoulder which rests against the rim and holds the tooth from dropping down into it when the wood shrinks or is compressed by the pressure due to transmitting power. A pin P is used to hold the tooth in place, instead

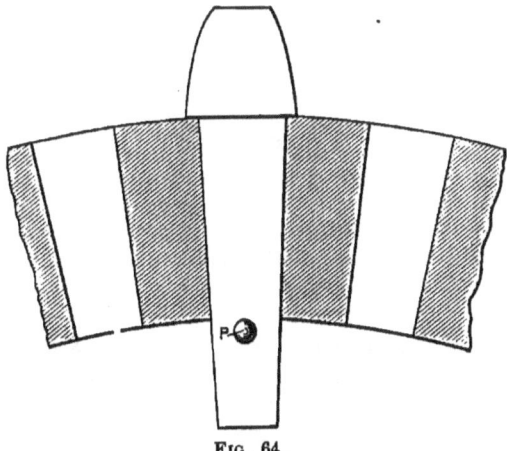

FIG. 64.

of the key of the preceding form; the key could be used in both forms equally well, however.

The working side of a tooth of the form of Fig. 64 is generally made with a deeper shoulder than on the opposite side, in order to allow for the wear due to service, the tooth retaining nearly its original strength till worn past the shoulder. Bevel- and spur-gears can both be made with wooden teeth when it is thought that the existing conditions to be met can be more satisfactorily fulfilled by them than with metal gears.

The more common practice is to mate a cast-iron and mortise gear, the smaller being of iron when they are of different diameters, the iron teeth having a thickness on the pitch-line of about 0.4 of the circular pitch, that of the wooden teeth being the remaining 0.6 of the pitch, due allowance being made for backlash and rough work by reducing these values slightly. The smaller gear is made of iron, because the work performed by any one tooth upon it is greater than for a tooth of the larger one; consequently greater durability is obtained by thus making the smaller gear of a material better able to withstand wear; also, the gears are stronger, since the teeth of the smaller, whose form makes them weaker than those of the same thickness on the pitch-line would be on the larger, are made of the stronger material. Sometimes, but very infrequently, both gears are made with wooden teeth.

Comparatively noiseless running at high speeds is one of the good qualities of mortise-gearing, this being especially marked when comparison is made with a pair of cast gears; the elastic quality of wood makes them able to resist sudden shocks that might break cast-iron gears of the same size and running at the same speed; they are very durable when run under moderate pressure with proper care in lubricating the teeth.

In practice, the wooden tooth-blanks are first keyed in the rim and then machined to proper form in the same manner as metal gears, the only difference in the two processes being that different cutting edges are required for shaping wood and metal, as, for instance, a saw, running at high speed, is used in a bevel-gear planing-machine when cutting wooden teeth, instead of the sharp-cornered planer-tool required for metal.

The woods more commonly used for teeth are maple, hickory, and locust. In order to prevent swelling and shrinking as the atmosphere changes its humidity, the blanks are thoroughly saturated with paraffin or some other oily substance before putting in place.

29. Rawhide, indurated vegetable fibre, etc., are frequently used for small gears where quiet running is desired. The gears are usually made of a number of thin disks placed side by side and held together by a pair of metallic disks at the ends. Most of them are durable under light service, but are not strong. Under heavy service they may wear rapidly, if they do not break. Rawhide and some of the other materials of a similar nature are liable to shrink considerably when used in a dry place. This may be a serious objection on account of reducing the diameter of the gear and causing the disks to become loose.

30. Factor of safety for tooth-gears.—While, as in most cases in machine-designing, it is impossible to fix a factor of safety, since its value must depend upon the conditions of each individual case, an examination of a few of the points to be considered may be of aid in selecting its value.

Thus, in a hand-driven crane, where the speed is slow and any unusual strain is not liable to occur, the factor used needs only to include an allowance for flaws, heterogeneous material, and internal stress, it not being necessary to include shocks and stresses that cannot be estimated, since neither of these last two really exist when proper safety-locking devices are used. If the change is made to power-driving, however, and high speeds are used, then an allowance must also be made for the shock due to the striking of the teeth together. The greatest stress that can come on the crane at any time is calculable within practical limits, being limited to the breaking strength of the chain, plus frictional resistances; hence no allowance for unknown stress need be made. Repeated stress certainly does occur in the teeth of a gear in service, but the allowance for flaws, heterogeneous material, and internal stresses is generally so large that the material does not regularly work near enough to its elastic strength to make necessary any allowance for this cause of fracture.

Rolling-mill machinery, stone- and ore-crushers, and other machines applied to similar purposes, are subjected to shocks and unknown stresses. In such cases the necessary factor of safety can be determined only by experience and some knowledge of the nature of the material to be operated upon.

31. The efficiency of spur-gearing depends very largely upon the frictional loss in the supporting bearings. The greater the loss of power in the bearings the lower the efficiency. The pressure between the teeth causes an equal amount of pressure upon the bearings supporting each gear if the bearings are on each side of the gear. The weight of the gear, as well as other weights and forces, must be taken into account when calculating the total pressure on the bearings.

The experiments made by Wm. Sellers & Co. upon a spur-gear 18.62 inches pitch diameter, having 39 teeth of $1\frac{1}{2}$ inches circular pitch, running on journals $2\frac{1}{8}$ inches diameter, and driven by a spur-pinion 5.73 inches pitch diameter, with 12 teeth of the same pitch, and running on one journal $2\frac{7}{16}$ inches and the other $1\frac{1}{8}$ inches diameter, both placed close to the hub of the pinion, show the average efficiencies given in Table X for pressures of 430, 700, 1100, 1600, and 2500 pounds pressure between the teeth.*

TABLE X.

EFFICIENCIES OF SPUR-GEARS.

Revolutions of pinion per minute	3	5	10	20	50	100	200
Efficiency, per cent	90	92	94	95.6	97.3	98.2	98.6

If the same amount of power were transmitted with the larger gear driving the smaller, the efficiency would be lower. This may be more readily seen by supposing that the journals of both gears are of the same size. Without friction the pressures on the journals of the two gears would be equal. When the resisting moment is applied to the pinion-shaft, the increase of pressure between the teeth, due to friction, over that which would be required were there no frictional resistance, is greater than that when the larger gear is the driver. The ratio of the increase in the two cases is inversely as the radii of the gears. The power required to drive a gear is proportional to the pressure against its teeth if the coefficient of friction remains constant. The change of pressure on the bearings would probably cause a slight change in this coefficient, but even if this should occur, the amount of driving power to be applied to the large gear when driving the small one, in order to deliver a given amount of power to an operating-machine, would be greater than that necessary if the driving gear were smaller than the driven.

BEVEL-GEARS.

32. Strength of bevel-gear teeth.—In order to investigate the nature of the pressure between the teeth of an accurately constructed and adjusted pair of bevel-gears, and of the fibre-stress in the material, let it first be assumed that they are not rotating, and that there is no pressure between their teeth. Then assume that one is locked so as to prevent its turning, and that a turning force is applied to the other, thus producing pressure between the engaging teeth.

On account of the elasticity of the material, a slight deflection

---

* "Experiments on the Transmission of Power by Gearing," by Wilfred Lewis. Trans. Am. Soc. Mech. Engrs., vol. VII., 1886, p. 273.

of the teeth will be caused by the pressure on them. The correspondingly slight rotation thus allowed in the other parts of the gear to which the turning force is applied, will cause a point at the top of the large end of a free tooth to move through a linear distance which bears the same ratio to the corresponding linear motion of a point similarly situated at the small end of the same tooth, as the ratio of the radial distances of the two points from the axis of the gear.

If the line of contact between a pair of engaging teeth is at the top of the working surface of a tooth on the locked gear, the top of the tooth will be deflected more at the large end of the gear than at the small, the deflection of any point along the top of the tooth being proportional to its radial distance from the apis of the gear. The linear dimensions of the tooth profile at the large and small ends of the tooth are proportional to the radii of the addendum-circles at the ends of the gear. The profiles are, of course, similar. Hence, in accordance with the property of cantilevers of similar profile that (referring to Fig. 56), when the breadth b remains constant, as well as the ratio of the length l to the height h, the linear deflection at the end is proportional to the load P, it can be seen that the distribution of pressure along the line of contact is in proportion to the radial distances from the axis of the gear, and also in proportion to the linear dimensions of the tooth sections. And, again in accordance with the property of cantilevers of similar profile, that, when the breadth b remains unchanged, the end deflection that will produce the same maximum fibre-stress S in each is proportional to the linear dimensions, it is evident that the maximum fibre-stress is the same in the tooth from end to end of the gear.

If H is the large, and h the small, addendum radius of the gear, then the resultant pressure against a tooth acts at a radial distance equal to $2(H^3 - h^3) \div 3(H^2 - h^2)$ from the axis of the gear.

The mean value of the pressure per unit length of the line of tooth contact equals the total or resultant pressure divided by the width of gear-face. In a tooth under pressure, this mean value is exerted at the middle of the gear-face. Therefore, a spur-gear having the same face width, and teeth of the same form and size as those of the bevel-gear at the middle of its face, is of the same

100 FORM, STRENGTH, AND PROPORTIONS OF PARTS.

strength as the bevel-gear. If m is taken as the mean addendum diameter of the bevel-gear, and a is the angle between the axis of the bevel-gear and the elements of the addendum-cone, the radius of the spur-gear must be $m \sec \alpha$, in order to agree with the system of gear-teeth used for the bevel-gear.

As in the case of spur-gears, the pitch-surface dimensions may be used without serious error for practical forms of gears. The dimensions of the pitch-cone will therefore be used hereafter instead of those of the addendum-cone.

Fig. 65 is a section of a bevel-gear. The dimensions given are:

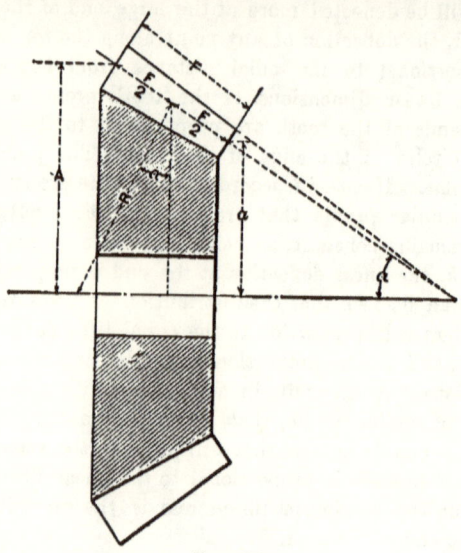

FIG. 65.

A = largest pitch radius of bevel-gear;
a = smallest pitch radius of bevel-gear;
$C = \dfrac{A + a}{2A} C'$ = circular pitch at middle of gear-face;
C' = circular pitch at large end of gear;
F = width of gear-face, measured on pitch-cone;
R = length of a normal to pitch-cone at middle of gear-face, measured between pitch surface and intersection with axis

of gear; this is the radius of the pitch-circle upon which the teeth are generated for the middle of the gear-face;
α = angle between gear-axis and pitch-cone elements.

The notation for the pressure is:

$P_{(\text{mean})}$ = mean value of pressure on bevel-gear tooth per inch of length of contact-line;
$P_{(\text{res})}$ = resultant of all elementary pressures on bevel-gear tooth;
P' = pressure which, if applied at the greatest pitch radius of the gear, would produce the same torsional moment on the gear-shaft as the actual acting pressure.

In accordance with the assumption just made, the resultant pressure $P_{(\text{res})}$ acts at a radial distance $2(A^3 - a^3) \div 3(A^2 - a^2)$ from the axis of the gear. This distance is inconvenient to use in designing, it being more satisfactory to use the pitch radius A of the large end of the gear. The relation between P' and $P_{(\text{res})}$ is expressed by the formula

$$P'A = P_{(\text{res})}\frac{2}{3}\frac{A^3 - a^3}{A^2 - a^2}.$$

Whence

$$P' = P_{(\text{res})}\frac{2}{3}\frac{A^3 - a^3}{A(A^2 - a^2)}.$$

The relation between $P_{(\text{mean})}$ and $P_{(\text{res})}$ is given by the formula

$$P_{(\text{res})} = FP_{(\text{mean})}.$$

Therefore

$$P' = \frac{2}{3}\frac{A^3 - a^3}{A(A^2 - a^2)}FP_{(\text{mean})}. \quad \ldots \quad (33)$$

The value of $P_{(\text{mean})}$ corresponds to those given on the scale of "pounds pressure at tooth-point" in Figs. 58 and 59. The value of $P_{(\text{mean})}$ therefore can be found on one of these diagrams by using: gear diameter $2R$; circular pitch C, taken at the middle of the bevel-gear face; and any given or assumed working fibre-stress S.

If the force P' is given, and the pitch C' required for a given value of S and a specified gear-blank, the value of $P_{(mean)}$ is first found by the equation

$$P_{(mean)} = \frac{3}{2} \frac{A(A^2 - a^2)}{A^3 - a^3} \frac{P'}{F}, \quad \ldots \quad (34)$$

and then, by the diagram, the value of C, from which C' can be determined.

Example.—Let $A = 10$ inches, $F = 4$ inches, $\alpha = 50°$, $C' = 1.0472$ inches (corresponding to 3 diametral pitch), and $S = 3000$ pounds per square inch. Then:

$$a = 10 - 4 \sin 50° = 10 - 4 \times .766 = 6.936 \text{ inches};$$

$$C = \frac{A + a}{2A} C' = (16.94 \times 1.047) \div 20 = 0.887 \text{ inch};$$

$$R = \frac{(A + a)}{2} \sec \alpha = \frac{16.94}{2} 1.556 = 13.18 \text{ inches};$$

$$2R = 26.36 \text{ inches}.$$

The value of $P_{(mean)}$ for 0.887 of an inch pitch, a pitch diameter of 26.36 inches, and 3000 pounds per square inch fibre-stress is found on the diagram Fig. 58 to be 260 pounds. Therefore, by equation (33),

$$P' = \frac{2}{3} \frac{(10)^3 - (6.94)^3}{10[(10)^2 - (6.94)^2]} 4 \times 260 = 900 \text{ pounds}.$$

The horse-power that would be transmitted by the gear at 100 revolutions per minute is

$$H.P. = \frac{900 \times 2\pi \times 10 \times 100}{33000 \times 12} = 9.52.$$

Example.—Take $P' = 1000$ pounds, $S = 2000$ pounds per square inch, and the dimensions of the gear-blank the same as those in the preceding example. The pitch C' is required.

SPUR- AND FRICTION-GEARS. 103

By equation (34)

$$P_{(mean)} = \frac{3}{2} \frac{10(10)^2 - (6.94)^2}{(10)^3 - (6.94)^3} \frac{1000}{4} = 292 \text{ pounds.}$$

In the diagram Fig. 58, $C = 1.65$ inches. Therefore

$$C' = \frac{2AC}{A+a} = \frac{2 \times 10 \times 1.65}{10 + 3.94} = 1.95 \text{ inches.} \quad \text{(End of example.)}$$

The value of $P_{(mean)}$ in equation (33) can also be determined by formula (11), P for spur-gears and $P_{(mean)}$ for bevel-gears being identical. Substituting, in equation (33), the value of $P_{(mean)}$ as given in equation (11) changes (33) to the form

$$P' = \frac{2}{3} \frac{A^3 - a^3}{A(A^2 - a^2)} CFS \left[0.106 - 0.215 \frac{C}{2R} \right]. \quad . \quad . \quad (35)$$

The value of $P_{(mean)}$ as given by formulas (21) and (25) can also be used for the system of gears that they represent.

The solution of the next to the last example by equation (35) gives

$$P' = \frac{2}{3} \frac{(10)^3 - (6.94)^3}{10(10)^2 - (6.94)^2} 0.887 \times 4 \times 3000 \left[0.106 - 0.215 \frac{0.887}{26.36} \right]$$
$$= 900 \text{ pounds.}$$

The side and end wear of the supporting bearings of a pair of bevel-gears, caused by the pressure between their teeth, both tend to localize the pressure between the teeth at the large ends. In allowing for such wear, when calculating the strength of the teeth, it is therefore correct to assume that the load is carried by a part of the larger end of the tooth. Probably a width of gear-face equal to the circular pitch at the large end is as great as can be safely taken when the apex angle a of the pitch-cone approaches near to 90°; when this angle is very small the gear becomes more like a spur-gear, and the width of gear-face may be increased to 1.5 times the circular pitch. Allowance for localization of pressure at the small end of the tooth may be made in the same manner as for the large. Such localization may be caused by foreign matter between the teeth, settling of supports, etc.

Mortise bevel-gears are very commonly used. Shrouding the teeth is also practised to a considerable extent. Teeth corresponding to the double helical teeth of spur-gears are less frequently adopted; these do not, of course, have an end sliding motion like that of screw-bevel and screw gears. It is believed that no machine has ever been put into practical use for cutting bevel-gears having teeth of any other form than those whose elements all intersect at the apex of the pitch-cone. The others are cast to form.

33. The efficiency of bevel-gearing is less than that of spur-gearing, for the reason that the pressure between the teeth causes an end thrust on the bearings, in addition to the side pressure corresponding to that of spur-gears. The frictional resistance due to this end thrust causes a reduction of efficiency.

FRICTION-GEARS.

34. Friction-gears, having smooth surfaces held in contact under pressure, and transmitting power by means of the frictional resistance between their surfaces, find a very considerable application in certain classes of machinery, notably that which is frequently stopped and started, and whose source of power is a constantly rotating shaft or pulley; they also afford a convenient method of obtaining different speeds when the primary shaft of a machine rotates uniformly. Less frequently they are used for transmitting power between two uniformly rotating shafts.

When a considerable amount of power is transmitted, and the pressure between the gears is heavy, one of a pair of gears which are in contact, is generally made of iron, and the surface of the other of some material such as wood, paper, leather, hard rubber, etc.

35. Cylindrical friction-gears.—The turning force P which can be exerted by one smooth cylindrical friction-gear upon another when they are pressed together with a force F, the coefficient of friction of the material being μ, is given by the formula

$$P = \mu F.$$

The coefficient of friction μ for paper friction-wheels, as determined by a series of experiments made by Prof. W. F. M. Goss,[*]

---

[*] Trans. Am. Soc. Mech. Engrs., vol. XVIII., 1897, p. 102.

is given below. In these experiments one of the pair of wheels in contact was iron, and the other of compressed straw-board in "thin disks cemented together under heavy pressure and strenghtened by iron side-plates, or fitted over iron centres." The edges of the disks were pressed against the iron wheel when power was transmitted.

The experiments were made upon paper friction-wheels approximately 5½, 8, 12, and 16 inches in diameter, all in contact with a 16-inch cast-iron wheel. "The contact pressure was varied from 75 pounds per inch of width to more than 400 pounds, and the speed limits gave a peripheral velocity varying from 450 to 2700 feet per minute."

It was found that:

The coefficient of friction μ increases as the rate of slipping between the gears increases;

When the slip is as great as 3%, there is apt to be a sudden increase in its value to 100%; i.e., motion ceases to be transmitted to the driven wheel;

"The coefficient of friction is apparently constant for all pressures of contact up to a limit which lies between 150 and 200 pounds per inch of width of wheel-face, beyond which limit its value apparently decreases";

"Friction-wheels of 8, 12, and 16 inches diameter give nearly the same value for the coefficient, while results from a 6-inch wheel are lower by about 10%";

"Variations in peripheral speed between 400 and 2800 feet per minute do not affect the coefficient of friction."

Fig. 66,* taken from Professor Goss's paper, shows the relation between the slip and the coefficient of friction which he found could be easily maintained with paper friction-wheels 8 inches or more in diameter.

It is probable that the coefficients of friction for the other materials that are most commonly used in contact with metal for friction-gears operating under comparatively high pressures, are lower than those generally found by the laboratory experiments where plane surfaces are moved over each other at unit pressure

---

* Trans. Am. Soc. Mech. Engrs., vol. XVIII., 1897, p. 103.

very much lower in comparison. It is believed that the following values of μ are as high as can be safely used for pressures of 100 pounds or more per inch of width of gear-face:

$$\begin{aligned}
\text{Metal on metal} &\ldots 0.2 \\
\text{Leather on metal} &\ldots 0.3 \\
\text{Wood on metal} &\ldots 0.3
\end{aligned}$$

A system of cylindrical friction-gears that is quite commonly used where it can be applied, and which has many advantages in

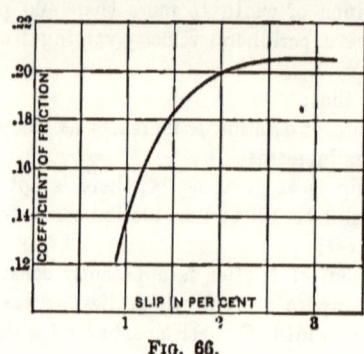

Fig. 66.

the way of economy of construction and ease of operation, is shown in Fig. 67. The larger gears, which are of iron, are placed on the driving and driven shafts, whose centres remain at a fixed distance apart. The small gear, intermediate between the other two, is so supported that it can be pressed against the others, or withdrawn from contact with them, at will. It is generally made of some of the so-called "friction materials," and thus affords a means of securing a high coefficient of friction, and a somewhat elastic surface in contact with the more rigid iron.

In a pair of friction-gears having different materials on their working faces, it is advisable to use the softer material on the driver; then, in case of excessive slipping, there is not so much danger that a flat place will be worn in the driven one.

Another system of cylindrical friction-pulleys that has been found satisfactory in some cases, where the distance between the centres of the shafts can be varied, consists of two pulleys, one on the driving shaft and the other on the driven, and an endless belt of leather fitting loosely around one, which is flanged to hold it in place. The face of the second pulley is slightly narrower than the space between the flanges of the other, so that when the two are drawn together the belt is pressed between the pulley faces and forms a cushion against which they work. The coefficient of fric-

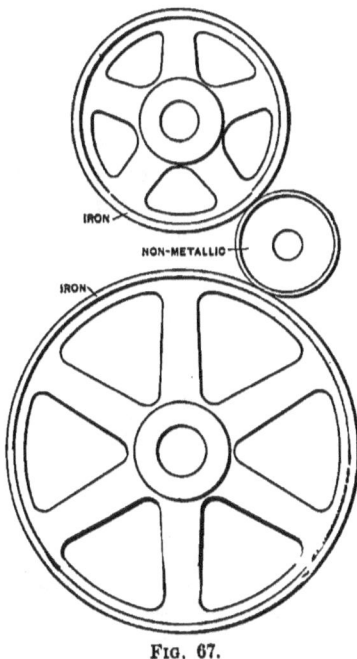

FIG. 67.

tion in such a system is, of course, that of leather on iron. This device has advantages in its simplicity and comparatively low cost of construction. It is hardly applicable for the transmission of large amounts of power.

The power that can be transmitted by a pair of cylindrical friction-gears in practical operation is about the same per inch of face width as that per inch of width of the kind of flat leather belt that would be used on the same machinery. It depends, of course, upon the pressure and coefficient of friction. When a given amount of horse-power H.P. is transmitted, the value of P can be found by equation (17), § 25.

The **efficiency of friction-gears** depends very largely upon the frictional resistance of their supporting journals. The heavy pressures between them, necessary to produce the required turning force, brings a correspondingly high pressure upon the journals, with its attendant frictional loss. While no efficiency tests upon this class of transmission machinery seem to have been made public, it would appear probable that the friction loss in it is much greater than in spur-gears or belt-connected pulleys constructed with the same care.

36. Grooved friction-gears, having their surfaces grooved circumferentially, are used when it is desired to have a tight grip between them without excessive pressure upon the bearings. Fig. 68 shows a section through the rims of such a pair of gears in con-

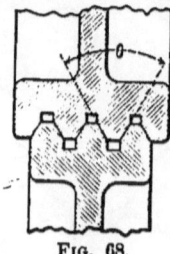

Fig. 68.

tact, cut by a plane passing through the axes of both gears. The wedge-like action gives them increased holding power.

There is a sort of rubbing or grinding action between a pair of grooved friction-gears in action which is detrimental to both their life and efficiency of service. The grinding action increases with the radial depth of the engaging surfaces; hence, if such gears are used, it is better to make them with a large number of shallow grooves than a few deep ones.

Both of a pair of grooved friction-wheels are frequently made of the same metal, cast iron being most used. Better service can be obtained by making them of such metals or alloys as work well together in journal- and step-bearings under heavy pressure.

SPUR- AND FRICTION-GEARS.   109

The angle θ between the sides of a groove should seldom be smaller than 30°, but the increase of grip over that of a pair of smooth cylinders is not great if the angle exceeds 40°.

The total pressure acting between the sides of the grooves of a pair of gears with parallel axes, when the gears are pressed together with a force F normal to their axes, the angle of the groove being θ, is $F \csc \dfrac{\theta}{2}$; and the tangential or turning force acting at their pitch-point is approximately $\mu F \csc \dfrac{\theta}{2}$, in which $\mu =$ coefficient of friction between the materials of the grooves.

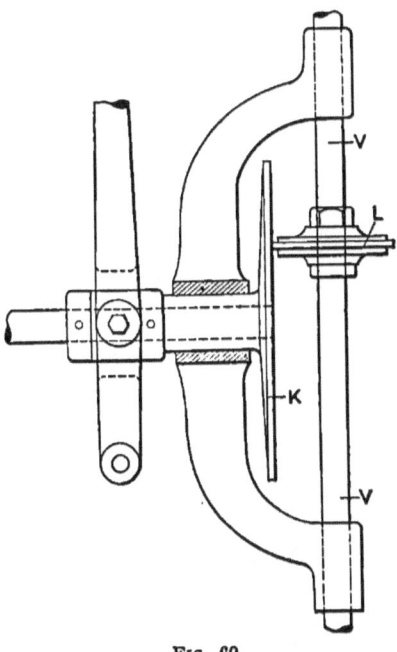

FIG. 69.

37. Friction bevel-gears are often used for connecting intersecting shafts where the same conditions of operation as have been mentioned for cylindrical friction-gears exist. In addition to the side pressure on the bearings, there is an end thrust which may be so great as to require special provisions, in the way of a step- or collar-bearing, to withstand it when the apex angle α of the gear-cone is large and the service is heavy. If F is the normal pressure between the gears, and α the angle between the axis of the gear and its face, then the value of the end thrust E may be expressed by the equation

$$E = F \sin \alpha.$$

38. Crown friction-gears are used on light machinery where it is desired to vary the speed of the driven shaft while the driver runs

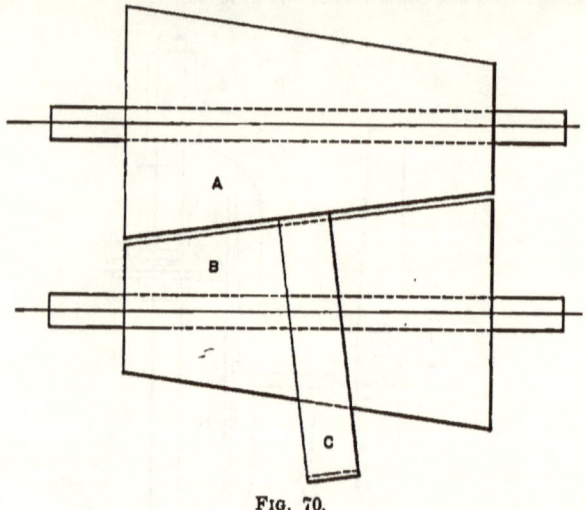

FIG. 70.

uniformly. Fig. 69 represents such a device. A ring L, of leather or other suitable material, is held between a pair of disks on the shaft V, which is in the same plane, and at right angles to the axis of the metal disk K, against which L presses.

If L is the driver, moving it across the face of K will change the speed of the latter, and the direction of K's rotation will be reversed by moving L across its centre. When possible L should be the driver, since it is faced with the softer material. When this method of driving is used on a drill-press, where the force required to turn the drill-spindle increases as the speed of rotation decreases, as is the case when changing from a small to a large drill-bit, L should always be the driver, since its lever-arm about the axis of K increases in the same proportion that the speed of K decreases, the speed of L remaining constant.

39. Double-cone, variable-speed friction-gears.—A special form of friction-gears for variable speeds is shown conventionally in Fig. 70. The mechanism consists of a pair of similar cones, A and B, with smooth surfaces, placed side by side on parallel axes, and separated a slight distance to allow a short endless belt C to pass between them. Their pressure against the belt serves to transmit power between them as they rotate. The belt is held in position by a shifter, not shown; by moving it along from end to end of the cones when they are running, their speed ratio can be varied.

CHAPTER III.

BELTS AND ROPES FOR POWER TRANSMISSION.

FLAT BELTS.

40. When a belt is transmitting power by its frictional resistance against the face of a pulley over which it passes, there must necessarily exist a difference of tension in it, at the points of tangency with the pulley, equal to the torsional force exerted upon the pulley. The tensile stress gradually increases from the point where the belt first comes in contact with a driven pulley to where it leaves it. The torsional force that can be transmitted to the pulley depends jointly upon the belt tensions, coefficient of friction between the belt and pulley face, radius of pulley, velocity at which the belt travels, and the weight of the belt per unit of length. The last three items must be included on account of the centrifugal force exerted on the belt as it passes around the pulley, which force reduces the pressure against its face, and is the principal factor which limits belt speed. The centrifugal force becomes greater as the speed increases, until, at high velocities, the belt is partly lifted from the pulley, and consequently but little turning force can be transmitted by it.

41. Equations for power transmission by flat belts.—By the use of the equations developed below, the torsional force that can be exerted by a belt, and the relations between all quantities involved, can be determined. The following notation is used:

$A =$ sectional area of belt, square inches;
$C =$ centrifugal force for 1 cu. in. of belt, pounds;
H.P. $=$ horse-power transmitted;
$P =$ turning force exerted by the belt against the pulley
$= T_n - T_o$, pounds;

BELTS AND ROPES FOR POWER TRANSMISSION.   113

R = radius of pulley, feet;
T = total tension in belt at any section, pounds;
T_n = total tension in belt on tight side, pounds;
T_o = total tension in belt on slack side, pounds;
T_r = total tension in belt when at rest, pounds;
V = velocity of belt, feet per minute;
c = centrifugal force per linear inch of belt, pounds;
g = acceleration due to gravity = 32.2 ft. per sec.;
p = pressure of pulley against 1 linear inch of belt, pounds;
r = radius of pulley or sheave, inches;
t = working tension of belt, pounds per square inch;
t_r = tension in belt when at rest, pounds per square inch;
v = velocity of belt, feet per second;
w = weight of belt per cubic inch, pounds;
$z = \dfrac{wV^2}{9660t}$, used for convenience only;
α = arc of belt contact, degrees;
$\theta = .001746\alpha$ = arc of contact between belt and pulley, circular measure (radians);
μ = coefficient of friction between belt and pulley;
$\epsilon = 2.71828 = 10^{0.4342945}$ is the base of the hyperbolic, Napierian, or natural logarithms.

The general equation for the centrifugal force acting on a body moving in a curved path is

$$(\text{centrifugal force}) = \frac{(\text{mass}) \times (\text{square of velocity})}{\text{radius of curvature}}.$$

The specific form for the present case is

$$C = \frac{w}{g}\frac{v^2}{R} = \frac{w(V \div 60)^2}{g(r \div 12)} = \frac{1}{9660}\frac{wV^2}{r},$$

or

$$c = A\frac{w}{g}\frac{v^2}{R} = \frac{T}{t}\frac{w}{g}\frac{v^2}{R} = \frac{1}{9660}\frac{T}{t}\frac{wV^2}{r}. \quad . \quad . \quad (36)$$

114 FORM, STRENGTH, AND PROPORTIONS OF PARTS.

By putting $z = \dfrac{1}{9660}\dfrac{wV^2}{t}$ for convenience, equation (36) reduces to

$$c = T'\dfrac{z}{r}. \qquad (37)$$

Fig. 71 shows a pulley with a belt passing around it, the angle of contact being θ. The nature of the forces that act along the

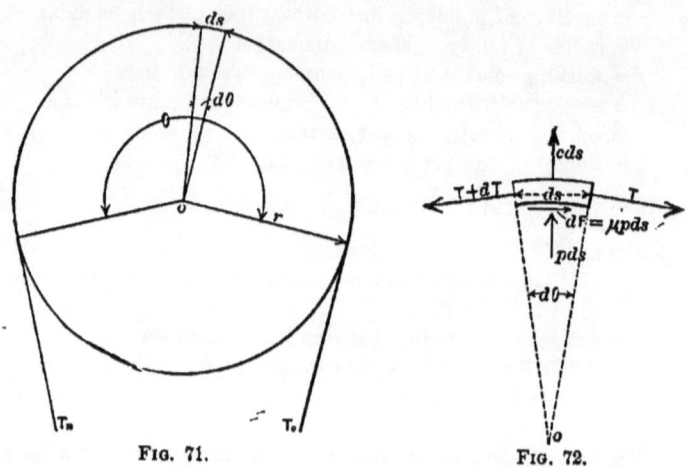

FIG. 71.   FIG. 72.

arc of contact, and the relation between the belt tensions T_n and T_o, when the belt is just beginning to slip, may be determined by taking an elementary length of the belt $= ds$, embracing an angle $d\theta$ on any part of the arc of contact, and considering it as a free body, as shown in Fig. 72.

The forces which hold this elementary body in equilibrium are, first, T and $(T + dT)$, normal to the end sections and making an angle $d\theta$ with each other, whose resultant force acts to press the belt against the pulley; second, cds, the centrifugal force; third, pds, the pressure of the pulley against it; and fourth, $dF = \mu pds$, the frictional resistance along its surface where it is in contact with the pulley.

The resultant of T and $(T + dT)$ can be taken as normal to the surface of the belt, since dT is small in comparison with T. This resultant is counterbalanced by pds and cds, dF not having any component normal to the belt surface, since it acts along the surface. The following equation for the value of pds can now be written

$$pds + cds = T \sin \frac{d\theta}{2} + (T + dT) \sin \frac{d\theta}{2},$$

in which $d\theta$ is so small that its sine can be taken as equal to the angle in circular measure, and $dT\left(\sin \frac{d\theta}{2}\right)$, being the product of two small quantities, can be neglected; the equation thus becomes

$$pds + cds = Td\theta,$$

or

$$pds = Td\theta - cds,$$

which, by substituting the value of c given in equation (37), takes the form

$$pds = Td\theta - T\frac{z}{r}ds;$$

but

$$ds = rd\theta,$$

therefore

$$pds = T(1-z)d\theta. \quad \ldots \ldots \quad (38)$$

By equality of moments about O,

$$T + dT = T + \mu pds;$$

whence

$$dT = \mu pds.$$

And by substituting the value of pds as shown by equation (38), this becomes

$$dT = \mu T(1-z)d\theta.$$

Further solution of this equation gives

$$\int_{T_0}^{T_n} \frac{dT}{T} = \mu(1-z) \int_0^\theta d\theta;$$

whence, by integration and further solution:

$$\text{hyp. log } \frac{T_n}{T_o} = \mu(1-z)\theta;$$

$$\frac{T_n}{T_o} = \epsilon^{\mu\theta(1-z)} = 10^{.00758\mu a(1-z)}; \quad \ldots \ldots \quad (39)$$

$$T_n = \epsilon^{\mu\theta(1-z)} T_o' = 10^{.00758\mu a(1-z)} T_o; \quad \ldots \ldots \quad (40)$$

$$P = T_n - T_o = T_n - \frac{T_n}{\epsilon^{\mu\theta(1-z)}} = T_n - \frac{T_n}{10^{.00758\mu a(1-z)}}; \quad (41)$$

or

$$P = T_n \frac{\epsilon^{\mu\theta(1-z)} - 1}{\epsilon^{\mu\theta(1-z)}} = T_n \frac{10^{.00758\mu a(1-z)} - 1}{10^{.00758\mu a(1-z)}}, \quad \ldots \quad (42)$$

and

$$T_n = P \frac{\epsilon^{\mu\theta(1-z)}}{\epsilon^{\mu\theta(1-z)} - 1} = P \frac{10^{.00758\mu a(1-z)}}{10^{.00758\mu a(1-z)} - 1}. \quad \ldots \quad (43)$$

For the values $\mu = 0.3$, and $\theta = \pi$, (i.e., $a = 180°$,) equations (42) and (43) reduce to

$$P = T_n \frac{\epsilon^{0.94248(1-z)} - 1}{\epsilon^{0.94248(1-z)}} = T_n \frac{10^{0.40932(1-z)} - 1}{10^{0.40932(1-z)}}, \quad \ldots \quad (44)$$

and

$$T_n = P \frac{\epsilon^{0.94248(1-z)}}{\epsilon^{0.94248(1-z)} - 1} = P \frac{10^{0.40932(1-z)}}{10^{0.40932(1-z)} - 1}. \quad \ldots \quad (45)$$

TABLE XI.

ANGLE OF CONTACT α ON THE SMALLER OF A PAIR OF PULLEYS DIRECTLY CONNECTED TOGETHER BY A BELT. NO ALLOWANCE FOR SAG OF BELT.

$D = $ diam. of large pulley; $d = $ diam. of small pulley; $C = $ distance between pulley-centres.

$\frac{D-d}{C}$	Angle of Contact. Degrees.	$\frac{D-d}{C}$	Angle of Contact. Degrees.	$\frac{D-d}{C}$	Angle of Contact. Degrees.
.05	177.13	.30	162.74	.55	148.07
.10	174.27	.35	159.84	.60	145.07
.15	171.37	.40	156.90	.65	142.07
.20	168.50	.45	153.67	.70	139.00
.25	165.64	.50	151.04	.75	135.34

Example.—Find the sectional area of a belt to transmit 200 H.P. when running at 4500 feet per minute, and working at 400 pounds tension per square inch on the tight side; the arc of contact on the working pulley having the smallest portion of its circumference embraced by the belt, being 160°.

The value of the turning force P which must be exerted by the belt is found by the equation

$$P = \frac{33000 \text{ H.P.}}{V}; \quad \quad \quad \quad (46)$$

which becomes, by the substitution of numerical values,

$$P = \frac{33000 \times 200}{4500} = 1467 \text{ pounds.}$$

Equation (43) is applicable to this problem. The value of z to be used in the equation for this problem, for belting weighing .035 pounds per cubic inch, is

$$z = \frac{.035(4500)^2}{9600 \times 400} = .184.$$

Substituting this value of z in equation (43), together with that of P found above, and taking $\mu = 0.3$, gives

$$T_n = 1467 \frac{10^{.00758 \times 0.3 \times 160(1-.184)}}{10^{.00758 \times 0.3 \times 160 \times .816} - 1} = 1467 \frac{10^{.2967}}{10^{.2967} - 1}$$

$$= 1467 \frac{1.981}{.981} = 2960 \text{ pounds.}$$

The sectional area of the belt is therefore

$$A = 2960 \div 400 = 7.4 \text{ square inches.}$$

If a belt $\frac{5}{16}$ of an inch thick is used, the width will be $7.4 \div \frac{5}{16} = 24$ inches (about).

The turning force per inch of width of belt is $1467 \div 24 = 61$ pounds.

A working strength of 400 pounds per square inch is doubtless higher than should be used for a belt that works nearly or quite up to its rated capacity continuously; from 200 to 300 pounds per square inch is more advisable.

When the belt is at rest, the tensions are practically equal in both stretches between pulleys. The sum of the tensions, $T'_n + T'_o$, is greater when a belt is transmitting power than when running idle or standing still. For a horizontal belt working at 400 pounds tension per square inch on the tight side, and having 2% slip (i.e., the surface of the driven pulley moving 2% slower than that of the driver) on cast-iron pulleys, the increase of the sum of the tensions may be taken, for speeds up to 1000 feet per minute, as an average of about $\frac{1}{2}$ of the value when the belt is idle. At high speeds the centrifugal action tends to reduce this increase, until, at the speed giving a centrifugal tension equal to the initial tension, they become equal on the two sides, and of the same value as when the belt is idle. It is on the safe side, so far as the stress in the belt is concerned, to allow for the full increase of tension, however, hence it will be so taken. Calling the tension in each side, when the belt is at rest, T_r, the relation between T_r and the sum of the working tensions is expressed by the equation

$$\tfrac{2}{3}(2T_r) = T_n + T_o;$$

whence

$$T_r = \tfrac{3}{8}(T_n + T_o).$$

The value of T_n has just been found above; that of T_o may be obtained by the equation

$$T_o = T_n - P.$$

Substituting this value of T_o in the last equation gives

$$T_r = \tfrac{3}{8}(2T_n - P) = \tfrac{3}{8}[(2 \times 2960) - 1467] = 1670 \text{ pounds}.$$

This is the tension that must be maintained in the belt, when at rest or running idle, in order that it shall not have more than 2% of slip when working under the full load for which it is designed. It corresponds to a tension t^r per unit area of section, when the belt is at rest, of

$$t_r = T_r \div A = 1670 \div 7.4 = 226 \text{ pounds per square inch}.$$

BELTS AND ROPES FOR POWER TRANSMISSION. 119

The required tension may be fairly well maintained by having the pulleys so that the distance between their centres can be adjusted, or by using an adjustable idle pulley as a tightener. When no means of pulley adjustment is used, the belt must be shortened at intervals in order to allow for the permanent stretch that continually occurs throughout its life of service. To allow for this stretch, the belt must, at each adjustment, be made shorter than is necessary to produce the required tension of rest. The tension, when shortening, may be weighed by a pair of belt-clamps, fitted with springs and a graduated scale, used to draw the ends together while the belt is in its working position.

If, in the example just given, μ is taken as 0.4 instead of 0.3, the following values are found:

$$T_n = 2460; \quad T_r = 1295; \quad \text{and} \quad t_r = 175.$$

42. The coefficient of friction μ, and slip of leather belting.—The coefficient of friction is an exceedingly variable quantity, generally lying between 0.25 and 1.0 for leather in good working order running on smooth iron pulleys; its value is even greater than this in some cases. The coefficient is somewhat higher for the same belt on wooden pulleys than on iron, and leather-covered pulleys give still higher values. A change in the intensity of pressure between the belt and pulley affects the value to a slight extent, decreasing it as the pressure increases; but, for ordinary working conditions, it may be considered as constant between the limits of pressure that can be used with reasonable durability of belting.

It has been found by several experimenters that the coefficient of friction increases as the slip of the belt over the pulley increases. The higher values therefore appertain to a high rate of slipping. If too much slipping occurs, there is danger that the heat generated will dry, or even burn, the surface of the belt, and thus not only weaken it, but at the same time reduce the coefficient of friction.

A slippage of 3% (i.e., a velocity of the driving-pulley face 3% faster than that of the driven) is probably as much as should be allowed in general practice, 2% being a good value. The rate of slipping, at which the higher values of the coefficient are obtained, varies from 10% to as much as 20% in some cases.

Belt-driven machinery that works against a high resistance for a short interval, as is customary with punching and shearing machines, should not have a maximum reduction of speed, and consequent rate of belt-slip, greater than 20% during the working period. Machines expending energy during a greater portion of each cycle should have a smaller variation of speed on account of the greater liability of the driving belt to slip from the pulleys.

With a belt in fair working condition, the coefficient of friction μ can safely be taken as 0.3 when the slippage is as much as 2%; probably 0.4 is allowable in a majority of cases. Fair working condition means that the surface next the pulley is not hard, dry, or cracked, or the belt stiff for the lack of belt dressing, but soft, pliable, clean, and running without excessive vibration.

When a belt becomes hard and dry, it can be softened, and the coefficient of friction increased, by the application of a suitable belt dressing, provided the belt has not been too long neglected.

43. Working strength of leather belting.—Tension causes a leather belt to elongate continuously throughout its life; upon removal of the tensile stress it will partly return to its original length, but a permanent elongation always remains. Both the total and permanent elongations increase much more rapidly during the earlier part of its life than later, if the belt is always used under the same conditions. During the time of its efficient service the elongation is very nearly uniformly distributed throughout the belt; but when it has elongated a certain percentage of its length, the stretching becomes uneven, and consequently the belt soon gets crooked and unfit for service.

It may be stated almost as an axiom that the elongation is more rapid the greater the stress. The life of a belt, therefore, grows shorter as the working stress increases. The permanent stretch that a belt will endure before becoming useless is probably a nearly constant percentage of its length, which, for good belts, may safely be taken at least as great as 6% for leather.*

Mr. F. W. Taylor, in a carefully kept record of the performance of belting in continuous use for nine years in the machine-shop of

---

* Notes on Belting, by F. W. Taylor. Trans. Am. Soc. of Mech. Engrs., vol. xv., p. 204.

the Midvale Steel Co., finds that double leather belting lasts well under a total load, t, of 240 pounds per square inch of sectional area, the annual cost of repairs, maintenance, and renewals amounting to not more than 14% of the first cost. When the stress is kept at 400 pounds per square inch of section, he finds that the annual expense for the same items becomes about 37% of the first cost. It should be kept in mind that these are the continuous working values and not, as is often the case, the maximum load under which a belt may work for a short time, the stress being much less most of the time, as, for instance, a belt driving a dynamo which carries its full load of lights or motors during only a small part of its run.

When a belt is laced together at the ends, the strength of the joint varies from 50 to 95% of that of the belt; the average efficiency of the laced joint is about 70%.* The average strength of joints made with metal fastenings is less than that for lacing. A carefully made cemented joint gives a strength about the same as that of the belt. This method of splicing, making what is commonly called an endless belt, is unquestionably the best for all cases of ordinary application, and becomes almost a necessity for high speeds, since any heavy place, such as a laced or metal-fastener joint, causes vibration.

By inspection of equation (43) it can be seen that, for a given belt working under uniform conditions as to velocity and turning force transmitted, the total tension, T_n, decreases as the coefficient of friction, μ, increases; hence if, in designing a belt, the value of μ is taken as small as it will probably ever be for that belt, it is reasonably safe to say that T_n will never exceed its calculated value as long as the belt works against a constant load.

An increased resistance of the pulley, and consequently of the turning force P, will increase T_n and the pressure against the pulley, at least on the part of the arc of contact next to the tight side of the belt; if there is no automatic tightener, and the distance between pulley centres is kept the same, the elongation of the tight side of the belt, due to its increased tension, will decrease the tension in the slack side; the coefficient of friction will decrease slightly on account of the increased pressure against the pulley over

---

* Digest of Physical Tests, Jan. 1896, p. 40.

the larger portion of the arc of contact, and the final result is that the belt slips. But the slipping increases the value of μ, and a greater turning force is exerted than when the resistance of the pulley was first increased.

From the above it can be seen that a long belt, keeping within practical limits, is better able to withstand suddenly increased loads than a short one, both on account of its greater capacity for elongating, and its greater ease of slipping. For these reasons it is not advisable to make belts short, even when they run at the speeds customary for transmitting power, if there is any probability that suddenly applied loads will come upon them.

In practice it is seldom possible to measure the tension in a belt when it is working. There is no reason, however, that the tension, practically equal in both stretches between pulleys, should not be known at the time of putting it in place or of tightening after it has become loose by service. This can be done by drawing the ends together with spring belt-clamps made to weigh the tension, as has been mentioned before. The belts used in Mr. Taylor's experiments, already cited, were adjusted in this manner. He finds that double leather belts, tightened to 240 pounds per square inch of section, or 71 pounds per inch of width, when at rest, and when made to exert a turning force on the pulley of 65 pounds per inch of width, will stretch so that the tension falls to 100 pounds per square inch, or 33 pounds per inch of width, in an average time of two and one half months of service, their average tension during this time being 150 pounds per square inch, or 46 pounds per inch of width.

Mr. Taylor concludes that, for continuous working, a double oak-tanned and fulled leather belt will give an effective pull of 35 pounds per inch of width on the face of the pulley when the arc of contact is 180°; and that other types of leather belts, and 6- to 7-ply rubber belts, will give an effective pull of 30 pounds per inch of width for the same arc of contact.

In iron-working machinery, such as lathes, planers, and drill-presses, 50 pounds per inch of width is quite commonly taken as the effective turning force or pull per inch of width for 180° of contact.

44. Velocity of leather belting.—Below 2500 to 3000 feet per minute, the velocity of a belt has but slight effect upon the length of its life. The most economical speed is probably from 4000 to 5000 feet per minute. Higher velocities, up to about 6000 feet per minute, will generally increase the driving power if the belt is worked at high tension, but the life of the belt is shortened, and vibration, commonly called "flapping," is liable to occur. It may also run from side to side of the pulley, an action known as "chasing"; this is more apt to occur in a thin, pliable belt than a thick, stiff one of the same width. At higher velocities the centrifugal force becomes an objectionable feature in addition to these, and, even if the belt runs smoothly, less power can be transmitted than at slower speeds.

When $z = 1$ in the equations for belting, the turning force $P = 0$, the tension in the belt due to centrifugal action being equal to the working stress. For a belt weighing .035 pounds per cubic inch, and adjusted to work at 400 pounds per square inch tension, $z = 1$ when the velocity is 10509 feet per minute; and for a working tension of 200 pounds per square inch $z = 1$ for a velocity of 7430 feet per minute. These are the velocities at which the turning force and the power transmitted become zero.

Belt speeds of 5000 feet per minute, and even more, are common in wood-working machinery having pulleys as small as 4 inches in diameter, or less. Thin leather belts are successfully used for such work.

When speeds as high as 5000 feet or more per minute are used in connection with pulleys as small as one inch in diameter, a woven linen web or tape has been found better than leather.[*]

45. Wear of leather belts.—Each time a belt is bent by passing over a pulley, the fibres farthest from the pulley are elongated, and those next to it compressed. The continuous repetition of the bending when the belt is in service has a tendency to crack the leather, particularly on the side away from the pulley. The smooth or hair side of a single belt is more easily cracked by this action than the flesh side. It is therefore advisable to run the hair side next the pulleys. Again, there is a slight wear on the surface of the belt that comes in contact with the pulleys. The flesh side is

---

[*] Mr. John T. Hawkins, in Trans. Am. Soc. of Mech. Engrs., vol. VII., 1886, p. 582. Belt used for stereotyper's routing-machine.

much stronger than the hair side. This is, therefore, another reason for running the hair side next the pulleys.

Double leather belts are made by cementing together the flesh sides of two thicknesses of leather, thus leaving the hair sides exposed to surface wear. Good double belts, properly cared for, are not subject to cracking on the side away from the pulley when working under proper conditions. Triple and quadruple thicknesses of leather are used for making very thick, heavy belts. It is possible, although there is no very definite proof, that a very thick belt, as the quadruple, when working up to the same stress per square inch that would ordinarily be used for lighter belts, may have so high a stress per inch of width, and consequent pressure against the pulley, that the heat generated by the slipping of the belt over the pulley face will dry and burn the surface of the leather so that it will become hard and crack on the surface. Such an action would, at the same time, reduce the coefficient of friction, and thus induce more slipping and injury to the belt.

The diameter of the pulley has a very considerable effect on the life of the belt; for, if a pulley is very small, the belt will be bent so short that there will be excessive wear between its particles, and the greater strain due to the short bend will have a greater tendency to form cracks. A common practice among engineers is to fix a minimum diameter of pulley for single, double, triple, and quadruple belts. This limitation will answer only for the ordinary thicknesses of such belts, however, and should be treated accordingly. To say that a belt is double does not fix its thickness by any means, on account of its great variation in common use. (See § 46.)

Mr. Taylor concludes, as a result of his experience, "that it is safe and advisable to use

"A double belt on a pulley 12 inches diameter or larger;

"A triple belt on a pulley 20 inches diameter or larger;

"A quadruple belt on a pulley 30 inches diameter or larger; and that it is inadvisable to use double, triple, and quadruple belts on pulleys respectively as small as 9 inches, 15 inches, and 24 inches diameter."

He also considers it safe to say "that the life of belting is doubled by spliceing and cementing the belt, instead of lacing, wiring, or using hooks of any kind."

A "quarter-turn" belt (i.e., one connecting a pair of pulleys whose axes are at right angles to each other, no intermediate or idle pulleys being used) is subjected to heavy stresses at its edges if the pulleys it connects are large in comparison with the distance between them. Belts used in this manner generally wear out rapidly, frequently becoming stretched and uneven on the edges. It is advisable to avoid such a device.

46. Weight and thickness of leather belting.—The weight of good leather belting varies from .03 to .04 of a pound per cubic inch when new. The various processes of tanning, and the difference between fulled and unfulled leather, are the principal causes of this very considerable variation of weight. Unfulled belts that are light when new, generally increase in weight when not allowed to become hard and dry in service; the application of belt dressing and, in many kinds of service, the accumulation of oil that gets upon them in various ways cause increased weight.

The thickness of belting cannot be given with any considerable degree of accuracy. The average thickness of single belts is about .22 of an inch, and often reaches .25 of an inch; it can readily be seen that such leather can be trimmed to any desired thickness. Double belts, such as are commonly used, vary in thickness at least from .22 to .35 of an inch, the majority lying between .30 and .35 of an inch. Unless specified as light or heavy, the thickness may ordinarily be taken as about .33 of an inch.

47. Rawhide, semi-rawhide, and rawhide with tanned-leather face, belts.—All of these classes of belts have a higher coefficient of friction than ordinary tanned leather belts, but do not seem to be so durable or satisfactory for service in dry places.

Rawhide seems to give better service in damp places than tanned leather, the moisture apparently not affecting its capacity to transmit power, or causing it to crack and distort as the tanned leather does, especially when alternately dry and wet. The weight of rawhide is about the same as that of tanned-and-fulled leather.

Semi-rawhide belting is made of hide that has only a thin surface layer tanned, the inner portion retaining the qualities of rawhide.

Rawhide belts with tanned-leather faces have the tanned leather pasted on the rawhide with sizing or glue. Under heavy service this facing is apt to come loose after a considerable time.

48. Cotton belts are generally made by folding together several thicknesses of cotton canvas or ducking, and fastening them by stitching or otherwise. They are sometimes woven solid so that no folding is necessary. Their strength is probably greater than that of any other form of belts commonly used. The coefficient of friction is rather low when the belt is used dry without any filling, sizing, or dressing. When properly sized or dressed, however, the coefficient of friction is equal to that of good leather belts. A weather-proof brand which is placed on the market seems to give excellent service for out-door work of the most trying kind. In it the interstices seem to be completely filled with the sizing, which, on the outside at least, is water-proof. It seems to be as durable as rubber belting, and has the advantage that there is no thin layer of rubber to rub or roll off, as is the case with the latter when excessive slipping occurs.

The weight of cotton belts depends largely upon the kind and amount of sizing that is used. Belts showing the following weights have been found in service: 0.026, 0.033, 0.037, 0.044, and 0.050 of a pound per cubic inch. The latter is the weight of the weather-proof belt mentioned above.

Cotton-leather belting is made by stitching a piece of thin leather to a cotton belt so as to make a leather facing on one side, which is used next the pulley. An unsized belt can be given a high coefficient of friction in this way. The facing is apt to tear off, especially in service where the belt must be shifted from step to step of a cone pulley.

49. Rubber belting is composed of a cotton web with a composition of rubber filling all its interstices and completely covering it. When of good material it is not injured by moisture, and is therefore excellent for damp places and out-of-door service. The coefficient of friction of good rubber belting is high. If the rubber compound is poor, the coefficient of friction may be low, and the compound will crack. When overloaded and caused to slip on the pulleys, the rubber is in danger of peeling loose and rolling up so as to tear off considerable areas, thus destroying the belt. The weight of rubber belting is about 0.045 pounds per cubic inch. The joints of rubber belting can be cemented by coating the surfaces with uncured rubber in solution and vulcanizing the splice with steam-heated clamps placed over it.

50. Leather-link belts.—Figs. 73 and 74. These are made of small pieces of leather, generally from one to two inches long and

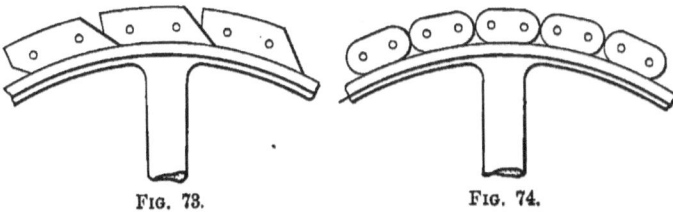

FIG. 73.   FIG. 74.

five-eighths to seven-eighths of an inch broad, perforated at each end at right angles to the natural surfaces of the leather. These are fastened together by pins through the holes so as to form a chain or belt of any desired width and length. The pins are generally of iron or steel, and of such a size as will fit the holes in the links. The edges of the links are exposed and form the broken surface which comes in contact with the pulley. In the better makes of this kind of belts each pin extends only half-way across its width, and the two half-belts thus formed are fastened side by side, to form the complete width of belt. By this means the belt is allowed to take the form of a "crowned" pulley whose face is made of two cone frusta placed base to base. A round piece of leather is sometimes used for uniting the links, instead of the steel pins. A belt thus made will adjust itself to a pulley whose crowning consists of a smooth curve, such as an arc of a circle.

The coefficient of friction of link belting seems to be about the same as that of good leather belts, possibly somewhat lower. The weight per cubic inch generally runs, for metal pins, from 0.035 to 0.050 pounds. On account of its great thickness it is much heavier per unit area of working surface than solid leather belting of the same capacity for transmission.

Leather-link belting is especially applicable to connecting a large and small pulley that are placed near together in the same horizontal plane, their axes being horizontal. This is because, on account of its weight, it can be very slack, and, the slack side being above, it sags down so as to make a large arc of contact on the pulleys, thus increasing the turning force without making the belt excessively tight, as would be necessary with any of the ordinary solid belts.

When a link belt breaks, it is generally without warning, and, on account of its weight, it is capable of doing serious damage, when running at high speed, by striking whatever may be in its path.

51. Effect of relative positions of pulleys.—When the axes of the driving and driven pulleys are in the same horizontal plane, and the loose side of the belt is uppermost, the arc of contact on each pulley is increased, both by the sagging of the loose side as the load is applied, and by the tightening of the lower side to a more nearly straight line on account of the increased tension in it. On account of this increase of the arcs of contact, the tension is smaller on the tight side of the belt than it would be if the rotation of the pulleys were reversed, thus bringing the slack side below and decreasing the arc of contact; hence it is always advisable, when possible, to run the pulleys so that the slack side of the belt will be uppermost.

Again, suppose that a large pulley is driving a comparatively small one placed vertically below it; the arc of contact on the small pulley will be much less than on the large one, and, in addition to this, the tension in either stretch of the belt at its point of tangency with the small pulley will be less than at the similar point on the large one. The difference of tension at the top and bottom of a stretch of the belt is the same as the weight of a portion of the belt, equal in length to the vertical distance between the points of tangency. Now suppose the small pulley is placed above the large one; the greater tension in the belt, still remaining at the upper pulley, will be applied so as to counteract, in a measure, the effect of the reduced arc of contact, while before, with the small pulley below, it was cumulative with it. No further investigation is necessary to show that it is advisable to place a small pulley above a large one, especially when the belt has a considerable length.

For inclined positions of a belt connecting pulleys running on horizontal axes, the two facts brought out above should be taken

BELTS AND ROPES FOR POWER TRANSMISSION. 129

into consideration. They indicate that the slack belt should always be kept above, and the small pulley higher than the large one, when they are not at the same level, and when other conditions will admit of such an arrangement.

Whatever the relative positions of a pair of pulleys, the tension, in a clear stretch of belt, at the point of tangency with the upper pulley, is greater than that at the similar point on the lower one, by an amount the same as the weight of a length of the belt equal to the vertical distance between the points of tangency.

52. A special system of flat-belt driving is illustrated in Fig. 75. It is especially applicable to close grouping of machinery.

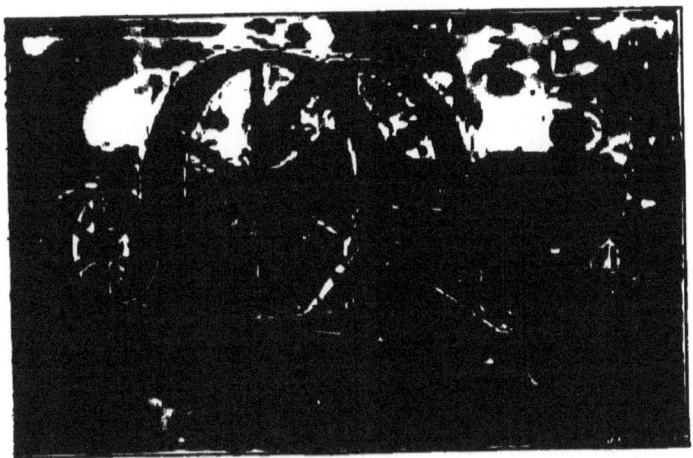

FIG. 75.

The belt is tightened by the mechanism shown near the engine-cylinder.

The driving capacity of this device, when used with ordinary double belts, and with a contact on the large pulley at least as great in linear measure as that on the small or driven pulley, can be safely taken as 1 H.P. per inch of belt width for a belt speed of 750 feet per minute.*

---

* Communication from A. L. Ide & Sons.

53. Efficiency of flat belting.—The power losses in flat leather belts are chiefly due to journal friction and belt slip. There is also a small loss caused by bending and straightening the belt as it runs on and leaves the pulley, but it does not seem to be great enough to need consideration. The slip is sometimes considered as made up of two elements, the creeping of the belt over the pulley, due to its elongation as it passes from the slack to the tight side, and the actual slip, which allows all parts of the belt to move at the same rate, faster than the driven pulley face or slower than the driving pulley. The effects of both are the same upon the efficiency, hence it does not seem necessary to separate them when dealing with this property.

If the journals are larger than necessary to withstand the pull of the belt, or the latter is working above or below its normal capacity, the efficiency is lowered.

The efficiency of flat leather belting is probably never greater than 97% in the grades of machinery that are generally used in practice. This refers to well-made machines with journal-bearings. If bearings having a very low frictional resistance, such as ball-bearings, are used, the efficiency might be increased to 98%, or possibly more. Probably 95% is a fair average for good practice when a belt is working at or near its most economical rate.

Crossed belts require less tension to transmit a given amount of power, on account of having greater arcs of contact on the pulleys than open ones; hence, if there is no rubbing together of the two stretches between the pulleys, the efficiency should be higher. Rubbing is almost certain to occur, however, thus adding an additional power loss. The rubbing may be of small consequence if the pulleys are of the same size; but if their diameters are greatly different, and especially if they are near together, a very considerable rubbing pressure is almost certain to be caused. The power loss under such a condition is proportionately large. It is seldom, possibly never, advisable to use a crossed belt when considered with regard to power transmission and belt economy.

Idle pulleys, used as tighteners or guides, lower the belt efficiency on account of the additional power that is required to drive them.

The efficiency of the other kinds of flat belting mentioned in the preceding paragraphs does not differ enough from that of leather to be practically appreciable.

ROPES.

54. Ropes, running on grooved pulleys or sheaves, are largely used for the transmission of power. The name "sheave" is applied to a pulley with but one groove; when there are two or more grooves the name "grooved pulley" is generally used.

While ropes are very frequently used for transmitting power between shafts that are parallel and at such a distance apart as is common for leather belts, their especial application seems to be that of connecting shafts that are not parallel, or are at a great distance apart. On account of their approximately circular form of section, they bend with equal ease in all directions, hence " quarter turns " and bends in various directions are not nearly so severe upon them as upon flat belts. Hemp, cotton, and wire ropes are the varieties almost exclusively used for power transmission, although leather, rawhide, and other materials are used to some extent, generally for light work. Manila hemp is generally called " manila," and the special make, which is largely used for power transmission, is called "Stevedore."

Fibrous Ropes.

55. Two systems of driving with fibrous ropes are in general use. They are commonly known as the Continuous or American system, and the Multiple or English system.

In the multiple system there are as many separate, endless ropes as there are grooves in each of the system of pulleys over which all the ropes run. Each rope always runs in the same groove of each pulley. The ropes are, of course, parallel to each other, practically speaking. The multiple system is generally used for transmitting very large amounts of power.

In the continuous system, shown in Fig. 76, there is a single endless rope, wound continuously over the pulleys. The winding is such that a point in the rope, starting at the groove at one end of any pulley, passes from this groove to the first groove of each of the other pulleys, then to the second groove of the first pulley and of the other pulleys in succession, thence to the third groove, and so on, until it has passed through all the grooves; it then passes over a single-groove guide-sheave, which leads it back to the first groove of the first pulley. The guide-sheave is often made to serve the double purpose of both guide and tightener. When used as a tightener it is supported on a carriage which is free to travel back and forth on a track or guide; weights are attached to the carriage to keep the tension of the rope uniform. The tension carriage must be free to have a considerable range of travel, in order that it may adjust itself for variation in the length of the rope, which may be caused by change of load, condition of the atmosphere, the gradual lengthening of the rope with service, and other influences. The tension-carriage should always be placed on the slack side of the rope. In Fig. 76 it can be seen in the upper middle portion of the figure. The tension-weight is shown alongside the column.

There are numerous modifications of the arrangement of the tension-carriage and guide-pulleys, to conform with local conditions, one or more idler sheaves frequently being added to guide the rope, but the principle is the same in all.

The method of transmitting power by ropes to the different floors of a building is shown in Fig. 77.

A varying load on a continuous-system rope-drive causes unequal tension in the stretches of rope between the pulleys. This can be seen by assuming that a drive which is running for some time without any load other than the journal-friction of the machinery immediately appertaining to the drive, has a load equal to the full capacity of the drive suddenly thrown upon it. While running light the tension in each stretch of rope, on both sides of the pulleys, is practically equal to that caused by the tension-carriage, i.e., equal to one half the effective weight on the tension-carriage. As the full load is suddenly applied, all the stretches of rope on the slack side are slackened somewhat more by the stretch of those on the tight side; the tension-carriage, however, maintains the same tension in the stretch between it and the driven pulley. Consequently when a length of rope approximately equal to one half of what would be required for connecting both pulleys with a single band has passed from the tension-carriage to the pulley, the tension on the tight side in the first stretch after leaving the

To face page 132.

76.

BELTS AND ROPES FOR POWER TRANSMISSION. 133

tension-carriage is greater than in any of the others, because the
slack side of the rope running on the driven pulley is tighter for

Fig. 77.

this stretch than for any other, and consequently there is less slip
in the first groove of the driven pulley than in any of the others of

the same pulley. As the rope continues to run over the pulley, the first stretch on the slack side becomes tighter than the others on the slack side, and the second stretch on the tight side increases its tension to a higher value than that of the stretches coming later on the tight side. After a time the tension becomes uniformly distributed among the stretches on the tight side.

On account of the inequality of tension in the stretches of the rope on the tight side, when working under varying loads, which may be so marked as to very appreciably shorten its life when there are many turns of a continuous rope around a pair of pulleys, it is advisable to limit the turns to a comparatively small number; the greater the variation of load, the smaller should be the number of turns of a continuous rope.

When, in order to fulfil the requirements of power transmission, a large number of turns of rope are required, and the continuous system is adopted, it is often advisable to use two or more continuous drives, operating side by side on the same pulleys if desired; the main pulleys in such a case would be the same as for a single continuous drive; an additional tension-carriage and guide sheave or sheaves, if the latter are used, must be supplied for each continuous rope. The additional cost of the latter may be more than counterbalanced by the increased life of the rope, however.

56. The equations for ropes transmitting power are similar to those for flat belts. On account of their approximately circular sectional form, it is more convenient to take a rope of unit diameter as the basis of calculations, instead of a unit area, as is done for flat belts. The coefficient of groove friction ϕ, which is generally used, is that of the rope in the groove, and is, of course, greater than that of the same rope on a flat surface of the same material as the pulley, on account of the wedge-like action of the rope in the groove; the sharper the angle between the two sides of the groove, the higher this coefficient of friction.

The following notation is applicable to ropes only; the symbols used in the equations, but not given in this notation, are the same as for flat belts:

W = weight of a rope 1 inch in diameter and 1 foot long;
τ = working strength of a rope 1 inch in diameter;

$\sigma = \dfrac{1}{115920}\dfrac{WV^2}{\tau}$, used for convenience only;

β = angle between sides of groove;

$\phi = \mu \csc \dfrac{\beta}{2}$ = coefficient of groove friction.

The equations, deduced in the same manner as for flat belts, take the form, by substituting σ for z in the equations for flat belts,

$$P = T_n \frac{\epsilon^{\phi\theta(1-\sigma)}-1}{\epsilon^{\phi\theta(1-\sigma)}} = T_n \frac{10^{.00758\phi a(1-\sigma)}-1}{10^{.00758\phi a(1-\sigma)}}, \quad . \quad . \quad (47)$$

and

$$T_n = P \frac{\epsilon^{\phi\theta(1-\sigma)}}{\epsilon^{\phi\theta(1-\sigma)}-1} = P\frac{10^{.00758\phi a(1-\sigma)}}{10^{.00758\phi a(1-\sigma)}-1}. \quad . \quad . \quad (48)$$

Example.—It is required to design a rope-drive to transmit 200 H.P. when running at 4500 feet per minute, and working at a tension on the tight side equivalent to 200 pounds for a rope of 1 inch diameter; the arc of contact on the working pulley having the smallest portion of its circumference embraced by the belt being 160°. Manila hemp rope to be used.

This problem is essentially the same as the example given under flat belting; the only difference being that ropes are used instead of a belt. As in that problem, the turning force

$$P = 1467 \text{ pounds.}$$

Equation (48) may be used for the remainder of the solution. The value of σ, as used in this equation, is, for a value of $W = 0.283$,

$$\sigma = \frac{0.283 \times (4500)^2}{115920 \times 200} = 0.247.$$

Taking the coefficient of friction $\phi = 0.3$, and substituting this value, as well as those of P and σ given above, in equation (48), gives

$$T_n = 1467 \frac{10^{.00758 \times 0.3 \times 160(1-0.247)}}{10^{.00758 \times 0.3 \times 160 \times 0.753}-1} = 1467 \frac{10^{.27397}}{10^{.27397}-1}$$

$$= 1467 \frac{1.8792}{.8792} = 3135 \text{ pounds.}$$

Assuming that rope 1¼ inches in diameter will be used, and that it will work under a tension equivalent to 200 pounds for a rope 1 inch in diameter, the working strengths being taken as proportional to the squares of the diameters, gives, for the working strength or the 1¼-inch rope,

$$\frac{25}{16} \times 200 = 312.5 \text{ pounds.}$$

The number of 1¼-inch ropes required is, therefore,

$$3135 \div 312.5 = 10 \text{ (about).}$$

The tension that must be put on a rope when at rest may be determined in the same manner as for a flat belt. It is not known, however, that any experiments have been made to determine whether there is the same increase of the sum of the tensions in the two sides of the rope when working, over that when at rest, as there is in flat belts, but it seems reasonable to assume that such is the condition. Upon the assumption that the increase is one third of the sum of the tensions of rest, the tension in each stretch of the rope when at rest is, as for the flat belt, taking into account the fact that there are ten stretches of rope between the pulleys on each side,

$$\frac{T_r}{10} = \frac{3}{8} \frac{2T_n - P}{10} = \frac{3}{8} \frac{(2 \times 3135) - 1467}{10} = 180 \text{ pounds.}$$

If a tension-carriage is used, as in the continuous system, the effective weight for producing the total tension T_o on the slack side of the rope must be $2T_o \div 10$, since there are 10 ropes in this particular drive. Therefore the effective weight of the tension-carriage must be

$$\frac{2T_o}{10} = \frac{2(T_n - P)}{10} = \frac{2(3135 - 1467)}{10} = 334 \text{ pounds.}$$

57. The grooves for non-metallic ropes found in practice are of numerous forms. The most common angle between the sides of the groove is about 45°, however. It varies from 30° to 60° in extreme cases. It is clear that the smaller this angle, the tighter the rope will wedge into it, and the less liable will it be to slip.

BELTS AND ROPES FOR POWER TRANSMISSION.   137

More power will be required to pull it out of a sharp-angled groove, however, thus causing more power loss; it is also possible that there will be more wear of the rope. In some forms the sides are straight, as in Fig. 78, while in others they are curved, as in Fig.

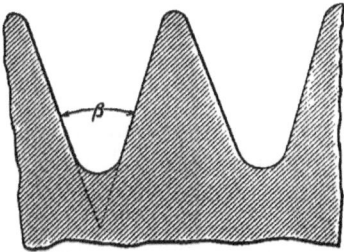

FIG. 78.

79,* where O is the centre of curvature for the left side of the groove; the right side has a similar curvature.

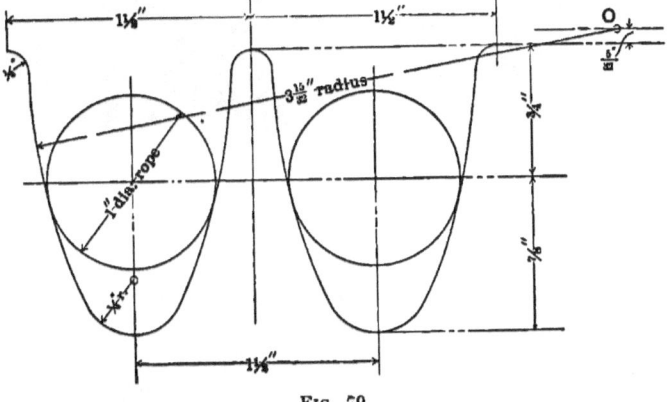

FIG. 79.

In a groove with curved sides, the angle between the sides where the rope comes in contact with them, when resting lightly in the groove, is smaller than would be used for straight sides; at the

---

* Redrawn from prints kindly furnished by the Walker Company, Cleveland, Ohio.

bottom it is larger. This arrangement secures a good grip of the rope when it is at or near the top of the working part of the groove, but allows it to slip more readily when, on account of wear, it is brought nearer to the bottom. Such a provision for more ease of slipping when near the bottom of the groove is of considerable importance in the working of a rope-drive, especially of the multiple system.

In order to show the action of a multiple-rope drive, let it be assumed that one has been in use until some of the ropes have become so worn that they must be replaced by new ones, and that the pulleys are of greatly different diameters, as is frequently the case when a line-shaft or some high-speed machine is driven directly from a pulley on the main shaft of an engine. The old ropes, being smaller than the new ones of the same nominal diameter, on

FIG. 80.

account of wear and stretching, will lie nearer the bottoms of the grooves than the new ones, and their effective radii will accordingly be less on each pulley, the whole system appearing as in Fig. 80. For convenience it may be assumed that the grooves are of the same form on both pulleys, this being very frequently the condition in practice. The reduction of the effective diameter will therefore be the same on both pulleys as the rope wears. By putting

D = effective diameter of the large pulley with new rope,
a = reduction of effective diameter of each pulley,
d = effective diameter of the small pulley with new rope,

the conditions of working can readily be expressed mathematically; it will be assumed that $D = 3d$.

If there were no slipping of the new ropes in the grooves, the small pulley would make three revolutions for one revolution of the engine; and if there were no slipping of the old ropes, the portion of length of one of them which would pass over the large pulley during one of its revolutions is $\pi(D - a)$. This length of rope, by passing over the small pulley at an effective diameter of $d - a$, would turn it through

$$\frac{\pi(D - a)}{\pi(d - a)} = \frac{\pi(3d - a)}{\pi(d - a)} = 3 + \frac{2a}{d - a} \text{ revolutions.}$$

This shows clearly that there is a tendency for the smaller, or old, ropes to drive the small pulley faster than the new ones. As a result, the entire load will be carried by the old ropes unless they slip in their grooves. A form of groove which will let a rope slip more easily as it becomes smaller with wear is better than one which does not allow such slippage.

If, instead of a large pulley driving a small one, the small one drives the large, there will be a tendency for the load to be thrown upon the larger or new ropes. The increased tendency to draw these ropes down into the grooves acts as a corrective, however, and thus prevents the excessive loading of the larger ropes.

In order to show the action of the ropes in a drive having curved grooves similar to that of Fig. 79, let it be assumed that an engine is driving a line-shaft through a rope-drive, the pulley on the engine being larger in diameter than the one on the line-shaft. It may be further assumed that the drive has been in service for some time, and that half the ropes have become so worn that they must be replaced by new ones. The old ropes left in place will, on account of wear and stretch, be smaller than the new ones of nominally the same diameter, and will therefore lie nearer the bottom of the groove. The effective diameter at which the new ropes work on each pulley will therefore be greater than that of the old ones. The difference of the effective pulley diameters, for the new and old ropes, is less than for straight-side grooves, provided the groove angle where the new rope comes in contact with the groove is the same as used for straight sides; at the bottom the angle is larger. The grip on the rope is therefore not as great when it approaches the bottom of the groove as it would be if the sides of the latter were straight. The curved sides of the grooves thus afford a means of preventing the excessive tension which would be thrown on the rope if it were running in V grooves. The life of the ropes is therefore lengthened.

A groove whose angle grows more flat at the bottom strains the rope less when an excessive load is thrown on the system than does one with straight sides making an angle which is the mean of that of a curved-side groove.

Another method of preventing unequal loading of the ropes in a multiple drive, when pulleys of different diameters are used, is to make the groove of the larger one with a smaller angle than those of the smaller. The decrease of the effective diameter of the larger pulley is thus made more rapid than for the smaller, as the rope becomes smaller. A case is cited where the engine driving pulley was about three times the diameter of the driven pulley. By making the angle of the groove 30° on the larger pulley, and 45° on the smaller, unequal pulling was obviated.*

There seems to be a difference of opinion among engineers as to what form of groove should be used for idler- or guide-pulleys. On the one hand it is maintained that a groove of circular section at the bottom, and large enough to let the rope run in it without wedging against the sides, is the best on account of causing no frictional loss as the rope enters and leaves the groove; the contrary argument is that the rope slips in the round-bottom groove, and thus causes as much loss as the V groove, together with more rapid wear of the rope. It would seem that the round-bottom groove would be at its best when the rope is in contact with a considerable portion of the circumference of the pulley, and the V groove when the arc of contact is small, the round-bottom groove being the more apt to permit slipping with a small arc of contact; but whether the round bottom is better, even for large arcs of contact, does not seem to be positively settled.

Cast iron and hard wood, such as maple, are materials which are almost universally used for the rope to run on; the grooves are generally turned in the rim of the pulley, no lining being used. Cast-iron and wooden grooves therefore correspond to pulley-rims of these materials.

---

* Kent's "Mechanical Engineers' Pocket-book," 1896, p. 927.

The sides of the groove should be perfectly smooth in order to prevent rapid external wear of the rope.

58. Coefficient of friction of non-metallic ropes.—There seem to be no experiments showing the coefficient of friction of ropes running in grooved pulleys. The experiments made on ropes running over flat pulleys are scarcely applicable to modern practice in rope-driving, for there is little probability that the ropes used in the two cases were in even approximately the same condition with regard to lubrication. The coefficients given below are not experimentally determined, but based upon modern practice. It is believed, however, that they are sufficiently correct for all practical applications.

The value of μ for well-lubricated manila ropes running in polished grooves generally lies between 0.1 and 0.3; 0.12 to 0.15 are safe values to use in designing. These latter two values give for 45° groove angles the corresponding values of the coefficient of groove friction

$$\phi = 0.12 \csc \frac{45°}{2} = 0.32; \text{ and } \phi = 0.15 \csc \frac{45°}{2} = 0.40,$$

and for 30° grooves,

$$\phi = 0.12 \csc \frac{30°}{2} = 0.46; \text{ and } \phi = 0.15 \csc \frac{30°}{2} = 0.58.$$

When the rope is dry, or lubricated with a somewhat sticky substance, the coefficient of friction is higher.

Other qualities of hemp rope have practically the same coefficient of friction as manila, when lubricated in the same manner.

Cotton rope has a higher coefficient of friction than hemp. This may be due to the fact that it does not require so much lubrication, on account of its containing a natural lubricant, as well as its softer texture. It is doubtless safe to take μ as high as 0.2 in designing; this gives for 45° grooves $\phi = 0.52$, and for 30° angles $\phi = 0.77$.

Rawhide rope does not seem to have been used extensively enough to give a very definite knowledge of its frictional qualities. It is doubtless safe to take μ as high as for cotton ropes when the rawhide is kept in good condition and not allowed to get moist; if there is any liability to moisture, μ may be taken the same as for well-lubricated hemp ropes.

59. Working strength of non-metallic ropes.—When manila ropes are used for power transmission they have been found to be durable when working under a stress of $200d^2$ pounds on the taut side, where $d =$ diameter of rope in inches. Cotton ropes have given good service under the same tension. Rawhide ropes can be successfully operated at a tension at least one quarter higher than that of hemp and cotton, i.e., $250d^2$ pounds or more.

The above values are for economical working and reasonable durability. As with flat belts, ropes can be operated at much higher stresses, for a short time, than those just given; double the above values, or even more, may be used for a short time.

60. The velocity of ropes for power transmission is limited by the action of centrifugal force in the same manner as for flat belts. Non-metallic ropes run much more steadily at high speeds, however, the flapping and chasing common to flat belts being practically absent. On account of this latter advantageous quality they are commonly run at higher speeds than are found satisfactory for belts; a speed of 5000 feet per minute, or more, is frequently used. The tension due to centrifugal force in a rope weighing 0.32 of a pound per foot, and running at 8493 feet per minute, is equal to 200 pounds; this is practically the weight and working strength of a 1-inch manila rope. Roughly speaking, therefore, no power can be transmitted by a well-lubricated hemp rope working at $200d^2$ pounds tension when the velocity reaches 8500 feet per minute. The speed at which maximum power can be transmitted with such a rope, working under $200d^2$ pounds tension, is about 5500 feet per minute for an arc of contact $\theta = 160°$ to $180°$. Cotton rope, being somewhat lighter, has its speeds of maximum power transmission and of theoretically no power transmission both somewhat higher.

The most economical speed for manila rope does not vary much from 4500 feet per minute, although there is no great change in economy for variations of 1000 feet per minute on either side of this value. The speed that is most economical, for the rope alone, is that which gives the minimum cost of rope per horse-power transmitted, which cost includes first cost and maintenance of the rope only.

61. Wear and lubrication of non-metallic ropes.—The wear of vegetable-fibre rope belts is of two kinds, external and internal. External wear is caused by the slipping of the rope in the groove, and the rubbing against the side of the groove as it winds on and is unwound from the pulley. The outer strands are gradually worn away by external wear, and the rope weakened accordingly. In order to prevent any considerable weakening of the rope by external wear, a covering, of some material weaker and cheaper than that of the body of the rope, is sometimes placed over it, thus forming a rope which maintains a nearly uniform strength until the covering is worn through, provided the rope is well lubricated internally.

Internal wear requires serious consideration, for, unless some suitable lubricant is used to reduce the friction between the strands and fibres, the life of the rope is apt to be short. When an unlubricated or improperly lubricated rope has been in service for some time, a fine dust is formed in it by the particles worn off by internal friction. This dust can be easily seen by opening the strands. A reverse bend in a rope is a cause of greatly increased internal wear. For this reason all sheaves and pulleys should be placed, as far as practicable, so that the belt will bend in the same direction in passing over them.

A large number of the lubricants used for rope are made of graphite mixed with some substance such as molasses, beeswax, or tallow. The best way of applying a lubricant is to saturate the strands as they are being laid up to form the rope. Hemp rope for power transmission is generally so treated during its manufacture. A lubricant that is applied externally must be of such a nature that it will penetrate the rope and act upon all the strands to reduce their frictional resistance to rubbing against each other.

Cotton fibres are covered with an oleaginous wax in their natural condition; this coating serves as a lubricant for the rope, and eliminates the necessity of artificial lubrication.

When a vegetable-fibre rope is exposed to the weather while in service, a dressing which forms a water-proof coating should be applied to it, care being taken first to have the interior lubricated. A mixture of beeswax and graphite is frequently used for waterproofing. Substances of an adhesive nature, such as tar, while answering excellently as a preservative for a rope that does not run over pulleys or sheaves, is not suitable for those used for power transmission. Such a substance cements the fibres of the rope together so that they cannot slip over each other freely when the rope is bent around the pulley, thus causing them to tear or break each other. It may also cause the rope to adhere to the pulleys so that the fibres will be picked off by the pulley, which, of course, causes rapid deterioration.

Roughness of the grooves is certain to cause rapid wear, and, as has already been stated in § 57, care should be taken to have them perfectly smooth and without flaws.

The diameter of the pulleys over which a rope runs has much to do with the wear upon it; the smaller the pulley, the more rapid the wear on account of the sharper bend. A diameter of the pulley equal to about 40 times the diameter of the rope represents a fair average of the size found in practice. A somewhat smaller size than is represented by this may be used for the smaller diameters of rope up to $\frac{3}{4}$ inch; but for ropes as large as 2 inches diameter the pulley may be a few inches larger than this indicates. At high velocities it is believed to be advantageous to use somewhat larger pulleys than are suitable for low speeds.

62. Weight of hemp and cotton ropes.—The weight of new Firmus manila rope is about $0.283d^2$ pounds per foot.* In service the rope generally becomes heavier per foot, notwithstanding it stretches some. The increase of weight is doubtless due to the accumulation of the lubricant, and probably some dust, upon it. The maximum weight of an old manila rope of modern manufacture for power transmission seems to be about $0.32d^2$ pounds per foot; a table given in Kent's "Mechanical Engineers' Pocket-book," 1896, shows weights considerably heavier for the smaller sizes.

The average weight of cotton transmission rope that has been in service for some time is about $0.26d^2$ pounds per foot.

63. The diameter of ropes used for power transmission depends very largely upon the size of the sheave that can be conveniently used. A rope 2 inches in diameter is the largest that is found

---

* Computed from data in the catalogue of the Rice Machinery Co.

in practice to any considerable extent. From 1½ to 2 inches in diameter are the limits of the sizes generally used where there is a large amount of power to be transmitted, and the pulleys can be made large without inconvenience, as in rope-drives connecting the main pulley of a large engine with a line-shaft, electric generator, or cable-drums of a cable railway. Smaller sizes are used for general transmission in buildings using power for manufacturing purposes.

64. The effect of the relative position of pulleys upon ropes used for power transmission is of the same nature as for flat belts. In practice, however, it is demonstrated that hemp and cotton ropes work more satisfactorily than flat leather belts when one pulley is at a considerable distance above the other. This is doubtless largely due to the fact that a tension-carriage is used to take up the slack in the rope, while the belt, ordinarily operated without any device to make adjustment for stretch, soon becomes loose on the lower pulley. The weight of a leather belt for such a drive is generally greater than that of a hemp or cotton rope, which is disadvantageous for the belt.

Hemp and cotton ropes are often used successfully, in factories, for directly connecting sheaves or pulleys three or four stories apart, one being nearly or quite vertically above the other.

65. The efficiency of rope belting seems to be about the same as that of flat belts when the distance of transmission is short, so that only a single stretch is necessary between the driving and the driven pulleys. For long distances of transmission the rope is more efficient than flat leather belting; this is partly owing to the fact that much longer stretches of rope can be used than of flat belting.

WIRE ROPES.

66. Wire ropes made of iron or steel, and generally with a hemp core, have been extensively used for transmitting power through long distances. They are being replaced by electrical transmission machinery for the greater distances, however, and by nonmetallic ropes for the lesser. The large diameter of the sheaves, generally from 100 to 140 times that of the rope, is an objectionable feature on account of the space required for them, their great weight and comparatively high cost. The rope itself is much more expensive than non-metallic ones for transmitting the same amount of power.

The sheaves for wire ropes are generally made of cast iron. The grooves are made much wider than the diameter of the rope, which runs against the bottom only; the flanges act only as guides to prevent it from leaving the sheave in case of excessive swaying. The bottom of the groove is lined with wood, leather, gutta percha, or some similar material, soft as compared with the iron.

Wire rope sometimes has a strong tendency to sway when running. One of the chief causes of this is lack of roundness in the sheave. After the lining is placed in the groove, it should be accurately turned to run true; a small depression, sufficient to fit possibly one quarter of the circumference of the rope, will hold it in the centre of a sheave rotating on a horizontal axis, when the rope is running without swaying.

The Richmond Manufacturing Company of Lockport, New York, have doubtless had the most extensive experience with wire-rope power transmission of any concern in the United States. As a result of an experience of thirty years, they have arrived at the following conclusions:

"First. The best motion for the pulleys is not less than 100 or more than 140 revolutions per minute. The power will often do well at a less or greater motion, but generally the result is not so satisfactory.

"Second. If a cable is run over level ground for a long distance a support will be needed every 400 or 500 feet, but where both ends of the cable are high enough so as to allow plenty of room for the sag in the middle, a support is not required for less distance than 1000 feet. Where supports are required, a stout post set firmly in the ground and extending about 20 feet above it, with two small pulleys upon the top, answers the purpose.

"Third. Where the cable is short, some method of tightening it should be provided, either by arranging the driving pulleys so that they can be pushed farther apart, or by using an extra pulley as an idler; where the cable is long, this is not required, as the weight of the cable itself will then prevent any slipping on the pulleys.

BELTS AND ROPES FOR POWER TRANSMISSION.

"Fourth. Large cables must never be used upon small pulleys, as the continual bending of coarse wires around too small pulleys would soon break them."

Table XII has been adopted by this company for its own practice, and is recommended for general work.

TABLE XII.

SHOWING THE NUMBER OF HORSE-POWER EACH SIZE OF CABLE WILL SAFELY TRANSMIT; THE DIAMETER OF DRIVING AND DRIVEN LEATHER-PACKED GROOVED PULLEYS NECESSARY TO BE USED WITH EACH SIZE; THE NUMBER OF REVOLUTIONS THEY SHOULD MAKE, ETC.

Diameter of Cable. Inches.	Diameter of Pulley. Feet.	Revolutions per Minute.	Horse-power Transmitted.
$\frac{1}{2}$	3	100	2
	3	140	3
$\frac{5}{8}$	4	100	4
	4	140	6
$\frac{7}{8}$	5	100	9
	5	140	13
1	6	100	14
	6	140	20
$1\frac{1}{8}$	7	100	23
	7	140	32
$1\frac{1}{4}$	8	100	33
	8	140	42

Their experience has also shown that 100 feet is the shortest distance between the driving and driven pulleys that will give satisfactory operation. They find that a cable working according to the conditions given above will last from five to six years when working ten hours a day.*

In the use of wire rope on cable railways, and where power is transmitted over a long distance by a single endless rope, it is desirable to have the tension on the tight side of the rope several times that on the slack. This is accomplished by using a pair of winding drums at the power plant of a railway, and at both ends of the line when power is transmitted between two points. The two drums of a pair are placed near together, with their axes parallel;

---

* Data and information kindly furnished by the Hon. William Richmond, president of the Richmond Manufacturing Co.

the tight side of the rope winds in the groove at one end of one drum, passing half-way around it, and then goes to the corresponding groove on the other, then back to the second groove of the first drum, and so on, winding consecutively in all the grooves of both drums. This device is most effective when the drums are geared together so that both act as drivers, or, if they are at the driven end of the line, power can be taken from both.

The wear on such a pair of winding drums, used as drivers, is most rapid in the groove where the rope first winds on, and gradually decreases in each successive groove. When the first groove has become worn to a smaller effective diameter than the last, there is a tendency toward unequal winding, and, as a consequence, heavy strains are produced in the rope on and between the two drums, which can only be relieved, in a measure, by the rope's slipping in the groove.

In order to prevent the rapid destruction of the rope and drums by this action, Mr. John Walker * has designed a differential pulley. A section of the rim of this pulley is shown in Fig. 81. Each

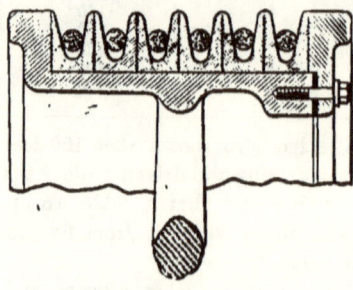

Fig. 81.

groove is cut in a ring, separate from the rest of the pulley, and a number of the rings are placed side by side on what corresponds to a flanged pulley. The rings have a free-running fit on the cylindrical part of this pulley, but their resistance to turning is regulated by one of the flanges, which is separate from the pulley and held in

---

* Formerly of the Walker Co. of Cleveland, Ohio; now Consulting Engineer located in Chicago.

place by bolts that can be adjusted to give the desired pressure against the sides of the rings. A cushion of some elastic material is placed between the loose flange and the main part of the pulley. The loose flange should be adjusted so that the rings will turn on the pulley a little more easily than the rope will slip in the grooves.

The differential pulley can be equally well applied to any system of rope-driving, using any kind of rope, where there are two or more grooves in a pulley.*

---

* A large amount of data on power transmission by wire ropes is given in Kent's "Mechanical Engineers' Pocket-book."

CHAPTER IV.

SCREWS FOR POWER TRANSMISSION.

67. In some classes of machines, screws are used for applying a great force acting through a small distance. Examples of such application may be seen in testing-machines for determining the strength of materials, and in presses for copying, baling cotton, etc. Again, in such machines as those for planing the edges of boiler-plates, a proportionately smaller force, exerted by the screw against the tool-carriage, acts through a greater distance.

The thread used on such screws is generally square, and in all cases, unless some special requirements make it necessary to have some other form, the surface of the driving side of the thread should be such as is generated by a line always remaining perpendicular to the axis of the thread. In other words, the working side should be the same as that of a square thread.

The following notation for screws will be used:

A = sectional area of screw at bottom of thread, square inches;
D = mean diameter of collar- or step-bearing, inches;
E = efficiency of screw and collar for forward motion;
E' = efficiency of screw and collar for backward motion or overhauling;
F = turning force applied to the screw for raising the load, pounds;
F' = turning force applied to the screw for lowering the load, pounds;
F'' = turning force exerted by screw when overhauling, pounds;
I_p = polar moment of inertia of the section of the screw, biquadratic inches;
T = tension or compression in screw, pounds;
d = mean diameter of thread, inches;

SCREWS FOR POWER TRANSMISSION. 151

e = efficiency of screw-thread alone for forward motion;
e' = efficiency of screw-thread alone for backward motion or overhauling;
l = length of lever-arm of the turning force, inches;
p = pitch of thread, inches;
r_1 = radius of screw at bottom of thread, inches;
s = shearing stress per unit area caused by the screw-thread resistance, pounds per square inch;
t = tensile or compressive stress per unit area caused by the axial force, pounds per square inch;
β = angle between surface of thread and a normal to the axis of the screw;
θ = angular pitch of thread, which is the angle between the mean helix and a plane perpendicular to axis of screw, degrees;
μ = tan ϕ = coefficient of friction between threads;
μ' = tan ϕ' = coefficient of friction between collar and supporting surface;
ϕ = angle of friction between threads, degrees;
ϕ' = angle of friction between collar and supporting surface, degrees.

The pitch p is the distance, parallel to the screw axis, between similar points of adjacent turns of the thread on a single-thread screw; in a screw having more than one thread p is the axial distance between similar points on successive turns of the same thread. The mean thread-diameter d is taken as that of a helix lying midway between the top and bottom of the thread, and, for convenience, all the pressure of the nut against the thread is considered as acting along this mean helix. Theoretically the diameter of this helix is slightly greater than this mean value, but the difference is so small as to be far within the necessary limits of accuracy for any practical requirements. The same is true of D. The mean angular pitch θ is found by laying out a right triangle, making one side equal p, and the other equal πd, θ being the angle between the latter and the hypothenuse. This gives

$$\tan \theta = \frac{p}{\pi d} \quad \ldots \quad \ldots \quad (49)$$

Square-thread Screws.

68. Relation between the turning moment and axial force in a square-thread screw.—When a screw is used for power transmission, as described above, it is often desirable to know the turning moment which must be applied to produce a given axial force, or *vice versa;* the axial force may be either tension or compression.

Fig. 82 may be taken to represent a portion of such a screw. As a convenient method of dealing with the problem, it may be assumed that a load is suspended from it by means of a nut (not shown) which fits on the thread; also, that the screw is supported by a bearing on which the collar near the top rests.

It is evident that friction must be considered. Since the nut and supporting surface under the collar may be of different materials, or differently lubricated, the coefficient of friction between the screw and nut, and that between the collar and the material against which it bears, may have different values; to make the case general, they will be considered as different.

If there were no friction between the threads of the nut and screw, the pressure between them would be normal to their working surfaces and make an angle θ with the axis of the screw. On account of friction, however, the direction of the force acting between them at any point makes the angle of friction ϕ with a normal to the thread at that point. When the screw is turning to raise the load, the direction of pressure between the threads is at an angle $(\theta + \phi)$ with the axis of the screw.

An elementary portion ab (see figure), of the tension T, is therefore held in equilibrium by the forces ac and cb; $cb = ab \tan(\theta + \phi)$. The latter is the external force, normal to the axis of the screw, which must be applied to turn the screw against the resistance due to the elementary axial force ab. The nut resistance to the turning of the screw, acting as it does at a distance $\frac{d}{2}$ from the screw's axis, is of a value $(cb) \times \frac{d}{2} = ab \tan(\theta+\phi) \times \frac{d}{2}$. The total nut resistance, for the total load T, is the sum of the resistances for all the elementary forces, each equal to ab, and is expressed by the equation

$$\text{Nut resistance} = T \tan(\theta + \phi) \times \frac{d}{2}.$$

SCREWS FOR POWER TRANSMISSION.

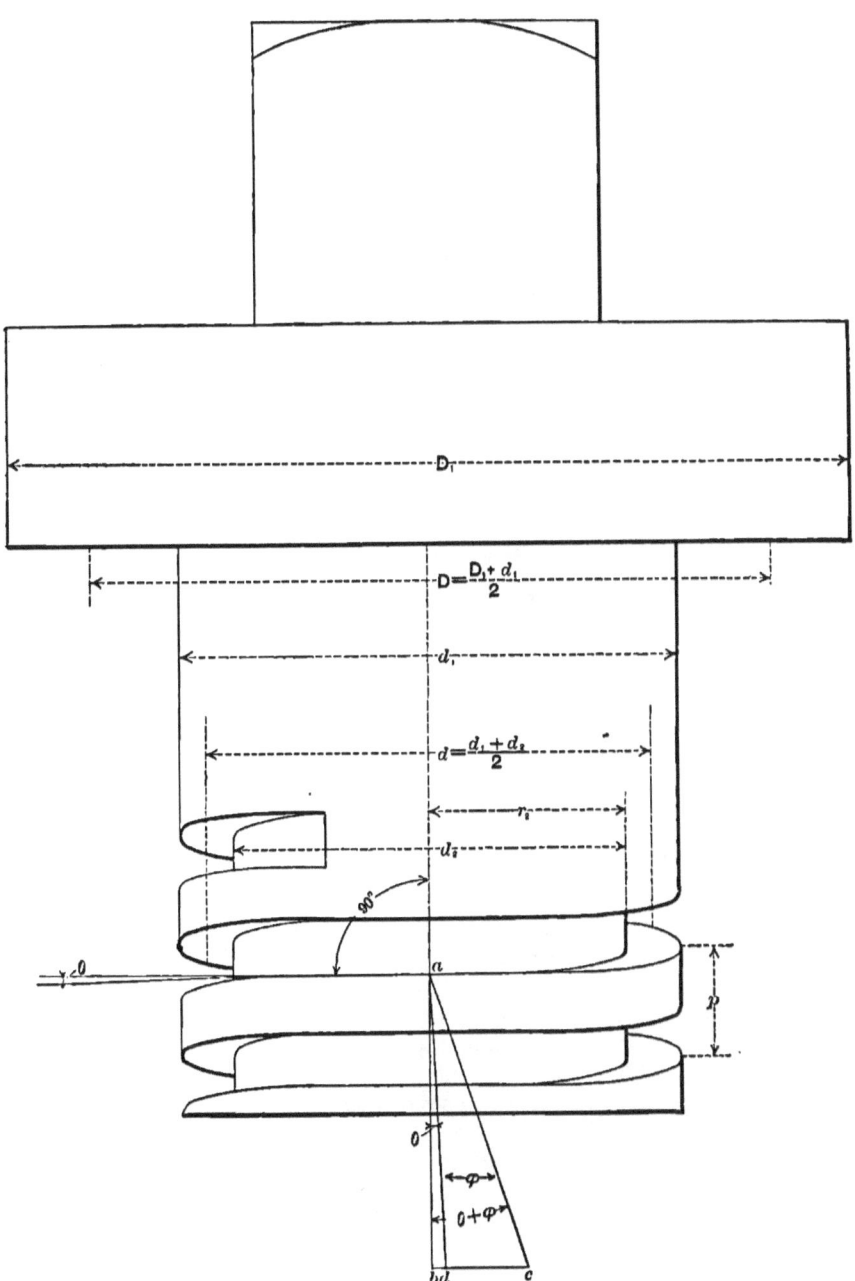

Fig. 82.

This represents the value of the turning moment which must be applied to overcome the resistance in the nut.

By the same method, the turning moment necessary to overcome the friction of the pressure T against the collar, is shown by the equation

$$\text{Collar friction} = T \tan \phi' \frac{d}{2} = T\mu'\frac{D}{2}.$$

In order to prevent side pressure of the body of the screw against the supporting frame, the torsional moment acting to turn the screw may be applied as a couple, each force having a lever-arm equal to $\frac{l}{2}$ about the axis of the screw, thus making the arm of the couple equal to l. Calling each of the external turning forces F, the driving torsional moment becomes Fl.

The following equation of turning and resisting moments can now be written for a square-thread screw with thrust collar when lifting a load:

Screw Resistance.   Collar Friction.

$$Fl = T \tan(\theta + \phi)\frac{d}{2} \quad + T\mu'\frac{D}{2}. \quad . \quad . \quad (50)$$

$$= T\frac{\tan \theta + \tan \phi}{1 - \tan \theta \tan \phi} \times \frac{d}{2} + T\mu'\frac{D}{2}$$

$$= T\frac{p \div \pi d + \mu}{1 - \mu(p \div \pi d)} \times \frac{d}{2} + T\mu'\frac{D}{2}$$

$$= T\frac{p + \mu \pi d}{\pi d - \mu p} \times \frac{d}{2} \quad + T\mu'\frac{D}{2}. \quad . \quad . \quad (51)$$

In equations (50) and (51), as well as those between them, the screw resistance is expressed by the first part of the right-hand members, and the collar friction by the last part of the right-hand members.

For lowering the load by turning the screw backward, the angle of friction ϕ must be taken on the opposite side of the normal ad to the thread from that shown in the figure, because the frictional resistance acts in a direction opposite that when the load is lifted.

The equation of turning moment and tension then becomes, for lowering the load, when ϕ is greater than θ,

Thread Resistance.   Collar Friction.

$$F'l = T\tan(\phi - \theta)\frac{d}{2} + T\mu'\frac{D}{2}, \quad \ldots \quad (52)$$

or

$$F'l = T\frac{\mu\pi d - p}{\pi d + \mu p} \times \frac{d}{2} + T\mu'\frac{D}{2}. \quad \ldots \quad (53)$$

And when θ is greater than ϕ,

Collar Friction.   Driving Moment of Screw-thread.

$$F'l = T\mu'\frac{D}{2} - T\tan(\theta - \phi)\frac{d}{2}, \quad \ldots \quad (54)$$

or

$$F'l = T\mu'\frac{D}{2} - T\frac{p - \mu\pi d}{\pi d + \mu p} \times \frac{d}{2}. \quad \ldots \quad (55)$$

The value of θ that will just hold the load at rest, or allow it to descend uniformly, when no external turning force is applied, is given by the equation

$$\tan(\theta - \phi) = \mu'\frac{D}{d}. \quad \ldots \quad (56)$$

Any larger value of θ than obtained by this equation will require a resisting moment applied to the screw to hold the load.

A screw that will allow a load to descend by its own weight is said to overhaul.

The turning moment $F''l$ that a screw will exert when overhauling is

Driving Moment of Screw-thread.   Collar Friction.

$$F''l = T\tan(\theta - \phi)\frac{d}{2} - T\mu'\frac{D}{2}, \quad \ldots \quad (57)$$

or

$$F''l = T\frac{p - \mu\pi d}{\pi d + \mu p} \times \frac{d}{2} - T\mu'\frac{D}{2}. \quad \ldots \quad (58)$$

If there were no collar friction, overhauling would occur for any value of θ greater than ϕ.

The principle of overhauling is applied to advantage in some mechanisms. Probably the most familiar example is the screw-driver having a handle which contains a nut that engages with a screw-thread of rapid pitch on the stock for holding the bit or blade: by forcing the handle down over the screw, the blade is rotated.

69. Efficiency of a square-thread screw and collar.—For forward motion, when the end thrust of the screw is resisted by a collar, the efficiency may be found by dividing the useful work accomplished during one revolution of the screw, $= Tp$, by the work, $F(2\pi l)$, applied to turn the screw through one revolution; the efficiency for forward motion is therefore expressed by the equation

$$E = \frac{Tp}{F(2\pi l)}. \qquad \ldots \ldots \ldots (59)$$

By substituting in this equation the value of p, as expressed in equation (49), and that of Fl as given in equation (50), it becomes

$$E = \frac{T\pi d \tan \theta}{T\pi d \tan(\theta + \phi) + T\pi D\mu'} = \frac{\tan \theta}{\tan(\theta + \phi) + \dfrac{D}{d}\mu'}. \quad (60)$$

If the coefficients of collar friction and thread friction are equal, which corresponds to $\mu' = \mu = \tan \phi$, the maximum value of E is obtained when

$$\tan \theta = \frac{\sqrt{1 + \dfrac{D}{d}}}{\sec \phi + \tan \phi \sqrt{1 + \dfrac{D}{d}}}. \quad \ldots \ldots (61)$$

The efficiency e of the screw-thread alone is obtained by dropping from the denominator of equation (60) the expression $T\pi D\mu'$ for the work done in overcoming collar friction, which gives

$$e = \frac{\tan \theta}{\tan(\theta + \phi)}. \qquad \ldots \ldots \ldots (62)$$

SCREWS FOR POWER TRANSMISSION. 157

The maximum value of e is obtained when

$$\tan \theta = \sec \phi - \tan \phi = \sqrt{1 + \mu^2} - \mu. \quad . \quad . \quad (63)$$

The value of the thread angle θ is less for maximum efficiency when there is no collar friction than when there is.

When a screw is overhauling, the points of application of the driving force and resistance are exchanged; the axial force or load becomes the driving force, and the external resistance is applied at the same point as the driving force when lifting the load.

The efficiency E' for overhauling is therefore expressed by the equation

$$E' = \frac{2\pi F'' l}{Tp}. \quad . \quad . \quad . \quad . \quad (64)$$

The value of $F''l$, in this expression, is given in equations (57) and (58); that in (57) will be substituted. Therefore,

$$E' = \frac{T\pi d \tan(\theta - \phi) - T\pi D\mu'}{T\pi d \tan \theta} = \frac{d \tan(\theta - \phi) - D\mu'}{d \tan \theta}. \quad (65)$$

Dropping the collar friction $T\pi D\mu'$ from this equation gives for the efficiency e' of the screw-thread alone, when overhauling,

$$e' = \frac{\tan(\theta - \phi)}{\tan \theta}. \quad . \quad . \quad . \quad . \quad (66)$$

When a driving force F' must be applied to a screw when lowering its load, there is no efficiency. The efficiency is zero when the load runs down uniformly of its own accord.

70. Coefficient of friction μ for square-thread screws.—This quantity has been very carefully determined by Prof. Albert Kingsbury for different materials and lubricants.* The dimensions of all the bolts and nuts tested were as follows:

Outside diameter of screw .... 1.426 inches
Inside diameter of nut .... 1.278 "
Mean diameter of thread .... 1.352 "
Pitch of thread..... .... $\frac{1}{3}$ "
Depth or height of nut (effective)...... $1\frac{1}{16}$ "
Area of rubbing surface of thread...... 1 sq. in. (about)

* Trans. Amer. Soc. Mech. Engrs., vol. XVII., 1896, p. 96.

158   FORM, STRENGTH, AND PROPORTIONS OF PARTS.

The nuts fit the screws loosely in order that there might be no friction other than that between the working surfaces. "The threads were cut carefully in the lathe and had been worn to good condition by trials previous to those here recorded. Screw No. 5 was not quite so smooth as the others." The lubricant was applied when the nut and screw were put together, and no more was added during the taking of the series of cards for them. The speed was slow; about one revolution in two minutes.

The results of the tests are given in Table XIII.

TABLE XIII.
COEFFICIENTS OF FRICTION FOR SQUARE-THREAD SCREWS.

Working surface of thread approximately 1 square inch in area; pressure per square inch of thread approximately the same as the total tension in screw-bolts.

Screws.	Coefficients of Friction μ. Average Values of Four Readings.				Value of μ for a Single Card.		Pressure and Lubricant.
	Nut 6, Mild Steel.	Nut 7, Wrought Iron.	Nut 8, Cast Iron.	Nut 9, Cast Brass.	Highest.	Lowest.	
1. Mild steel....	.141	.16	.136	.136	Screw 5 Nut 9 $\mu = .20$	Screw 3 Nut 8 $\mu = .11$	Pressure 10000 lbs. per sq. inch. Machinery-oil.
2. Wrought iron.	.139	.14	.138	.147			
3. Cast iron.....	.125	.139	.119	.171			
4. Cast bronze...	.124	.135	.172	.132			
5. Mild steel, case-hardened.	.133	.143	.13	.193			
1.........	.12	.105	.10	.11	Screw 4 Nut 9 $\mu = .25$	Screw 3 Nut 8 $\mu = .09$	Pressure 10000 lbs. per sq. inch. Lard-oil.
2.........	.1125	.1075	.10	.12			
3.........	.10	.10	.095	.11			
4.........	.1150	.10	.11	.1325			
5.........	.1175	.0975	.105	.1375			
1.........	.111	.0675	.065	.04	Screw 5 Nut 6 $\mu = .15$	Screw 5 Nut 9 $\mu = .03$	Pressure 10000 lbs. per sq. inch. Machinery-oil and graphite.
2.........	.089	.07	.075	.055			
3.........	.1075	.071	.105	.059			
4.........	.071	.045	.044	.036			
5.........	.1275	.055	.07	.035			
1.........	.147	.156	.132	.127	Screw 5 Nut 7 $\mu = .19$	Screw 2 Nut 9 $\mu = .11$	Pressure 3000 lbs. per sq. inch. Machinery-oil.
2.........	.15	.16	.15	.117			
3.........	.15	.157	.14	.12			
4.........	.127	.13	.13	.14			
5.........	.155	.1775	.1675	.1325			

- The coefficient of friction was low enough, in some cases, to allow the screw to "overhaul"; i.e., a pressure against the screw, parallel to its axis, caused it to rotate and pass through the nut.

Following out the fairly well established fact that, in oil-lubricated journals, the coefficient of friction decreases rapidly as the velocity of rubbing increases from slightly above zero to 10 or 100 feet per minute, according to whether the pressure is high or low, it would seem that lower values of the coefficients than given in the table might be found in screws running at the higher speeds that are frequently used for power transmission.

71. Problem.—Design a screw to lift 7 tons = 14000 pounds; end thrust of screw to be taken by collar-bearing; screw must not overhaul; driving gear attached to screw to be 18 inches in diameter.

For finding the thread angle to prevent overhauling, the lowest values of the screw and collar friction that may occur with the materials used should be adopted. It will be assumed that the screw is to be of mild steel; the nut and collar-bearing may be of cast brass. The lowest values of the coefficients of friction may be taken as

$$\mu = .03 = \tan \phi, \text{ whence } \phi = 1° \, 43'; \text{ and } \mu' = .025.$$

The ratio of the mean diameter of the collar to that of the thread will be taken as $1.6 = D \div d$.

The value of the thread angle θ is found by equation (56), by which

$$\tan (\theta - \phi) = \mu' \frac{D}{d} = .025 \times 1.6 = .04,$$

whence

$$\theta - \phi = 2° \, 18';$$
$$\theta - 1° \, 43' = 2° \, 18';$$
$$\theta = 4° \, 1'.$$

Since the pitch must generally be such that the screw can be cut in an ordinary lathe, it will be assumed that $p = .5$ inch, and the mean diameter d of the screw, to give the required angle θ, calculated.

By equation (49),

$$\tan \theta = \frac{p}{\pi d},$$

whence

$$d = \frac{p}{\pi \tan \theta} = \frac{.5}{\pi \tan 4°} = \frac{.5}{.06993\pi} = 2.27.$$

Taking the outside diameter of the screw as $0.45p$ greater than the mean diameter d gives

Outside diameter of screw $= 2.27 + .45 \times .5 = 2.5''$.

The mean diameter of the collar is

$$D = 1.6d = 1.6 \times 2.27 = 3.63 \text{ inches.}$$

The turning moment Fl which is necessary to raise the load is found by equation (51). The greatest values of the coefficients of friction, μ and μ', that are probable to occur at any time when the lubrication is poor, should be used in this equation in order that the maximum value of the turning moment may be found.

The values $\mu = .15$ and $\mu' = 0.1$ are probably the highest that occur in well-made machines operated with reasonable care and cleanliness.

Applying equation (51),

$$Fl = T\frac{p + \mu\pi d}{\pi d - \mu p} \times \frac{d}{2} + T\mu'\frac{D}{2}$$

$$= 14000 \frac{.5 + .15\pi 2.27}{\pi 2.27 - .15 \times .5} \times \frac{2.27}{2} + 14000 \times 0.1 \frac{3.63}{2}$$

$$\begin{array}{ccc} \text{Screw Resistance.} & \text{Collar Friction.} & \\ = \quad 3532 & + \quad 2541 & = 6073 \text{ pounds.} \end{array}$$

And the force F acting tangent to the pitch circle of a gear whose pitch radius is 9 inches is

$$F = 6073 \div 9 = 675 \text{ pounds.}$$

No allowance has been made for journal friction in the above calculations. This makes the value of F greater than 675 pounds. The pressure between the teeth of the driving gears is greater than F on account of their obliquity of action. Assuming that the actual pressure between the teeth, taking into account both journal friction and obliquity of action, is 800 pounds, and that the coefficient of journal friction is .06, then the corresponding value of the tangential force is

$$F = 675 + (.06 \times 800) = 723 \text{ pounds.}$$

The efficiency of the screw when working with the highest assumed coefficients of friction, and including journal friction, is, by equation (59),

$$E = \frac{Tp}{F(2\pi l)} = \frac{14000 \times 0.5}{723 \times 2\pi \times 9} = .174 = 17.4\%.$$

72. Maximum stress in a screw.—A screw that is turning and lifting a load is subjected to an axial tensile stress equal to the weight of the load, together with a torsional moment equal to the turning moment that must be applied to overcome the resistance of the screw-thread. The combination of these two stresses produces what may be called a maximum tensile stress and a maximum shearing stress. The values of these maximum stresses should not exceed the working strengths, tensile and shearing, of the material.

If the axial force acting on the screw is compression instead of tension, and the distance between the nut and collar is small enough to allow the part in compression to be considered as a short column, not liable to bend or buckle, the maximum compressive stress will be equal to the maximum tensile stress for the same load, and the maximum shear will be the same in both cases.

The formulas for maximum stress in a rod of circular section, as given in works on the "mechanics of materials," are:

$$\text{Maximum tension or compression} = \frac{t}{2} + \sqrt{\frac{t^2}{4} + s^2}; \quad (67)$$

$$\text{Maximum shear} = \sqrt{\frac{t^2}{4} + s^2}. \quad (68)$$

In dealing with the strength of a screw of ordinary construction, the thread may be neglected, since it adds but little to the strength. The screw may therefore be considered as a cylindrical bar of a diameter the same as that of the bottom of the thread. Under this assumption the value of t is

$$t = T \div A = T \div \pi r_2^2,$$

and that of s, for square-thread screws, is

$$s = \frac{Fl}{I_p \div r_2} = \frac{T \tan(\theta + \phi) \frac{d}{2}}{\frac{1}{2}\pi r_2^4 \div r_2} = \frac{T \tan(\theta + \phi) d}{\pi r_2^3}.$$

By substituting these values of t and s in equations (67) and (68), the following equations are obtained for square-thread screws:

$$\left.\begin{array}{l}\text{Max. tension or}\\ \text{compression per}\\ \text{sq. in.}\end{array}\right\} = \frac{T}{2A} + \sqrt{\frac{T^2}{4A^2} + \frac{T^2 \tan^2(\theta + \phi) d^2}{\pi^2 r_2^6}},$$

which reduces to

$$\left.\begin{array}{l}\text{Max. tension or}\\ \text{compression per}\\ \text{sq. in.}\end{array}\right\} = \frac{1}{2}\frac{T}{A}\left[1 + \sqrt{1 + 4\tan^2(\theta + \phi)\left(\frac{d}{r_2}\right)^2}\right]. \quad (69)$$

In the same manner,

$$\left.\begin{array}{l}\text{Maximum shear}\\ \text{per sq. in.}\end{array}\right\} = \frac{1}{2}\frac{T}{A}\sqrt{1 + 4\tan^2(\theta + \phi)\left(\frac{d}{r_2}\right)^2}. \quad . \quad . \quad (70)$$

Example of application of formulas for maximum stress in a screw.—In the problem solved in § 71 the following data were assumed, or else determined for the highest coefficients of friction: $T = 14000$ pounds; $\tan \phi = .15$; $\phi = 8° 31' 10''$; $\theta = 4° 1'$; $\theta + \phi = 12° 32' 10''$; and $d = 2.27$ inches. In addition to these data, the radius at the bottom of the thread may be taken as $r_2 = 1$ inch; whence $A = 3.1416$ square inches.

By the substitution of these quantities in equation (69),

$$\text{Maximum tension} = \frac{1}{2}\frac{T}{A}\left[1 + \sqrt{1 + 4\tan^2(12°\,32'\,10'')\left(\frac{2.27}{1}\right)^2}\right]$$

$$= \frac{1}{2}\frac{T}{A}[1 + \sqrt{1 + 4(0.22236)^2(2.27)^2}]$$

$$= \frac{1}{2}\frac{14000}{3.1416}(1 + 1.4211) = 5390 \text{ lbs. per sq. in.};$$

and, by equation (70),

$$\text{Maximum shear} = \frac{1}{2}\frac{14000}{3.1416} \times 1.4211 = 3180 \text{ lbs. per sq. in.}$$

Angular-thread Screws.

73. Relation between turning moment and axial force in an angular-thread screw.—When the working surface of a screw thread makes an angle β with a normal to the axis of the screw, as in Fig. 83, the pressure between the threads of the nut and screw is

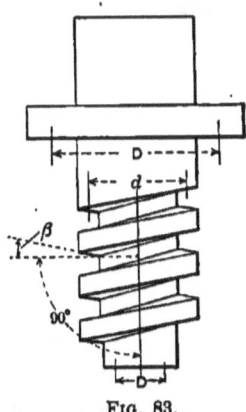

Fig. 83.

increased by the wedge-like action of the thread.

The following equations apply to an angular-thread screw when lifting a load.* (Notation given in § 67.)

---

* The development of the equation for thread resistance is given in § A of the Appendix.

164 FORM, STRENGTH, AND PROPORTIONS OF PARTS.

$$Fl = T\frac{\tan\theta + \tan\phi \sec\beta}{1 - \tan\theta \tan\phi \cos\beta} \times \frac{d}{2} + T\mu'\frac{D}{2} \quad . \quad . \quad (71)$$

Thread Resistance.   Collar Friction.

$$= T\frac{(p \div \pi d) + \mu \sec\beta}{1 - (p \div \pi d)\mu \cos\beta} \times \frac{d}{2} + T\mu'\frac{D}{2}$$

$$= T\frac{p + \mu\pi d \sec\beta}{\pi d - \mu p \cos\beta} \times \frac{d}{2} + T\mu'\frac{D}{2}; \quad . \quad . \quad . \quad (72)$$

$$E = \frac{Tp}{F(2\pi l)} = \frac{Td \tan\theta}{2Fl}; \quad . \quad . \quad . \quad . \quad . \quad . \quad (73)$$

$$e = \frac{\tan\theta(1 - \tan\theta \tan\phi \cos\beta)}{\tan\theta + \tan\phi \sec\beta}. \quad . \quad . \quad . \quad . \quad (74)$$

For the backward motion of an angular-thread screw when a driving force is applied to lower the load:

Thread Resistance.   Collar Friction.

$$F'l = T\frac{\tan\phi \sec\beta - \tan\theta}{1 + \tan\theta \tan\phi \cos\beta} \times \frac{d}{2} + T\mu'\frac{D}{2}. \quad . \quad (75)$$

If $\tan\theta$ is greater than $\tan\phi \sec\beta$ in equation (75), then the thread resistance is negative, and must be subtracted from the collar friction.

The thread resistance for backward motion becomes zero when

$$\tan\theta = \tan\phi \sec\beta,$$

or

$$\tan\phi = \tan\theta \cos\beta.$$

The last two equations indicate the values of θ and ϕ which would just allow the screw to overhaul if there were no collar friction.

When overhauling, the driving moment, exerted by the nut, is

Driving Moment of Screw-thread.   Collar Friction.

$$F''l = T\frac{\tan\theta - \tan\phi \sec\beta}{1 + \tan\theta \tan\phi \cos\beta} \times \frac{d}{2} - T\mu'\frac{D}{2}. \quad . \quad (76)$$

A series of experiments upon a screw-bolt were made by James McBride to determine its efficiency.* An ordinary V-thread screw-bolt was purchased on the market, and was not specially prepared for the experiments; the nut fitted the thread freely, so that it could be run along the thread easily with the hand; the flat end of the nut rested on a washer which formed the bearing-surface; both nut and washer were of malleable iron with bearing-surfaces left rough or unfaced. Lard-oil was used as a lubricant. The bolt was suspended vertically by the nut, which rested on a support 4 feet 6 inches high; a load was hung on the lower end of the bolt. The torsional moment was applied by different men pulling on a horizontal, single-ended wrench which spanned the nut. The following are the more important data:

Diameter of bolt = 2 inches;
Pitch of screw   = 0.22 of an inch = 4.5 threads per inch;
Standard V thread;
Load suspended = 7500 pounds.

The highest efficiency found was 12.29%, the lowest 9.71%, the average being 10.19%.

At the average efficiency of 10.19% a torsional moment of 1 inch-pound, applied to the wrench, would produce an axial stress $T = (2\pi \div 0.22)0.1019 = 2.91$ pounds in the bolt.

---

*Trans. Amer. Soc. Mech. Engrs., vol. XII., 1891, p. 781.

CHAPTER V.

SCREW-GEARING.

74. The most common form of screw-gearing that is used, when there is to be any considerable amount of power transmitted, is the worm and worm-wheel; and, of this mechanism, that having the axes of the worm and wheel at right angles is most generally adopted, doubtless because it is only when the shafts are at a right angle that the face of the wheel can be concaved to embrace a portion of the worm. In the worm and worm-wheel there is generally a great difference of angular velocities, the worm making many revolutions to give one turn to the wheel. The diameter of the wheel is ordinarily much greater than that of the worm.

When the axes are at any other angle than 90°, the face of the wheel cannot be concaved, and its teeth, instead of being curved so as to secure line contact with the worm, must be portions of a many-threaded screw, or else similar to those of a spur-gear, both of which give only point contact theoretically.

When the worm and wheel do not have greatly different angular velocities, and the teeth on both are short lengths of the threads of a many-threaded screw, the name "screw-gears" is commonly applied. This term will be used to designate all such mechanisms in which the wheel does not have a concave face. The axes may be at any angle.

WORM AND WORM-WHEELS.*

75. The strength of a worm and worm-wheel seldom needs consideration, for, under ordinary conditions, the allowable pressure

---

* The coefficient of friction for worm-gearing is probably about the same as that of screw-gears (Table XXI) or square-thread screws (Table XIII), according to the velocity of rubbing.

between them, that will not produce abrasion, is much less than would be required to break the teeth when made of any of the materials common to engineering.

Since the teeth are cut on a wheel concaved to fit the worm, they are much stronger than those of the same proportions that are straight. Their strength cannot be satisfactorily calculated, and it is scarcely desirable to make such a calculation.

The strength of the worm-thread cannot be calculated, but when of the same material and thickness at the pitch surface as the teeth on the wheel, it is the stronger of the two.

76. Equations for turning force and efficiency of worm and worm-wheel.—The equations of the relation between the turning moment and pressure against the teeth in a direction parallel to the axis of the worm are essentially the same for a worm-wheel as those of an angular-thread screw, given in § 73. This assumes that the thread is angular, not square, for it is not known that a square-thread worm is ever used when it must perform a service that demands even moderately high speeds and pressures.

If a worm works in engagement with two wheels that are tangent to its pitch surface at diametrically opposite points, and whose axes are parallel to each other and at right angles to that of the worm, the equations for an angular-thread screw apply to it without modification. For the more common mechanism, in which the worm engages with a single wheel, there is a side pressure on the supporting journal-bearings which may materially increase the turning moment required to rotate the worm against a given pressure of the wheel against it. This is especially true if the journal is large or has a high coefficient of friction.

The thrust-bearing of a worm can be obviated by using two worms, one right-hand and the other left-hand, on the same shaft, each engaging with a suitable worm-wheel; the shafts of the worm-wheels must be geared together so as to give them both the same rate of rotation in opposite directions, provided each worm and wheel forms a part of the mechanism differing from the other only in the direction of the thread. It can readily be seen that the thrust of one worm will be annulled by that of the other, thus eliminating the need of a thrust-bearing.

For convenience of reference the notation used in the equations

168 FORM, STRENGTH, AND PROPORTIONS OF PARTS.

for worm-gearing will be given in full, although most of the symbols are the same as for screws:

D' = mean diameter of thrust-bearing, inches;
D'' = mean diameter of journal, inches;
E = efficiency of mechanism for forward motion;
E' = efficiency of mechanism for overhauling;
F = force acting to rotate the worm, pounds;
F'' = turning force exerted by the worm when overhauling, pounds;
P = turning force acting on driven wheel, pounds;
T = thrust of worm or screw-gear parallel to its axis, pounds;
d = mean diameter of worm-thread, inches;
e = efficiency of worm-thread alone, journal and thrust-bearing friction not included;
e' = efficiency of worm-thread alone when overhauling;
l = length of lever-arm of turning force, inches;
p = pitch of worm-thread, inches;
β = angle between the surface of the thread and a plane normal to the axis of the screw;
θ = angular pitch of screw-thread;
μ = tan ϕ = coefficient of friction between worm and wheel;
μ' = tan ϕ' = coefficient of thrust-bearing friction;
μ'' = coefficient of journal-friction;
ϕ = angle of friction between worm and wheel;
ϕ' = angle of friction for thrust-bearing.

In the following discussion it is assumed that the axes of the worm and worm-wheel are at right angles.

The pressure, due to the forces acting between the teeth, against the journals supporting a worm that engages with but one worm-wheel, assuming that there is a journal at each end of the worm, is equal to the resultant of two forces, one approximately equal to the force resisting the rotation of the worm when driving the wheel, acting parallel to the axis of the wheel, and at a distance d from the axis of the worm, and the other equal to the component, normal to and intersecting the axis of the worm, of the pressure T tangent to the mean diameter of the wheel and parallel to the axis of the worm.

SCREW-GEARING. 169

The first of these forces, acting parallel to the axis of the wheel, is approximately $T\dfrac{\tan\theta + \tan\phi \sec\beta}{1 - \tan\theta \tan\phi \cos\beta}$*; the second, normal to and intersecting the axis of the worm, is $T\tan\beta$. Since these two forces act at right angles, their resultant, which is the

$$\text{Pressure on worm-journals} = T\sqrt{\left(\dfrac{\tan\theta + \tan\phi \sec\beta}{1 - \tan\theta \tan\phi \cos\beta}\right)^2 + \tan^2\beta}.$$

The journal friction, assuming that both journals are of the same diameter D'' and have the same coefficient of friction μ'', is

$$\text{Journal friction of worm-shaft} = \mu'' T \dfrac{D''}{2}\sqrt{\left(\dfrac{\tan\theta + \tan\phi \sec\beta}{1 - \tan\theta \tan\phi \cos\beta}\right)^2 + \tan^2\beta}. \quad (77)$$

If the journals supporting the worm are of different diameters or have different coefficients of friction, the pressure on each journal may be found and the friction of each determined. The total pressure is divided between the two journals in amounts inversely proportional to their distances from the pitch point of the worm and wheel. Such a refinement as this would seldom, if ever, be worth applying in practice.

There is an end thrust on the worm-wheel shaft approximately equal to $T\dfrac{\tan\theta + \tan\phi \sec\beta}{1 - \tan\theta \tan\phi \cos\beta}$; also a journal pressure having a value $T'\sqrt{1 + \tan^2\beta}$.

The following equations may now be written by referring to those for an angular-thread screw. In them the frictional losses in the bearings supporting the worm-wheel are not taken into account as producing part of the thrusting force T. The equations of efficiency, therefore, do not exactly represent the efficiency of a complete mechanism, on account of the friction of the worm-wheel bearings being neglected, but the speed of the worm-wheel is so slow that the friction losses in its bearings are comparatively small ordinarily. The efficiencies given by the following equations are

---

* See Appendix, § A.

170 FORM, STRENGTH, AND PROPORTIONS OF PARTS.

far more accurate than the assumptions that can be made for the coefficients of friction.

The value of the "journal friction" in the following equations may be obtained by equation (77).

When the worm drives the wheel:

$$Fl = T\frac{\tan \theta + \tan \phi \sec \beta}{1 - \tan \theta \tan \phi \cos \beta} \times \frac{d}{2} + T\mu'\frac{D'}{2} + \text{Journal friction}, \quad (78)$$

(Thread Resistance. Thrust-bearing Friction.)

or

$$Fl = T\frac{p + \mu\pi d \sec \beta}{\pi d - \mu p \cos \beta} \times \frac{d}{2} + T\mu'\frac{D'}{2} + \text{Journal friction}, \quad (79)$$

and

$$E = \frac{Tp}{F(2\pi l)} = \frac{Td \tan \theta}{2Fl}; \quad \ldots \ldots \quad (80)$$

$$e = \frac{\tan \theta(1 - \tan \theta \tan \phi \cos \beta)}{\tan \theta + \tan \phi \sec \beta}. \quad \ldots \quad (81)$$

If it is desired to neglect the effect of the angle β, then, for the approximate turning moment,

$$Fl = T \tan (\theta + \phi)\frac{d}{2} + T\mu'\frac{D'}{2} + \text{Journal friction}, \quad \cdot \cdot \quad (82)$$

(Approximate Thread Resistance. Thrust-bearing Friction.)

or

$$Fl = T\frac{p + \mu\pi d}{\pi d - \mu p} \times \frac{d}{2} + T\mu'\frac{D'}{2} + \text{Journal friction}, \quad \cdot \quad (83)$$

And, for the approximate worm-thread efficiency, neglecting β,

$$e = \frac{\tan \theta}{\tan (\theta + \phi)} \text{ (approximately)}. \quad \ldots \quad (84)$$

When the wheel drives the worm, which action corresponds to overhauling in a screw:

$$F''l = T\frac{\tan \theta - \tan \phi \sec \beta}{1 + \tan \theta \tan \phi \cos \beta} \times \frac{d}{2} - T\mu'\frac{D'}{2} - \text{Journal friction}, \quad (85)$$

(Turning Force due to Pressure of Wheel against Worm. Thrust-bearing Friction.)

or

$$F''l = T\frac{p - \mu\pi d \sec \beta}{\pi d + \mu p \cos \beta} \times \frac{d}{2} \qquad - T\mu'\frac{D'}{2} - \text{Journal friction,} \quad (86)$$

and

$$E' = \frac{F''2\pi l}{Tp} = \frac{2F''l}{Td \tan \theta}; \quad \ldots \ldots (87)$$

$$e' = \frac{\tan \theta - \tan \phi \sec \beta}{\tan \theta (1 + \tan \theta \tan \phi \cos \beta)}. \quad \ldots \ldots (88)$$

As is indicated by the above equations, the efficiency of worm-gearing increases with the angular pitch up to a certain limiting value, which depends upon the coefficient of friction of the worm and the frictional resistance of the thrust-bearing. This limiting angle is greater than has apparently ever been found satisfactory in practice. There seems to be a general tendency, however, to use a greater angular pitch than formerly, when the primary function of the worm-gearing is to transmit power in considerable amounts. A pitch angle of 20°, or even more, is quite commonly used for worm-driven machine-tools.

77. Tests of worm-gearing.—In 1885 Wilfred Lewis published the results of an extensive series of tests made by Wm. Sellers & Co. on worm and spiral gearing. The result of each individual test on a particular mechanism, at different speeds and pressures, was plotted on a diagram, which also had a curve of the mean efficiency of the mechanism for the entire range of speed covered. From these curves of mean efficiencies the readings given in Table XIV were taken. The teeth were approximately of the involute system.

In the experiments it was found that abrasion, or cutting, began between the surfaces of the worm-thread and wheel-teeth at certain limiting pressures and speeds. The mechanism could still be run after abrading, but the efficiency was materially reduced. This is shown very clearly in the third and fourth columns of Table XIV, which give the efficiency both before and after cutting; the drop in efficiency in the last column, at 100 revolutions, also shows the effect of cutting.

TABLE XIV.

EFFICIENCY OF CAST-IRON WORM-GEARING.*

Efficiency = ratio of power delivered by worm-wheel shaft to that applied to worm-shaft.

All worms 4" pitch diameter; all worm-wheels 18.62" pitch diameter, 39 teeth, 1½-inch pitch. Worm ran in oil-bath. When two values are given for the efficiency in the same column, the lower one was obtained after abrasion began. The drop from 69 to 65 in the last column is due to abrasion.

Revolutions per minute, approximately equals the velocity of sliding in feet per minute.	Efficiency, per cent.			
	2 threads, 3 in. pitch. Angular pitch 13° 51'. Cast thread and teeth.		1 thread, 1½ in. pitch. Angular pitch 6° 40'. Step-bearing 1.6 in. mean diam.	
	Collar thrust-bearing 4" mean diam. Pressure 200 to 6000 lbs.	Step-bearing 1.0" mean diam. Pressure 1200 to 5500 lbs. (After abrasion.)	Cast thread and teeth. Pressure 450 to 5600 lbs. (After abrasion.)	Machine-cut thread and teeth. Pressure 1200 to 5600 lbs.
3	55	49	....	42
5	59	52	....	45
7	62	54	46	48
10	65	57	49	51
15	68	59	52	55
20	70	61	53	57
30	72	64	57	61
40	74	66	59	63
60	75	68	64 and 46	67
80	76	70 and 59	67 and 48	69
100	....	71 and 60	69 and 50	65 (abraded)
120	....	72 and 61	71 and 52	67 (abraded)
150	....	73 and 62	73 and 53	68 (abraded)
200	....	74 and 63	54	70 (abraded)
300	....	65	56	....
400	....	66	57	....
500	....	67	57.5	....
600	....	....	58	....
900	....	....	59	....

* Readings of efficiency taken from diagrams by Wilfred Lewis in Trans. Amer. Soc. Mech. Eng., vol. VII., p. 273.

The second column gives the efficiency of a worm whose thrust was taken by a collar-bearing formed by turning down the end of the thread to a flat surface; this did not form a complete ring of metal to bear against the bearing, about half the material being removed on account of the space between the threads. In all the other tests the thrust was carried by a step-bearing made of two hardened-steel disks, carefully ground, with a hard-brass washer

SCREW-GEARING. 173

interposed. The step-bearing was $2\frac{1}{8}$ inches outside diameter. The shaft at one end of the worm was $2\frac{7}{8}$ inches diameter, and $1\frac{5}{8}$ inches at the other.

An examination of the table shows that the efficiency increased with the speed, in all cases, until abrasion began. The greater efficiency of the double-thread worm may be seen clearly for the lower speeds, but is not so decided as the speed increases.

A series of experiments were also made on the single-thread worm with cast teeth, whose efficiency is given in the fourth column of Table XIV, to determine the speeds and pressures liable to produce cutting during a ten-minute run. The results of these experiments are given in Table XV, taken from Mr. Lewis's paper.

TABLE XV.

SPEEDS AND PRESSURES LIABLE TO PRODUCE CUTTING IN CAST-IRON WORM-GEARING.

Velocity of sliding, feet per minute.	Pressure on teeth, pounds.	Temperature.		Efficiency.		Duration of run, minutes.	Ft. lbs. per min. consumed in friction before cutting began.
		Initial.	Final.	Initial.	Final.		
800	1785	106°	140°	.609	.387	6	117,600
880	1780	118	132	.607	.462	3	129,300
880	1205	137	150	.575	.360	3	97,000
800	448	118	133	.594	.445	10	29,400
480	2822	144	167	.591	.450	7	117 800
400	3481	170	180	.639	.415	3	98,300
360	4837	138	166	.641	.473	6	122,400
*306	5558	163	186	.677	.677	10	102,000

* No cutting at 306 feet per minute.

In the last of these experiments, at a speed of rubbing of 306 feet per minute, and a thrust of 5558 pounds on the worm, no abrasion occurred. The temperatures given are those of the oil-bath in which the worm ran; the initial efficiency was obtained before abrasion, and the final after cutting began.

In 1883–84 Dr. R. H. Thurston made a series of experiments on worm-gearing for the Yale & Towne Mfg. Co. The results are given by Henry R. Towne in the Transactions of the American Society of Mechanical Engineers.† The data show that the worm

---

† Vol. VII., 1886, p. 300.

and wheel were both of cast iron, with machine-cut threads and teeth; the worm was $6\frac{1}{16}$ inches pitch diameter, double-threaded, and 4 inches long on the thread; the worm-wheel was $15\frac{1}{8}$ inches pitch diameter, with 50 teeth and 2-inch face. The pitch, calculated from these data, is 2 inches linear, or 5° 59′ angular.

During the experiments three forms of thrust-bearings for taking the thrust of the worm were used. They were: First, a collar thrust-bearing having a collar 1 inch wide and $2\frac{1}{2}$ inches mean diameter. The rubbing surfaces were the faced ends of the worm-hub and of the cast-iron supporting frame of the mechanism. Second, a button thrust-bearing, or step, made by capping the end of the worm-shaft with a piece of hardened steel, having its exposed face slightly convex, and letting it run against the hardened end of an adjusting set-screw. Third, a roller thrust-bearing, consisting "of 12 chilled cast-iron coned rollers of $\frac{7}{16}$ inch mean diameter, contained within a brass cage having a separate pocket for each cone, the cones travelling at a mean radius of $1\frac{3}{8}$ inches from the axis of the shaft, between two steel collars or rings, one bearing against the hub of the worm, and the other against the face of the frame-bearing, the faces of these collars being coned to the shape of the rollers. The centrifugal thrust of the cones was resisted by a wrought-iron ring surrounding the cage, the ends of the cones being convex."

Table XVI shows a comparison between the button-bearing and

TABLE XVI.
COMPARATIVE EFFICIENCIES OF THE SAME WORM AND WHEEL WITH BUTTON THRUST-BEARING AND ROLLER STEP-BEARING.

Horse-power per 100 revolutions of worm per minute.	Efficiency, per cent.		Horse-power per 100 revolutions of worm per minute.	Efficiency, per cent.	
	Button thrust-bearing.	Roller thrust-bearing.		Button thrust-bearing.	Roller thrust-bearing.
.25	6	9	4.0	38	43
.50	10	12	5.0	44	47
.75	13	18	6.0	49	50
1.00	16	22	7.0	54	52
1.5	21	27	8.0	59	54
2.0	25	31	8.28 *	60	....
2.5	29	35	9.00	....	56
3.0	32	38	10.5 †	....	57

* Highest power given for button-bearing.   † Highest power given for roller-bearing.

SCREW-GEARING. 175

roller-bearing for different rates of working, and Table XVII between the collar-bearing and roller-bearing. The readings in the tables are taken from diagrams showing the results of the experiments.

TABLE XVII.

COMPARATIVE EFFICIENCIES OF THE SAME WORM AND WHEEL WITH COLLAR THRUST-BEARING AND WITH ROLLER STEP-BEARING.

Revolutions of driver per minute.	Efficiency, per cent.		Revolutions of driver per minute.	Efficiency, per cent.	
	Collar thrust-bearing.	Roller thrust-bearing.		Collar thrust-bearing.	Roller thrust-bearing.
60	38	48	250	40	60
100	42	52	300 *	38	63
150	48	55	350	....	65
200	42	58	400 †	....	67

* Highest speed for collar-bearing.   † Highest speed given for roller-bearing.

The difference in the efficiencies obtained with these three forms of thrust-bearings shows clearly how much power may be lost in this part of the mechanism. It should be noted that while the efficiency of the roller-bearing is higher than that of the button-bearing for rates of working up to 6 horse-power, it becomes lower at 7 horse-power. The original diagram shows equal efficiencies at about 6.5 horse-power.

The results of a number of experiments on three tool-steel worms, hardened, running against cast-iron worm-wheels, are given by Bertram P. Flint.* Two of the worms were of the same diameter, but one was of practically twice as great lead or pitch as the other. The results of the experiments are given in Table XVIII. That the limit of pressure is lower at the highest speeds of rubbing is clearly shown.

The Sprague Elevator Co. found the efficiencies given in Table XIX by experimenting on a pair of Hindley worms such as are used by them for running their elevators.† In the elevator mechanism, two winding-drums carry the cable attached to the elevator-cage. These drums are geared together with spur-gears, so that they must

---

* *Engineering News*, April 9, 1892, p. 348.
† *American Machinist*, Jan. 21, 1897, p. 45.

TABLE XVIII.
EFFICIENCY OF WORM-GEARING.*

Efficiency = ratio of power delivered by worm-wheel shaft to that received by the worm-shaft.

Worm of tool-steel, hardened ; wheel of cast-iron ; axes of worm and wheel both horizontal ; worm on top of wheel ; bottom of wheel dipped in oil-bath. Thrust of worm taken by phosphor-bronze plate against which the end of the shaft turned. The limit of pressure indicates the highest pressure that did not cause abrasion.

Thrust of Worm. Pounds.	Efficiency, per cent.							
	Single-thread Worm. Angular Pitch 6° 20′.			Double-thread Worm. Angular Pitch 12° 30′.			Double-thread Worm. Angular Pitch 10°.	
	128 rev. per min. 96 ft. per min. Sliding velocity.	201 rev. per min. 151 ft. per min. Sliding velocity.	272 rev. per min. 204 ft. per min. Sliding velocity.	128 rev. per min. 96 ft. per min. Sliding velocity.	201 rev. per min. 151 ft. per min. Sliding velocity.	272 rev. per min. 204 ft. per min. Sliding velocity.	200 rev. per min. 236 ft. per min. Sliding velocity.	272 rev. per min. 320 ft. per min. Sliding velocity.
200	......	37	30	43	49	47.5		
300	39.5	39	32	47	53	49.5	50	53
400	uni-	41	33	51	56	51	53	54
500	formly	43	34.5	53	57	51.5	54	54
600	in-	44	36	54	54	51	54	52
700	creases	44.5	37	54	52	49	53	51
800	to	44	37	53	51.5	48	52	
1000	41	43	36.8	51	49.5	47	51	
1200		42	36	50	48			
1400		39	......	48.5				
1600								
Limit of pressure, pounds.	1400	1225	1470	1260	1060	1275	710	

* Readings of efficiency taken from diagram by Bertram P. Flint, *Engineering News*, April 9, 1892, p. 348.

rotate at the same angular speed in opposite directions. A worm-wheel is rigidly attached to each drum or its shaft, one having a right-hand, and the other a left-hand, pitch to its teeth. One right-hand and one left-hand worm are attached to the same shaft, so as to engage properly with the two wheels. By this arrangement no thrust-bearing is required. The experiments were conducted by driving the worms with an electric motor whose armature was direct-connected to the worm-shaft, and hoisting a weight correspond-

SCREW-GEARING. 177

TABLE XIX.
EFFICIENCY OF HINDLEY WORM AND WORM-WHEEL. DRIVEN ELEVATOR.[*]

$$\text{Efficiency} = \frac{\text{Load lifted} \times \text{distance traversed.}}{\text{Electrical energy received by motor.}}$$

Two Hindley worms, right- and left-hand, on same shaft. No thrust-bearing or step. Pitch diam. of worm, 5.47 inches; pitch diam. of gear, 26.89 inches; double-thread worm; lead or pitch of each thread, 2.86 inches nearly; angular pitch at pitch line 9° 39'; velocity ratio 29¼ to 1; speed of worm, 500 rev. per min.; velocity of rubbing at pitch line, 718 ft. per min.; speed of travel of worm-wheel at pitch line, 112 ft. per min.

Pressure in Pounds on		Efficiency, per cent.	Pressure in Pounds on		Efficiency, per cent.
Both Worms.	One Worm.		Both Worms.	One Worm.	
1000	500	44	3000	1500	68
1500	750	54	4000	2000	72
2000	1000	60	6000	3000	76
2500	1250	65	8500	......	78

[*] Readings taken from diagram in *American Machinist* of Jan. 21, 1897, p. 45.

ing to an elevator-cage, which was lifted by means of a rope running from the winding-drums on the worm-wheel shafts. Measurements of the electrical energy received by the motor, and of the work done upon the weight, were taken. The ratio of the latter to the former corresponds to the efficiency given in the table. This, of course, does not represent the efficiency of the worm-gearing alone, as has been given in the preceding tables, since the motor losses, as well as those of the machinery between the worm-wheels and weight lifted, are included. The experimenters estimate that the efficiency of the worm-gearing alone "can scarcely be less than 90%." Attention is also called to the fact that the percentage motor loss increases as the load becomes lighter; hence the efficiency-curve drops more rapidly than the real efficiency of worm-gearing.

A parallel series of tests on a pair of worms of the ordinary form, and working under the same conditions as the Hindley, gave an efficiency 2% lower at 2000 pounds pressure on both worms, which gradually decreased to 10% lower at 6000 pounds.

The requirement of the Hindley worm for accurate adjustment should be kept in mind when applying it where the service is heavy.

SCREW-GEARS.

78. Screw-gears seem to be rapidly increasing in favor for use where the service is light. Numerous examples are found in the feed mechanism of both light and heavy machine tools. In a great many cases they are now used where heretofore bevel-gears were almost universally applied. The objection against them, that the teeth cannot be accurately cut in a milling-machine, but must be finished by hand, is not sufficiently strong to prohibit their use.

79. Strength of screw-gear teeth.—When the velocity ratio is great, and the teeth of the gear corresponding to a worm-wheel run nearly or quite straight across its face, they may be dealt with for strength as those of a spur-gear having pressure applied at one point. While the point of application may be at the top of the tooth, it should never come at the end of its length across the gear face; therefore it is doubtless safe to consider the effective width of the gear face for resisting fracture at least as great as 1.5 times the circular pitch. When the teeth run at a considerable angle across the face of the gear, they are somewhat stronger than those of a spur-gear of the same diameter and thickness. When the angle between the axis of the gears is not so great as to prevent cutting the teeth of only such a length, in the middle of the gear face, as is necessary to allow the engagement of other gears, the teeth are much stronger on account of the supporting material at their ends; such a gear corresponds in a way to a worm-wheel or one that is shrouded.

80. Equations for turning force and efficiency of screw-gears.—When the axes of the gears are at right angles, the equations for worm-gearing apply equally well to screw-gears. If the angle between the axes does not differ greatly from 90°, the same equations will give results more accurate than our knowledge of the coefficiencies of friction entering into them.

The slight practical use to which equations for smaller angles between the gear-axes can be put does not seem to warrant their presentation.

It is worth noting, however, that, as the angle between the axes decreases, the amount of sliding between the teeth of the gears also decreases, and in consequence of this, together with the reduction of thrust on the spiral pinion, the efficiency increases uniformly, reaching that of spur-gears when the axes become parallel.

In the tests already referred to as being made by Mr. Lewis, he found the screw-gear efficiencies given in Table XX. The larger

TABLE XX.

EFFICIENCY OF CAST-IRON SCREW-GEARS.*

Efficiency = ratio of power delivered by spur-gear shaft to that applied to spiral-pinion shaft; or, ratio of output to input.

All spiral pinions 4" pitch diameter; all spur-wheels 18.02" pitch diameter, 39 teeth, 1¼ inch pitch. Spiral pinion ran in oil-bath.

Cycloidal teeth, accurately cut.

Revolutions of Screw-pinion per minute.	Efficiency, per cent.			
	1 thread. 1.511" pitch. 6° 51' angular pitch.	2 threads. 3.086" pitch. 13° 49' angular pitch.	4 threads. 6.828" pitch. 28° 31' angular pitch.	6 threads. 12.894" pitch. 45° 44' angular pitch.
3	....	60	70	81
5	....	63	73	83
7	45	66	75	84.5
10	46	68.5	78	86
15	48	72	80	87.5
20	49	74	82	89
30	51	77.5	84.5	91
40	53	80	86	92
60	65.5 and 55.5	83	89	94
80	68 and 57.5	85	90	94.8
100	70 and 59	86	91	95.3
120	71.5 and 60.5	87	92	96
150	73.5 and 62	88.5	92.3	96.5
200	75.5	89	92.3	96.6
275	77.5	....	92	96.4

* Readings of efficiency taken from diagrams by Wilfred Lewis in Trans. Amer. Soc. Mech. Eng., vol. vii., p. 273.

wheel in each test was a spur-gear with accurately cut cycloidal teeth; the screw-pinion was also accurately cut to fit the gear. The angle between the gear-axes was $90° - \theta$, the pinion-shaft being set at the pitch angle θ with a normal to the axis of the spur-gear, in order to obtain accurate intermeshing of the gears. The two efficiencies of the single-thread pinion, at speeds from 60 to 150 revolutions, are due to different conditions of the rubbing surfaces; the higher efficiencies correspond to an improvement of these surfaces, secured by running the mechanism for some time under a light load.

81. Coefficient of friction of screw-gears.—By determining the journal friction as accurately as possible, and allowing for it, Mr. Lewis found, by calculation, the coefficient of friction μ for the two-, four-, and six-thread pinions given in Table XX. These coefficients are given in Table XXI, together with their average.

TABLE XXI.

COEFFICIENT OF FRICTION OF SCREW-GEARS.

For Three of the Gears given in Table XX.

Revolutions of Screw-pinion per minute.	Coefficient of Friction, μ.			
	2 threads 13° 49′, angular pitch.	4 threads 28° 31′, angular pitch.	6 threads 45° 44′, angular pitch.	Average Values.
3	.086	.105	.094	.095
5	.078	.097	.089	.088
10	.064	.081	.076	.074
20	.050	.065	.061	.059
50	.035	.042	.038	.038
100	.025	.030	.024	.026
200	.018	.026	.015	.020

values for various speeds. In view of the low values of μ for square-thread screws, given in Table XIII, when lubricated with graphite and oil, it would seem that such a lubricant might be excellent for worm- and screw-gears. In fact it is recommended by a leading concern building large worm-driven metal-working planers.

Valuable data could doubtless be obtained by a series of experiments on worm- and screw-gears of different materials and with different lubricants.

CHAPTER VI.

SCREW-FASTENINGS.

82. In machine construction it is frequently desirable to fasten parts together in such a manner that they can readily be separated and put together again without injuring the fastening or destroying its usefulness. In order to accomplish this, fastenings having screw-threads cut upon them are used.

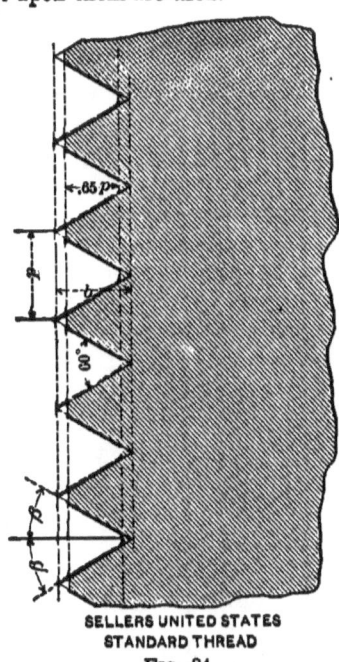

SELLERS UNITED STATES STANDARD THREAD
FIG. 84.

The Sellers United States Standard thread, shown in sectional outline in Fig. 84, has the sides inclined at an angle $= 60° = 2\beta$.

The proportions are obtained by extending the lines of the sides until they intersect, then cutting off one eighth of the distance b from the top, and filling in to the same extent at the bottom, of the angles thus formed; this makes the height of the thread $= h = 0.65$ of the pitch.

Table XXII gives the proportions of U. S. Standard threads.

Table XXII.

U. S. OR SELLERS SYSTEM OF SCREW-THREADS.

Diameter of Bolt, Inches.	Threads per Inch.	Diameter of Root of Thread, Inches.	Width of Flat, Inches.	Area of Bolt-body in Square Inches.	Area at Root of Thread in Square Inches.
1/4	20	.185	.0062	.049	.027
5/16	18	.240	.0074	.077	.045
3/8	16	.294	.0078	.110	.068
7/16	14	.344	.0089	.150	.093
1/2	13	.400	.0096	.196	.126
9/16	12	.454	.0104	.249	.162
5/8	11	.507	.0113	.307	.202
3/4	10	.620	.0125	.442	.302
7/8	9	.731	.0138	.601	.420
1	8	.837	.0156	.785	.550
1 1/8	7	.940	.0178	.994	.694
1 1/4	7	1.065	.0178	1.227	.893
1 3/8	6	1.160	.0208	1.485	1.057
1 1/2	6	1.284	.0208	1.767	1.295
1 5/8	5 1/2	1.389	.0227	2.074	1.515
1 3/4	5	1.491	.0250	2.405	1.746
1 7/8	5	1.616	.0250	2.761	2.051
2	4 1/2	1.712	.0277	3.142	2.302
2 1/4	4 1/2	1.962	.0277	3.976	3.023
2 1/2	4	2.176	.0312	4.909	3.719
2 3/4	4	2.426	.0312	5.940	4.620
3	3 1/2	2.629	.0357	7.069	5.428
3 1/4	3 1/2	2.879	.0357	8.297	6.510
3 1/2	3 1/4	3.100	.0384	9.621	7.548
3 3/4	3	3.317	.0413	11.045	8.641
4	3	3.567	.0413	12.566	9.963
4 1/4	2 7/8	3.798	.0435	14.186	11.329
4 1/2	2 3/4	4.028	.0454	15.904	12.753
4 3/4	2 5/8	4.256	.0476	17.721	14.226
5	2 1/2	4.480	.0500	19.635	15.763
5 1/4	2 1/2	4.730	.0500	21.648	17.572
5 1/2	2 3/8	4.953	.0526	23.758	19.267
5 3/4	2 3/8	5.203	.0526	25.967	21.263
6	2 1/4	5.423	.0555	28.274	23.098

The proportions of heads and nuts, as made by the leading manufacturers, do not generally conform with the U. S. Standard, and not always with each other. It is advisable to consult the catalogue of the manufacturer whose bolts are to be used.

Table XXIII gives the proportions of bolt-heads and nuts adopted by some of the leading makers.

TABLE XXIII.*

PROPORTIONS OF BOLT-HEADS ADOPTED BY DIFFERENT MANUFACTURERS.

The dimensions are the same whether finished or rough.

Diam. of Screw.	Dimensions of Square and Hexagon Bolt-heads.†							
	Hoopes & Townsend.		Rhode Island Tool Co.		Wm. H. Haskell & Co.		J. H. Sternbergh & Son.	
	Width.	Thickness.	Width.	Thickness.	Width.	Thickness.	Width.	Thickness.
1/4	7/16	3/16	3/8	3/16	7/16	3/16	7/16	3/16
5/16	1/2	1/4	15/32	1/4	1/2	1/4	17/32	1/4
3/8	19/32	9/32	9/16	5/16	5/8	9/32	5/8	5/16
7/16	11/16	3/8	21/32	3/8	23/32	11/32	23/32	3/8
1/2	3/4	7/16	3/4	7/16	13/16	3/8	13/16	13/32
9/16	27/32	1/2	27/32	1/2	15/16	7/16	29/32	15/32
5/8	15/16	17/32	15/16	17/32	1	1/2	1	1/2
3/4	1 1/8	5/8	1 1/8	5/8	1 3/16	5/8	1 3/16	9/16
7/8	1 5/16	3/4	1 5/16	3/4	1 3/8	3/4	1 3/8	11/16
1	1 1/2	7/8	1 1/2	7/8	1 9/16	13/16	1 9/16	3/4
1 1/8	1 5/8	1	....	....	....	....	1 3/4	7/8
1 1/4	1 7/8	1 1/8	....	....	....	....	1 15/16	1

* *Machinery*, April 1897, page 242.

† The width of the nut is the same as that of the head; its thickness is commonly equal to the diameter of the thread.

The Whitworth English Standard thread, Fig. 85, has an angle of 55° between the sides, and the top and bottom are rounded for a distance equal to one sixth of the depth of the corresponding V thread; this leaves the height of the thread $= h = 0.64$ of the pitch.

By cutting off the sharp points of the older forms of V threads, so as to form the standard threads just mentioned, they become less subject to bruising by accidental blows, and the taps, dies, and turning tools used to form them are more durable than those with sharp threads.

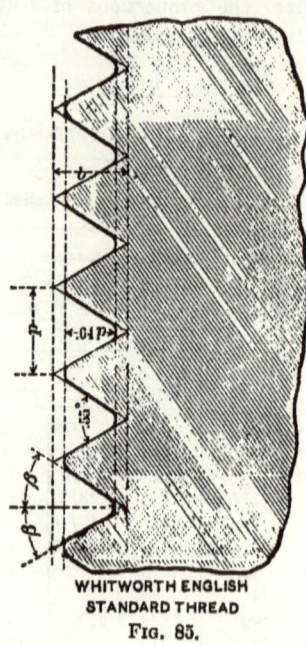

WHITWORTH ENGLISH STANDARD THREAD
FIG. 85.

The square thread, Fig. 86, is commonly used when there is considerable wear due to relative motion of the threaded parts engaging together; it has the advantage of presenting a surface almost at right angles to the line of pressure, which is ordinarily parallel to the axis of the thread.

The buttress thread, Fig. 87, is sometimes useful when the pressure against the thread is all, or nearly all, in one direction. The surface taking the thrust is made perpendicular to the axis of the thread, the other having any convenient angle with it. By this means a strong thread is obtained.

The forms of screw-fastenings are almost infinite in number; a few of the more common ones, together with their ordinary functions, are given below, chiefly in order that they may be referred to with a clear understanding of what each is.

A machine bolt consists of a bar of metal, forming the body of the bolt, having a thread and nut at one end, and a head, forming an integral part with the body, at the other. Ordinarily it is used

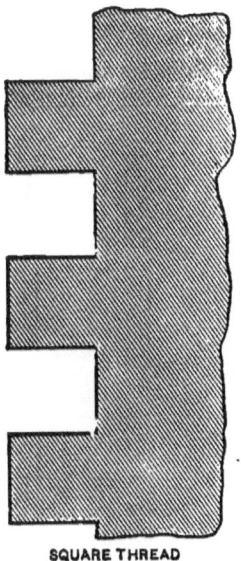

SQUARE THREAD
Fig. 86.

to clamp together parts of machinery by passing through them so that the inner surfaces of the head and nut press against the pieces held together by the bolt, as shown in Fig. 88. A washer is frequently placed between the nut and the clamped piece, to prevent marring the latter, and to give a better bearing surface for the nut.

A cap-screw, Fig. 89, is of the same form as a bolt with the nut removed. It is commonly used to fasten parts together by screwing into a threaded hole in one of them. The depth of the threaded hole depends largely upon the material; in all cases where the parts are drawn tightly together, its depth should be as great as the

186   FORM, STRENGTH, AND PROPORTIONS OF PARTS.

diameter of the screw. When the material tapped into is much weaker than that of the screw, as when a machine-steel cap-screw is screwed into cast iron, the threaded depth of the hole should be made considerably greater; twice the diameter of the screw, or more, is advisable under such conditions.

A stud, Fig. 90, consists of a bar threaded at both ends. One end is screwed into a tapped hole in one of the pieces to be held together, and the other receives a nut which presses against the

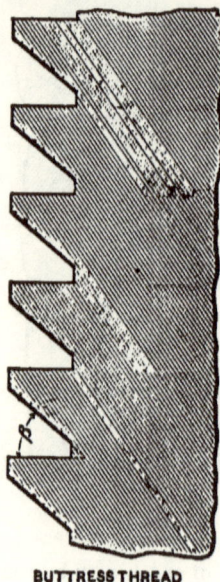

BUTTRESS THREAD
FIG. 87.

clamped piece. A stud performs the same service as a cap-screw. It is useful where the parts are to be separated frequently. If a cap-screw is drawn down hard and removed several times from cast iron, there is danger that the threads of the hole will break and crumble away, unless the material is close-grained and of good quality. By using a stud this is obviated, for it can be screwed tightly into the hole and left there. The nut and stud can both be

SCREW-FASTENINGS. 187

made of a strong, durable material, and in case of injury to the threads, can be replaced readily and at small expense. A stud does

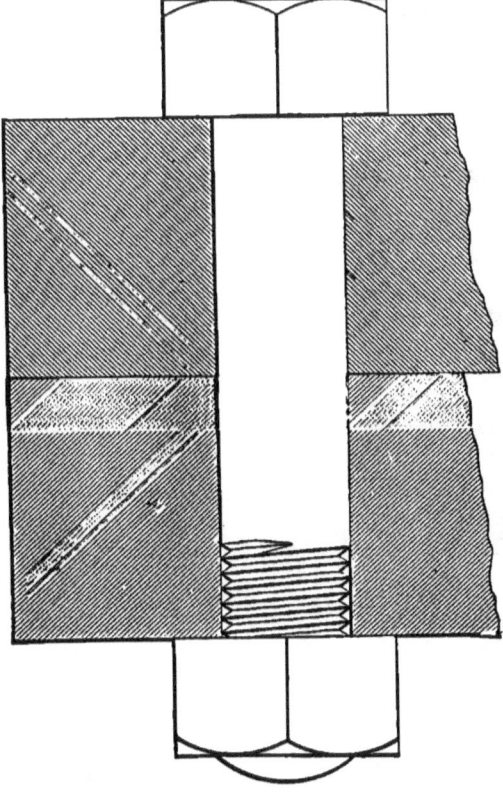

Fig. 88.

not need as long a thread in the tapped hole as a cap-screw, for the reason that it does not turn when under stress.

Fig. 91 shows a stud that can be used, when only a shallow hole can be made in a weak material, by screwing the larger end into the threaded hole and putting the nut on the smaller. Fig. 92 shows another form that can be used in the same way, or it can be used

188 FORM, STRENGTH, AND PROPORTIONS OF PARTS.

when the stud is screwed into a thin piece, through which the threaded hole passes. For such a purpose the small end is screwed in until the shoulder prevents it from entering farther; the nut is, of course, placed on the large end in such a case.

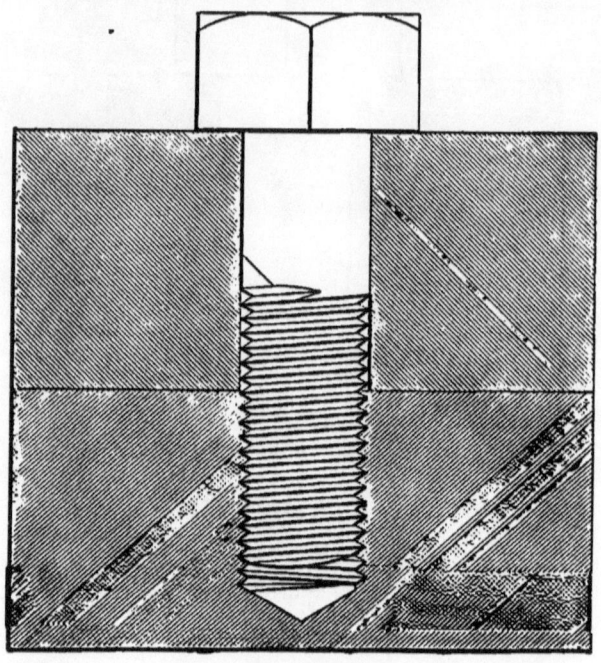

FIG. 89.

A set-screw is used to prevent sliding or rotation between parts that fit together; an ordinary application is that of fastening pulleys in place upon shafting. A common form of set-screw is that of Fig. 93; the point shown is round; other points, largely used, are the cup, cone, and pivot, shown in Figs. 94, 95, 96, and 97.

The method of using a set-screw is shown in Fig. 98, which represents a collar or pulley-hub held in place on a shaft by the set-screw. When the piece through which the set-screw passes must be held very firmly in position, so as to resist forces of considerable magnitude, tending to displace it, a recess, generally conical, is drilled in the shaft to receive the end of the screw; for light work, the screw is tightened against the smooth shaft so that the point is set into it far enough to hold.

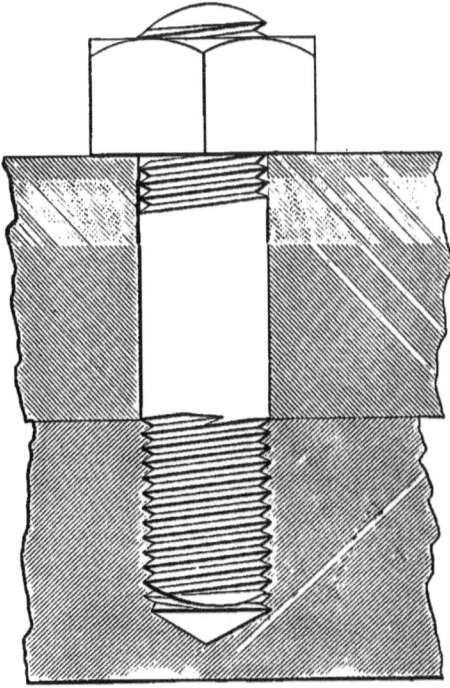

FIG. 90.

The conical and cup points hold more firmly than the round point when set lightly against the smooth shaft, but at the same time they mar it to a greater extent, thus making it more difficult to separate the parts.

Prof. Gaetano Lanza gives the results of a series of tests on the holding power of set-screws made on screws having different kinds of points.* The experiments were made on a pulley held in place

---

* Trans. Amer. Soc. Mech. Eng., vol. x., p. 230.

190   FORM, STRENGTH, AND PROPORTIONS OF PARTS.

on a ¼⅛-inch shaft by two ⅜-inch set-screws having ten threads to the inch. The screws were tightened with a force of 75 pounds applied to the end of a 10-inch wrench. All screws were of wrought

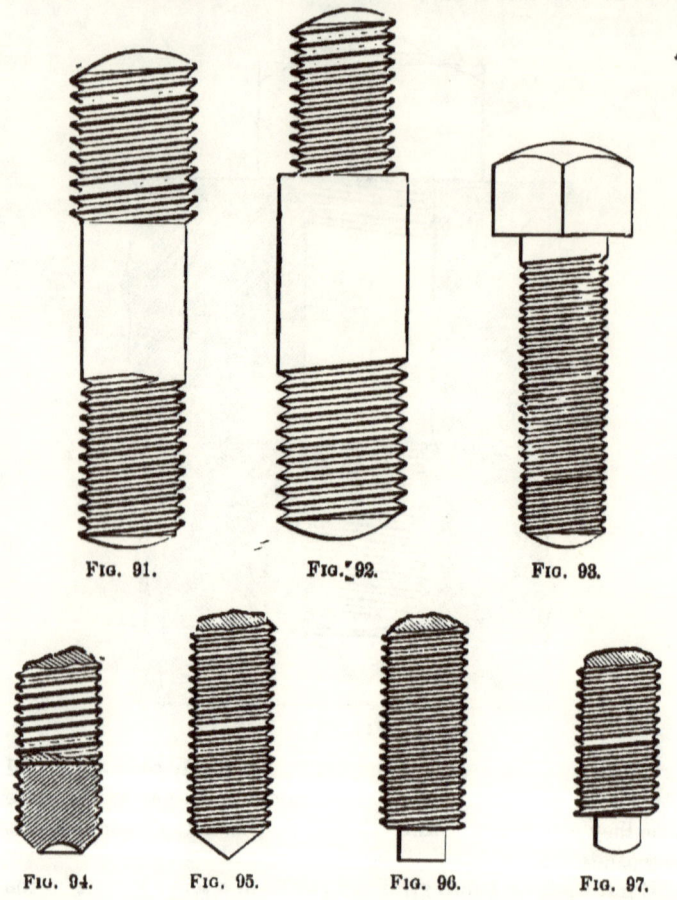

FIG. 91.   FIG. 92.   FIG. 93.

FIG. 94.   FIG. 95.   FIG. 96.   FIG. 97.

iron, and only one was case-hardened at the point. The shaft was of rather hard steel, and the set-screws made but little impression upon it. Six tests were made on each of the points A, B, and D; four on C. A summary of the results is given in Table XXIV.

SCREW-FASTENINGS.

TABLE XXIV.
HOLDING POWER OF SET-SCREWS.

	A Flat ends ⅜″ diam.	B End rounded to ¼″ radius.	C End rounded to ¼″ radius.	D End cupped and case-hardened.
Lowest holding power, pounds..	1412	2747	1902	1962
Highest " " "	2294	3079	3079	2958
Mean " " "	2064	2912	2573	2470

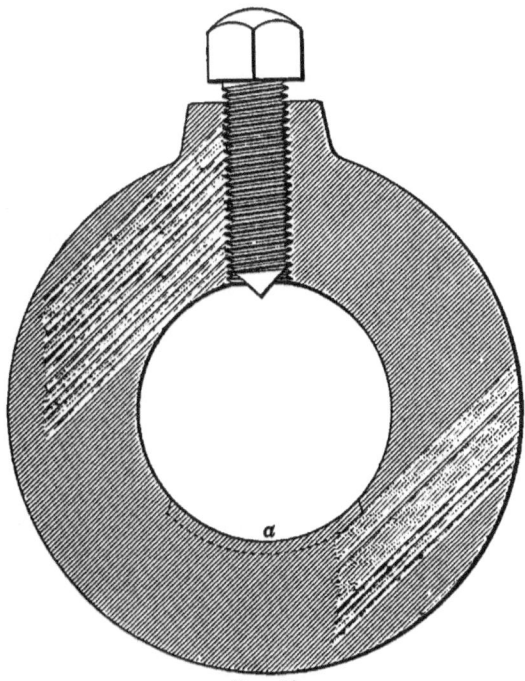

FIG. 98.

The following remarks are made by Professor Lanza regarding the screws:

"A. The set-screws were not entirely normal to the shaft;

hence they bore less in the earlier trials before they had become flattened by wear.

"B. The ends of these set-screws, after the first two trials, were found to be flattened, the flattened area having a diameter of about ¼ of an inch.

"C. The ends were found, after the first two trials, to be flattened as in B.

"D. The first test held well because the edges were sharp, then the holding power fell off till they had become flattened in a manner similar to B, when the holding power increased again."

A pulley that does not fit its shaft is difficult to hold in place by a single set-screw, or even by two or more placed on the same side of the shaft. In either case, the screw-points, pressing against the shaft on one side, cause contact between the hub and shaft on the opposite side, as at a, Fig. 98, while the remaining parts of the bore and shaft do not touch. If a torsional moment acts, tending to rotate the pulley clockwise around the shaft, the set-screw, holding firmly to the shaft, will cause the hub to slide toward the left over the shaft at a; a reversal of the turning force will allow the surfaces at a to slide back over each other. Numerous repetitions of this action will cause the point of the set-screw to enlarge the indentation in the shaft, and, finally, the point may wear away, or the screw work backward in the hub, so that the parts can rotate relatively to each other. Such a repeated application and removal of the turning force or resistance is of common occurrence in practice.

There are two methods of obtaining a more reliable set-screw fastening than the above, when there is not a close fit between the hub and shaft. One is to place two set-screws at an angle of about 90° with each other, in the same cross-sectional plane of the shaft and hub, both screws being radial; the other is to cut away a portion of the hub, as shown by the dotted line below a, Fig. 98, still retaining the single set-screw, or more, as the case may be, on one side of the shaft. Both these methods give three-point contact in a plane normal to the axis of the shaft, and passing through the axis of the screw. This prevents slipping between the hub and shaft, thus giving a firmer fastening.

In all the above fastenings the form of the head or nut, whether designed for a wrench or a screw-driver, does not change the name of the fastening. Very small screws, with heads slotted for a screw-driver, are, however, commonly called machine screws.

83. Locking devices for nuts and screws.—Since the constant jarring to which many kinds of machinery are subjected, frequently causes nuts and screws to work loose, and even to fall from their places, the necessity of securing them by some safety locking appliance has brought forward numerous devices for this purpose. A few will be described.

One of the most common is the lock- or jam-nut shown in Fig. 99. The ordinary proportions are given. They are locked together

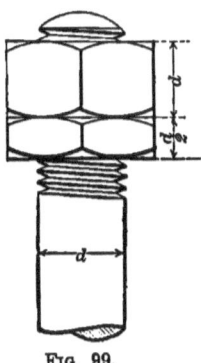

FIG. 99.

by screwing them tightly against each other. When they are also tightened against the piece held in place, a greater stress is brought upon the nut nearest the end of the bolt, which therefore should be thicker than is necessary for the one next the clamped piece. In practice the thin nut is frequently placed on top, but this is wrong. It is probably due to the fact that ordinary wrenches are not thin enough to turn the thin nut, when it is under, without catching against the top one.

Coiled-spring nut-locks, Figs. 100 and 101, are used by placing one just under the nut in the place ordinarily occupied by a washer; those in the figures are intended for bolts with right-hand threads. When the nut is tightened by turning it clockwise, its bearing surface slips freely over the nut-lock, and presses it down against the

194 FORM, STRENGTH, AND PROPORTIONS OF PARTS.

clamped piece; as soon as the nut is turned backward, however, the sharp corners of the lock cut into the surfaces of both the nut and the piece under it, thus preventing the loosening of the former. It can easily be seen that such a nut-lock can be used only where the nut is not to be removed, or where the cutting of the material by the lock is not objectionable.

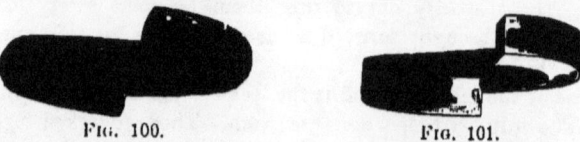

FIG. 100.   FIG. 101.

Another device, in which the thread is made so as to lock the nut, is shown in Fig. 102. The thread of the bolt is undercut so

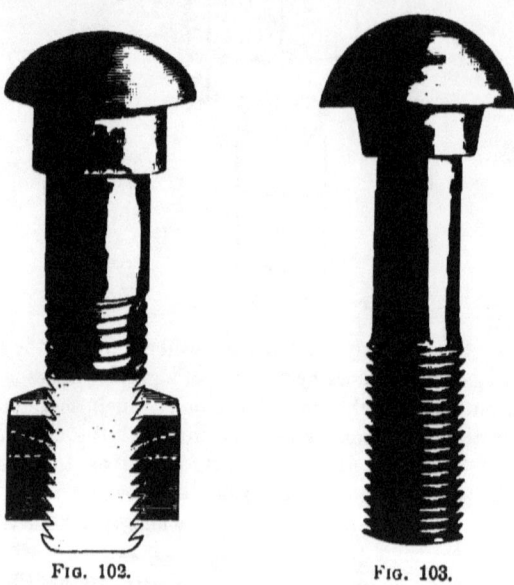

FIG. 102.   FIG. 103.

that its working side makes about five degrees less than a right angle with the axis of the bolt, and the apex of the thread is cut to a knife-edge. The nut has a thread cut so that its working side makes about five degrees more than a right angle with the axis of the nut. This leaves a cavity of about ten degrees between the bolt- and nut-threads, so that when the nut is tightened the thin edges of the bolt-thread are forced out into the nut, thus locking them together. It is claimed that this device is serviceable even where the nut is to be removed as often as a dozen times during its use.

Fig. 103 illustrates a bolt with a cold-rolled or "undercut" thread as made by J. H. Sternbergh & Son, of Reading, Penn.*

84. Strength of screw-bolts.—In many cases a bolt must have sufficient tensile strength to resist a known or estimated stress, as, for example, when used to attach cylinder-heads to engines, pumps, etc.; to fasten together flanged pipes carrying some substance under pressure; or to support a suspended load.

If, as is ordinarily the case, the nut is to be screwed up tightly enough to prevent even the slightest separation of the clamped parts when under stress, the initial tension in the bolt, due to the tightening of the nut, must be at least as great as the service tension, in order to prevent lengthening of the bolt under the service tension, and the consequent separation of the parts clamped together by it. This assumes that the parts are rigid and large in comparison with the bolt.

For example, suppose that, in Fig. 104, the part B is to be clamped against A by the bolt, so that they will not separate when a given load T is suspended from B, the line of action of T being coincident with the axis of the bolt. The initial tension in the bolt must therefore be at least as great as T. In order to allow for a factor of safety, assume the working strength of the material of the bolt as 6000 pounds per square inch. The load T may be taken as 5000 pounds. The required area of the bolt to resist tension only, in a section across the roots of the thread, is therefore

$$\frac{5000}{6000} = 0.833 \text{ sq. in.}$$

The sectional area at the roots of a U. S. Standard thread, on a bolt having a body 1¼ inches in diameter, is 0.89 of a square inch,

---

* Other designs of nut-locks are illustrated in Rose's "Modern Machine-shop Practice," vol. I., pp. 118-121; Unwin's "Elements of Machine Design," Part I., pp. 159-163; Reuleaux's "Constructor," p. 56; Klein's "Elements of Machine Design," p. 9 and Plate II.

196   FORM, STRENGTH, AND PROPORTIONS OF PARTS.

the root diameter being 1.065 inches. Since this is the nearest standard size having an area as large as that required, it would be used in practice.

The turning moment necessary to give the tension of 5000

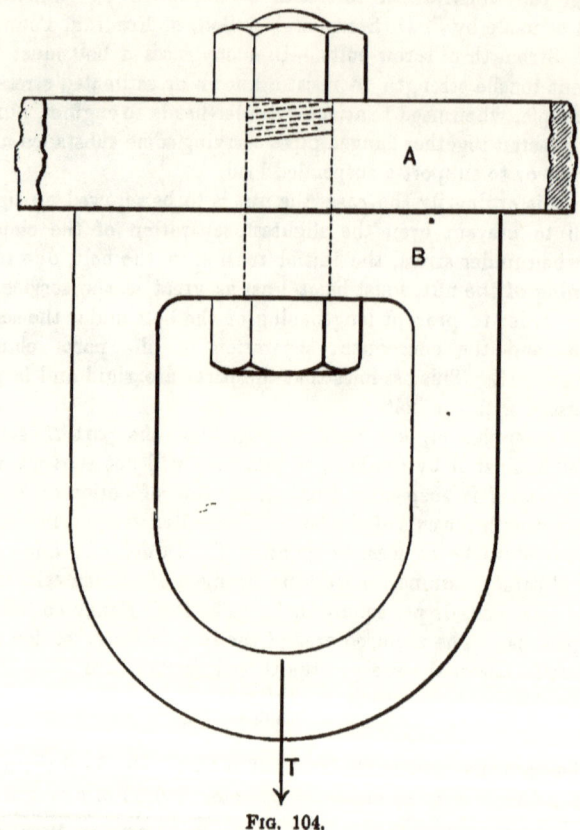

Fig. 104.

pounds can be found by equation (71) or those following it, in which, for this case: $p = \frac{1}{4}$ inch; $d = \dfrac{1.25 + 1.065}{2} = 1.16$ inches; $D = 1.7$ inches, about; $\beta = 30°$; and, since the coefficient of friction is uncertain and should be taken large enough to cover poor lubrication and rough bearing-surfaces, it is admissible to take $\mu = 0.15$, and $\mu' = 0.12$.

Substituting in equation (72),

$$Fl = 5000 \; \frac{\frac{1}{7} + .15 \times \pi \times 1.16 \times 1.15}{\pi 1.16 - 1.5 \times 1.16 \times .866} \frac{1.16}{2} + 5000 \times .12 \times \frac{1.7}{2}$$
$$= 622 + 510 = 1132 \text{ inch-pounds.}$$

Assuming $l = 16d = 16 \times 1.25 = 20$ inches, this being about the value used in practice for a solid wrench, gives

$$F = \frac{1132}{20} = 57 \text{ pounds,}$$

which is 1 pound of torsional force for every $5000 \div 57 = 88$ pounds tension in the bolt.

The value of Fl can also be obtained by assuming an efficiency E for the bolt, and substituting in equation (59), which can be applied to angular- as well as square-thread screws. Taking $E = 0.1$ and substituting,

$$0.1 = \frac{5000 \times 1/7}{F(2\pi l)}; \quad \text{or} \quad Fl = \frac{10 \times 5000}{2\pi \times 7} = 1140 \text{ inch-pounds.}$$

The maximum tensile and shearing stresses can be found by equations (67) and (68); the values of t and s to be used are $t = T \div A = T \div \pi r_2^{\,2}$, and $s = Fl \div \frac{1}{2}\pi r_2^{\,3}$. These maximum stresses do not need consideration, however, unless the nut is tightened while under the stress of the load.

If the material through which the bolt passes in Fig. 104 were totally without elasticity, and the bolt tightened to produce a tension T in it, then, for any suspended load not exceeding T, the tension in the bolt will always be equal to T; and when the load equals T there would be no pressure between the surfaces clamped together, for the bolt would elongate enough to relieve the pressure.

If a spring were interposed between the surfaces before tightening the bolt and put in compression, without clamping it solid, by screwing down the nut till a tension T is produced, then a suspended load would make the tension in the bolt equal to T plus the load. This assumes that the elongation in the bolt is so slight as to be inappreciable compared with the capacity of the spring for elongation.

Since all material is elastic, there is always something of the spring action on the bolt. This, of course, is exceedingly slight when heavy parts of rigid material are bolted together; when short bolts are used to clamp together pieces separated by a springy substance, such as rubber packing, the spring action may be great enough to require consideration.

Experiments show that grooves used for standard bolt-threads very materially increase the tensile strength per unit area, although the total strength may be made less than that of the original bar. This is due to the fact that fracture is made to occur at the groove, which, on account of having a large amount of resisting material on each side, requires a greater force to fracture it than a bar of the same diameter as a section at the bottom of the groove.

The effect of cutting a V thread upon a bolt is nearly the same as for a single circumferential groove of the same form as the thread-groove. It has been found by experimental investigation that, on account of this strengthening effect of the thread, with fairly good rubbing surfaces reasonably well lubricated, the axial tensile stress per square inch which will cause rupture in a U. S. Standard screw-bolt, while tightening the nut, is practically the same as the breaking strength per square inch of the body of the bolt.* This means that the twisting effect of the nut may be neglected.

The sectional area of U. S. Standard screw-bolts at the bottom of the thread is roughly 0.7 of the area of the bar on which the thread is cut, for diameters up to 2 inches. It therefore seems safe to say that the axial tensile stress which will rupture such a screw-bolt, when tightening the nut, is 0.7 as great as will fracture the body of the bolt, it being assumed that the body and the top of the thread are of the same diameter.

Major Wm. R. King found that by doubling the number of threads per inch on a screw-bolt, the total tensile strength was increased about 20%, and the resilience or shock-resisting power to a much greater extent. The gain was somewhat greater when the

---

* Zeitschrift des Vereines deutscher Ingenieure, April 27, 1895, p. 505.

SCREW-FASTENINGS. 199

number of threads per inch was tripled.* The average results of several tests were as follows:

Threads per inch	6	12	18
Relative tensile strength	1.	1.21	1.23
Elongation	.025	.06	.08
Relative work or resilience	.025	.0726	.0984

The stress was applied at the nut and head. Stripping of the thread did not occur in any of the experiments, but the reduction of the diameter of the screw by elongation was so great as to let a portion of the threads of the nut and bolt slip past each other.

It has been shown by numerous tests upon screw-bolts having the thread made by the cold-rolling process, which forces the metal up so that the top of the thread is somewhat larger in diameter than the body of the bolt, that, when subjected to tensile stress applied to the working faces of the head and nut, the bolt invariably fractures in the body.† These tests show very clearly the strengthening effects of the thread, for the sectional diameter across the bottom of the thread is, of course, considerably less than that of the body of the bolt.

85. Endurance of screw-bolts.—Repeated stresses in a screw-bolt, such as occur in the fastenings of a trip-hammer, steam or pneumatic rock-drill, connecting-rod ends of pumps, engines, etc., frequently cause bolts to fracture across the threads between the nut and body. This is caused by the slight temporary elongation of the bolt every time the shock or stress occurs. Such an elongation may occur without allowing the surface of the parts held together to separate even to the slightest extent; for the elasticity of any two parts clamped together with a bolt will, when an additional stress is applied to separate them, especially if it is a shock, cause them to increase their dimensions in the direction of the length of the bolt, and thus elongate it.

By reducing the sectional area of the body of the bolt so that it is not greater than that across the bottom of the thread, the bolt is made more durable. This is due to the fact that, by reducing the

---

* Trans. Amer. Inst. Mining. Engrs., 1885, p. 90.
† Catalogue of J. H. Sternbergh & Co.

sectional area of the body, the bolt elongates more readily, the slight elongation that occurs with each shock distributing itself nearly uniformly throughout the length of the bolt, thus reducing the actual stress which occurs in it with each shock. When the body of the bolt is of the same diameter as the top of the thread, the slight elongation which occurs with each repetition of stress is largely localized at the bottom of the thread between the nut and

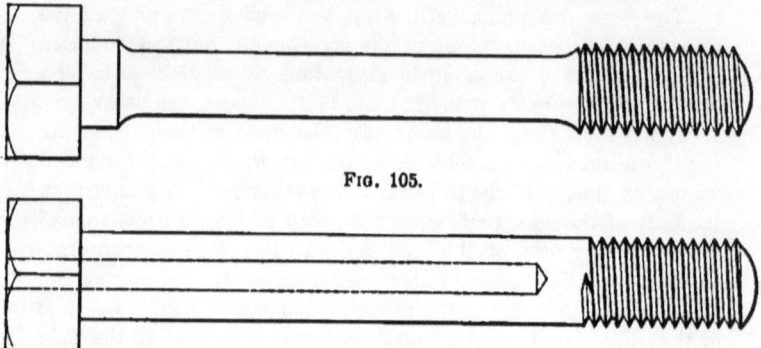

Fig. 105.

Fig. 106.

body of the bolt, and causes the metal to gradually give way by fatigue.

The reduction of the sectional area of the body of a bolt may be accomplished either by reducing its diameter, as in Fig. 105, or by drilling a hole in the centre, as in Fig. 106. The hollow bolt is the stronger torsionally, and will therefore withstand a greater tensile stress while the nut is being screwed on. The increased endurance of bolts so made has been demonstrated in practice.

CHAPTER VII.

MACHINE KEYS, PINS, FORCED AND SHRINKAGE FITS.

MACHINE KEYS AND PINS.

86. The principal function of machine keys is to prevent rotary motion of one part about another, as of a pulley about a shaft on which it fits; less frequently they are used to prevent lateral motion also, when the tendency to such motion is comparatively small. They are in a general way divided into three classes, commonly called flat, square, and feather or sliding keys.

There are no accepted standards of keys and key-way proportions. Tables XXV, XXVI, and XXVII, by John Richards,* give the average proportions of general practice, however.

The flat key, Fig. 107, is most suitable for heavy machinery, such as is used for mill-work, when both the relative rotation of the parts and the lateral motion incidental to the vibration of the machinery, and, in some cases, the weight of the parts, are to be resisted. A flat key should completely fill the key-way; the pressure against the sides should be greater than at the top and bottom, where a light pressure is all that is necessary. A heavy pressure against the top

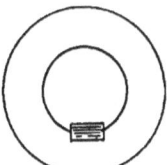
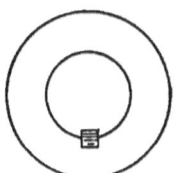

FIG. 107.   FIG. 108.

and bottom of the grooves has a tendency to spring the parts out of true and fracture the hub. The top of a key is the side remaining

---

* Richards' "Manual of Machine Construction." *Cassier's Magazine*, April, 1893, p. 416.

completely exposed after the key has been put in position in one of the parts to be keyed.

"**Square**" **keys**, Fig. 108, which are generally only approximately square in section, are generally used for the lighter classes of work, and especially for machine-tool construction. They should fit tightly on the sides, only merely touching at the top and bottom so as to prevent their tipping over under a heavy load, or not touching at all at the top.

TABLE XXV.*

DIMENSIONS OF FLAT MACHINE KEYS, INCHES.

Diam. of shaft	1	$1\frac{1}{4}$	$1\frac{1}{2}$	$1\frac{3}{4}$	2	$2\frac{1}{2}$	3	$3\frac{1}{2}$	4	5	6	7	8
Breadth of keys	$\frac{1}{4}$	$\frac{5}{16}$	$\frac{3}{8}$	$\frac{7}{16}$	$\frac{1}{2}$	$\frac{5}{8}$	$\frac{3}{4}$	$\frac{7}{8}$	1	$1\frac{1}{4}$	$1\frac{3}{8}$	$1\frac{1}{2}$	$1\frac{3}{4}$
Depth of keys	$\frac{3}{16}$	$\frac{3}{16}$	$\frac{1}{4}$	$\frac{5}{16}$	$\frac{5}{16}$	$\frac{3}{8}$	$\frac{7}{16}$	$\frac{1}{2}$	$\frac{1}{2}$	$\frac{11}{16}$	$\frac{13}{16}$	$\frac{7}{8}$	1

TABLE XXVI.*

DIMENSIONS OF APPROXIMATELY SQUARE MACHINE KEYS, INCHES.

Diameter of shaft	1	$1\frac{1}{4}$	$1\frac{1}{2}$	$1\frac{3}{4}$	2	$2\frac{1}{4}$	3	$3\frac{1}{2}$	4
Breadth of keys	$\frac{5}{16}$	$\frac{3}{8}$	$\frac{7}{16}$	$\frac{9}{16}$	$\frac{11}{16}$	$\frac{13}{16}$	$\frac{15}{16}$	$1\frac{1}{8}$	$1\frac{1}{4}$
Depth of keys	$\frac{7}{16}$	$\frac{1}{2}$	$\frac{9}{16}$	$\frac{11}{16}$	$\frac{13}{16}$	$\frac{15}{16}$	1	$1\frac{1}{8}$	$1\frac{1}{4}$

TABLE XXVII.*

DIMENSIONS OF FEATHER OR SLIDING MACHINE KEYS, INCHES.

Diameter of shaft	$1\frac{1}{4}$	$1\frac{1}{2}$	$1\frac{3}{4}$	2	$2\frac{1}{4}$	$2\frac{1}{2}$	3	$3\frac{1}{2}$	4	$4\frac{1}{2}$
Breadth of keys	$\frac{1}{4}$	$\frac{1}{4}$	$\frac{5}{16}$	$\frac{3}{8}$	$\frac{3}{8}$	$\frac{7}{16}$	$\frac{1}{2}$	$\frac{9}{16}$	$\frac{5}{8}$	$\frac{5}{8}$
Depth of keys	$\frac{3}{8}$	$\frac{3}{8}$	$\frac{7}{16}$	$\frac{1}{2}$	$\frac{1}{2}$	$\frac{9}{16}$	$\frac{5}{8}$	$\frac{11}{16}$	$\frac{3}{4}$	$\frac{3}{4}$

* Taken from *Cassier's Magazine*, April, 1893, p. 416.

If only a small portion of the power that the shaft is able to transmit is taken off through the key, the latter may be made smaller than given in the tables. The side pressure on the key is practically inversely as the diameter of the shaft.

A feather or sliding key is adapted to a service requiring lateral motion of the parts over each other, but not relative rotation. On account of the wear on the sides, due to the lateral motion, and in order to give the key a firm hold upon the part to which it is attached, its radial height is greater than for "square" keys. Such a key is fastened to one of the parts, either tightly or loosely, as is most suitable for other requirements, and moves laterally with regard to the other. Probably the simplest method of fastening such a key to a shaft is to dovetail it slightly on the sides and ends near the bottom, then force it into place and rivet the edge of the key-way down against it with a key-set. Another method of attaching a short feather to a shaft is shown in Fig. 109. The key

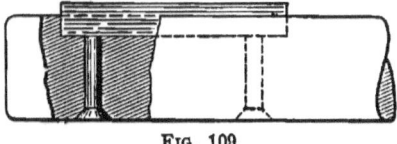

FIG. 109.

is made with a couple of round lugs or pins which pass through holes drilled in the shaft and are riveted on the side opposite the key. The same device is applied to a feather attached to a hub in

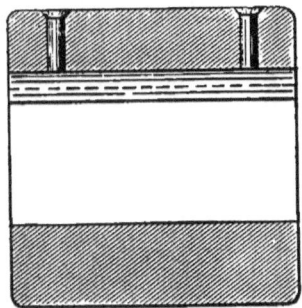

FIG. 110.

Fig. 110. Another method is shown in Fig. 111; the feather may be made a loose fit in this form, and can be removed by slipping the hub off the end of the shaft.

A method of using a square key which is not very generally applied in practice, but which appears to possess an advantage in the ease with which it can be fitted, is shown in Fig. 112; the key

204 FORM, STRENGTH, AND PROPORTIONS OF PARTS.

is square, and is placed so that one diagonal is radial. A heavy load would exert a strong bursting pressure on the hub. For this reason the key is hardly suitable for heavy service.

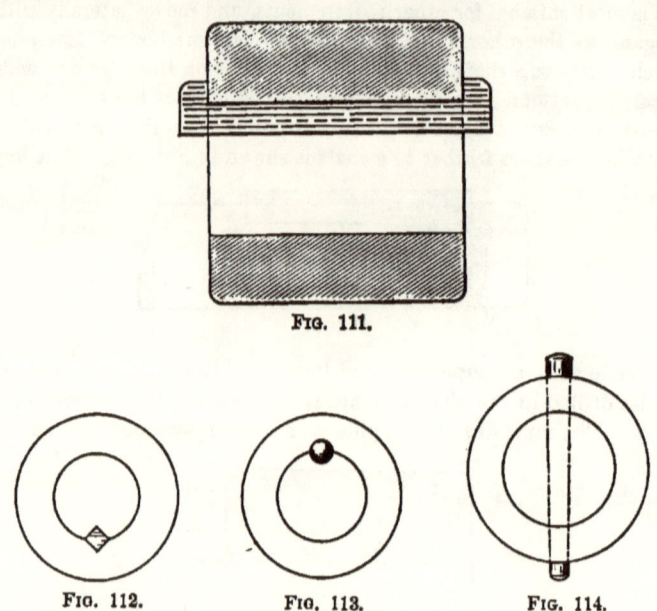

FIG. 111.

FIG. 112.　　FIG. 113.　　FIG. 114.

A cylindrical pin can be used, as in Fig. 113, when the shaft and hub are of the same material, or near enough alike to make it practicable to drill a hole for the pin as shown. Such a key or pin is hardly suitable for heavy machinery.

A taper pin, Fig. 114, can be used to advantage when there is considerable end pressure to be resisted, as well as turning force. In practice the smallest size of such a pin is ordinarily determined by the smallest taper reamer that is long enough to ream out the hole. Morse standard taper pins are generally used. The torsional moment, which a pin that is not large enough to weaken the shaft unduly will resist, is much less than for a key of the proportions given in the tables, if the length of the key is as great as the diameter of the shaft.

87. Roller keys are used to some extent for fastening small pulleys, etc., to shafting. The hub of the pulley is bored a short distance at each end to fit the shaft; the centre is bored eccentric with the ends, and larger in diameter by something more than the diameter of the roller key to be used. A hardened cylindrical key of about the same length as the larger part of the bore is placed in it, and the shaft slipped through the hub. A slight rotation of the latter causes the key to bind between it and the shaft so as to prevent further rotation in that direction; a slight rotation in the opposite direction loosens the connection. It is self-evident that this device can be used only when the tendency to turn the parts over each other is always in the same direction.

88. Eccentric keys or fastenings of the form shown in Fig. 115,

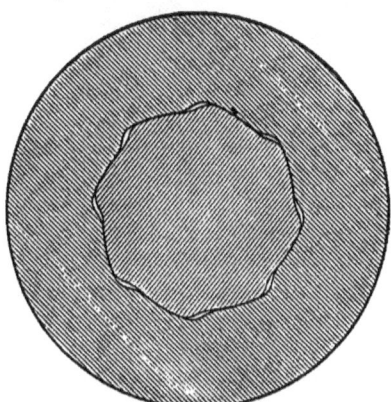

Fig. 115.

known as Blanton Patent Fastenings, are used to good advantage in certain classes of machinery. The surface of the shaft is turned into a series of corrugations as shown, and the hub is bored to the same form, being made large enough to slip along the shaft freely when the parts are put in the loose position. A slight angular rotation of one causes it to lock with the other and drive it. By turning them in the opposite direction relatively to each other they are loosened.

This fastening has been found especially applicable to the lifting-cams of ore stamp-mills, largely on account of the ease with which the cams can be removed and new ones substituted. When

206   FORM, STRENGTH, AND PROPORTIONS OF PARTS.

there is danger that the fastening may become loose on account of the momentum of the parts, slight variations of speed, reversals of motion, etc., a key is applied, as in Fig. 116, to prevent the

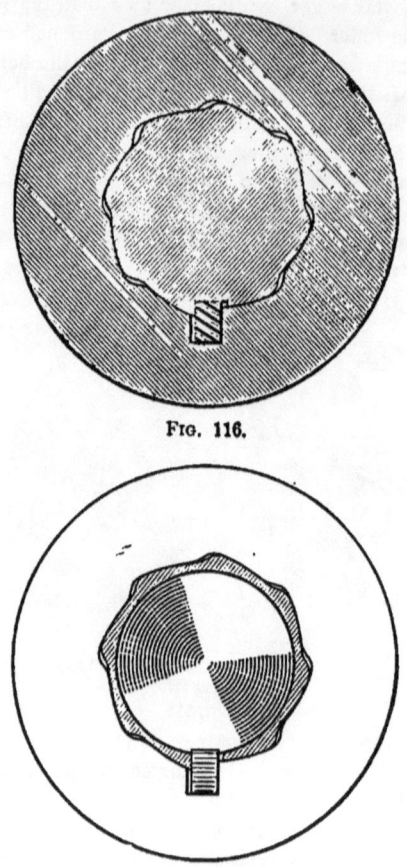

FIG. 116.

FIG. 117.

loosening.   The fastening, although thus modified, is still adapted to driving in only one direction.

The Blanton fastening is adapted to round shafts as shown in Fig. 117. The turning moment is largely resisted by the friction between the shaft and corrugated sleeve, thus relieving the rectangular key of much of the pressure that would come upon it if preventing rotation between two round parts.*

SHRINKAGE AND FORCED FITS.

89. Shrinkage and forced fits are adopted frequently when it is desired to have parts fit together very tightly; the parts are almost invariably either cylindrical or slightly conical on the surfaces in contact. For a shrinkage fit, the outer member, of two parts that are to be fastened together, is finished to a diameter slightly smaller than the inner. It is then heated so as to expand it enough to pass over the inner member, put into place and cooled. The contraction of cooling causes it to grip the inner part firmly.

For a forced fit the parts are prepared in the same manner as for shrinkage fits, but are forced together cold instead of the outer one being expanded by heat and then shrunk into place.

90. Tension in and pressure against a thin ring fitted by shrinking or forcing.—The tension in a ring that is thin radially in comparison with its diameter, as well as the pressure between the ring and body it encircles, can be calculated if the modulus of elasticity E of the material is known. This assumes that the ring is either forced into place, or is heated to a uniform temperature throughout and all parts cooled at the same rate. While such heating and cooling can probably never be attained in practice, they can be nearly enough approximated to warrant the above assumption. If cooled unevenly, the greatest tension will be at the place cooled last.

The following notation will be used for shrinkage and forced fits:

$A =$ sectional area of ring, square inches;
$D =$ diameter of ring before heating, inches;
$D_1 =$ diameter of ring after putting into place, inches;
$E =$ tensile modulus of elasticity of material of ring, pounds per square inch;

---

* The Blanton Patent Fastening is used in the United States by Fraser & Chalmers of Chicago, who control its use on stamp-mills. (Statement from pamphlet issued by the Blanton Patent Syndicate, Ltd., London.)

F = force required to slip the ring from place, pounds;
P = total pressure between surfaces of ring and inner part, pounds;
T = total tension in ring, pounds;
h = radial thickness of ring, inches;
p = unit pressure between surfaces of ring and inner member, pounds per square inch;
r = radius of ring, inches;
t = unit tension in ring, pounds per square inch;
w = width of bearing surface of ring, measured parallel to axis of bore, inches;
μ = coefficient of friction between surfaces bearing together.

Remembering that E = (stress in pounds per square inch) ÷ (elongation per inch of length), the two following equations may be written:

$$t = E\frac{D_1 - D}{D}; \qquad \ldots \ldots \quad (89)$$

$$T = At = AE\frac{D_1 - D}{D}. \qquad \ldots \ldots \quad (90)$$

The equation for the unit pressure p between the surfaces may be obtained by assuming that the ring is cut in two on a diametral

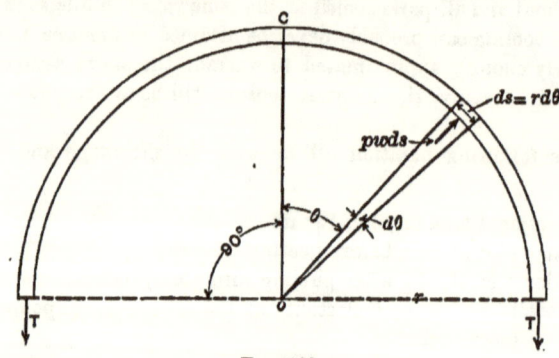

Fig. 118.

plane, and one half considered as a free body with the forces necessary to hold it in equilibrium acting upon it, as shown in Fig. 118.

SHRINKAGE AND FORCED FITS.

These forces consist, first, of the two tensions T and T which the other half of the ring exerted when in place, and, second, of the elementary forces normal to the inner surface of the ring and uniformly distributed over it. T and T are both normal to the plane of the section of the ring. The forces acting normal to the inner cylindrical surface of the ring may be taken as having a unit value of p pounds per square inch.

Take an elementary portion ds of the circumference of the ring. The corresponding inner surface of the ring is wds. The total pressure acting on this elementary area is $pwds$. This pressure may be resolved into two components, parallel and perpendicular to T and T. The one which is normal to T and T is annulled by the equal and opposite component of the pressure acting on an equal area symmetrically situated with regard to the median line OC. This component may therefore be neglected. The other component, parallel to T and T, and acting in the opposite direction from them, must be resisted by a portion of the tensile forces T and T acting at the ends of the half-ring. The value of this component is $pw(ds) \cos \theta$, in which θ is the angle between the median line OC and a radial line drawn to the elementary area $w(ds)$. The sum of the components parallel to T and T for all the elementary areas of the half-ring must equal $2T$, since T and T are the forces which hold these components in equilibrium. Therefore

$$\int pw(ds) \cos \theta = 2T$$

Since r is the radius of the ring, and $d\theta$ is the angle between the two radial lines drawn to the ends of the elementary length ds of the circumference, $ds = rd\theta$. The last equation may therefore be written

$$pwr \int_{-\frac{\pi}{2}}^{\frac{\pi}{2}} \cos \theta \, d\theta = 2T,$$

which gives

$$2pwr = 2T.$$

210   FORM, STRENGTH, AND PROPORTIONS OF PARTS.

Transposing and substituting the value of T, as given in equation (90), gives

$$p = \frac{T}{wr} = \frac{At}{wr} = AE\frac{D_1 - D}{Dwr}. \quad \ldots \quad (91)$$

The total pressure P between the surfaces bearing together is

$$P = pw\pi D = \frac{T\pi D}{r} = 2T\pi = 2At\pi. \quad . \quad (92)$$

The force F necessary to slip the ring laterally along the surface against which it presses is

$$F = \mu P = 2\mu T\pi = 2\mu At\pi. \quad \ldots \quad (93)$$

Example.—If a steel ring 4 inches wide, 0.7 of an inch thick radially, and 60 inches in diameter before putting in place, is shrunk or forced on another part which increases its diameter .06 of an inch, the tension per square inch is, by equation (89), putting $E = 30,000,000$,

$$t = 30,000,000 \frac{.06}{60} = 30000 \text{ lbs. per sq. in.}$$

The total tension is, by equation (90),

$$T = At = (4 \times 0.7) 30000 = 84000 \text{ lbs.}$$

The pressure per square inch between the cylindrical surfaces is, by equation (91),

$$p = \frac{T}{wr} = \frac{84000}{4 \times 30} = 700 \text{ lbs. per sq. in.}$$

The total pressure between the bearing surfaces, determined by equation (92), is

$$P = 2T\pi = 2 \times 84000\pi = 527780 \text{ pounds.}$$

The force required to slip the ring laterally, taking $\mu = .15$, is, by equation (93),

$$F = .15 \times 527780 = 79170 \text{ pounds}$$
$$= 19.6 \text{ tons (about).}$$

91. Shrinkage and forced fits for thick rings and heavy parts. —When the ring or outer member is very thick or heavy, the material in it is not subjected to even an approximately uniform stress when two parts are shrunk or forced together. The layers near the smaller diameter of the outer member are put into greater tension than those more remote, on account of the elasticity of the material and the supporting action of each successive annular layer against the one inside of it. On account of this uneven distribution of stress, it is not probable that any mathematical formula can be deduced that is of sufficient practical value for very thick pieces to warrant its adoption. This applies to such a member as a crank shrunk or forced on its shaft, or having a pin forced or shrunk into it.

Again, when a heavy part is shrunk or forced over another, the great pressure exerted on the inner piece decreases the diameter of the latter on account of its elasticity. This has been perceptibly shown when the steel tires are shrunk on locomotive drive-wheel centres. The increase of diameter of the tire, due to putting it in place, is less than the difference between the bore of the tire and the outer diameter of the wheel centre before putting the tire on. The extent of the reduction of the inner part depends upon the form and strength of the parts. It is therefore a quantity that cannot well be introduced into a formula, except empirically.

92. Allowance for shrinkage and forced fits.—For locomotive drive-wheels and steel-tired car-wheels, the bore of the tire is quite commonly made .001 of the diameter smaller than the wheel centre. The Midvale Steel Co. makes a greater difference of diameter than this, their rule being to bore the tire $\frac{1}{750}$ of the nominal diameter smaller than the wheel centre.

The collection of data, Tables XXVIII, XXIX, XXX, and XXXI,* shows the practice of several machine-builders in allowing for shrinkage and forced fits. The force necessary to press the parts together is given in nearly all cases.

It is the practice of some leading concerns to slightly taper the parts that are forced together. This obviates the necessity of slowly forcing the parts together throughout the entire length of the bore, and reduces the liability to abrasion and cutting between the surfaces. The parts are readily separated, since they become loose

---

* Taken from *Machinery*, May, 1897.

when given a small relative lateral movement. The difficulty of securing tapers accurately the same on both parts is, of course, detrimental to their general application.

Table XXXII indicates the practice of the Chicago, Milwaukee, and St. Paul Railway in the Milwaukee shops, for both taper and cylindrical fits.

Table XXVIII.*
EXTRACT FROM LANE & BODLEY CO.'S RECORD BOOK OF FORCED FITS.

No.	Mean Diameter of Pin, inches.	Length of Fit, inches.	Mean Diameter of Hole, inches.	Total Allowance.	Allowance per inch.	Area of Fitted Surface, square inches.	Volume under Fitted Surface, cubic inches.	Pressure to enter pin, tons.	Pressure at Mid-position, tons.	Maximum Pressure, tons.
1	1.8798	6.125	1.8767	.0031	.0017	36	16.7	2	10	20
2	1.8819	6.125	1.877	.0042	.0022	36	16.7	2	15	23
3	1.8774	4.375	1.8764	.001	.00052	24.4	13.7	.5	1	1
4	2.7455	4.5	2.7387	.0068	.00247	38.7	26.5	3	12	25
5	2.7465	4.5	2.7437	.0028	.001	38.7	26.5	5	12	23
6	3.261	5	3.2542	.0068	.0021	51	41.5	5	20	45
7	3.2625	5	3.2555	.007	.002	51	41.5	5	15	30
8	3.267	5	3.261	.006	.0018	51	41.5	5	15	20
9	4.2505	6.	4.2402	.0103	.0024	79.8	85.1	5	22	44
10	4.2388	6.625	4.2478	.0091	.0021	78.1	93.4	12	30	60
†11	4.2303	6.5	4.2224	.0079	.0019	95.8	91.	10	60	125
12	5.9343	4.0625	5.9216	.0127	.0022	75.7	112.2	6	16	25
13	5.9381	4.	5.9252	.0129	.0022	74.4	110.4	3	18	35
14	5.9294	4.125	5.9194	.01	.0017	76.7	113.8	5	15	25
15	6.8829	5.125	6.8697	.0132	.002	110.7	190.1	8	20	42
16	6.889	5.	6.8785	.0105	.0015	108.	18.9	5	22	45
17	6.8692	4.875	6.855	.0142	.0021	104.8	180.4	5	35	65
18	7.8884	5.5	7.873	.0154	.002	135.9	267.3	5	32	64
19	7.8715	6.5	7.8575	.014	.0018	160.5	315.9	5	25	50
20	7.862	5.625	7.846	.016	.002	138.2	272.8	8	40	80
21	8.924	6.125	8.905	.019	.0021	170.8	378.9	20	45	68
22	8.9	6.75	8.8848	.0152	.0017	188.4	419.9	5	47	96
23	8.878	6.5	8.8669	.0112	.0013	180.7	401.	10	45	92

* Taken from *Machinery*, vol. III., No. 9, May, 1897.   † No. 11 was a cast-steel crank-disk.

SHRINKAGE AND FORCED FITS. 213

TABLE XXIX.*

DATA FURNISHED BY C. O. HEGGEM, SUPT. RUSSELL & CO., MASSILLON, OHIO.

Allowance is for both press and shrinkage fits for steel cranks.

FOR PRESS FITS—CAST-IRON CRANKS.

```
Diameter of hole   4  to   5  inches.    Allow .0090 inch
    "    "   "     5   "  7¼    "          "   .0060  "
    "    "   "    7¼   "   9    "          "   .0055  "
    "    "   "    10   "  12    "          "   .0050  "
    "    "   "    12   "  16    "          "   .0040  "
    "    "   "    16   "  18    "          "   .0030  "
```

FOR SHRINK FITS—CAST-IRON CRANKS.

```
Diameter of hole   4  to   5  inches.    Allow .0045 inch
    "    "   "     5   "  7¼    "          "   .0030  "
    "    "   "    7¼   "   9    "          "   .0027  "
    "    "   "    10   "  12    "          "   .0025  "
    "    "   "    12   "  16    "          "   .0020  "
    "    "   "    16   "  18    "          "   .0015  "
```

TABLE XXX.*

FROM DATA FURNISHED BY H. BOLLINCKX, BRUSSELS, BELGIUM.

Diameter, inches.	Length, inches.	Total Allowance, inches.	Allowance per inch, inches.	Total Pressure, pounds.
CRANK-PINS.				
2.2	2.76	.0039	.0018	43300
2.79	3.78	.0049	.0018	49500
3.145	4.32	.0058	.0018	55500
LEVER-PINS.				
.865	1.57	.0023	.0027	6800
1.1	1.77	.0025	.0023	6800
1.18	1.965	.0027	.0023	6800
1.375	2.36	.0031	.0023	9060
CYLINDER-CASES.				
11.4		.0098	.00086	37100
17.7		.0098	.00055	37100
PISTONS.				
1.49	4.32	.0050	.004	37100
1.46	4.32	.0059	.004	49500
1.77	4.76	.0070	.004	55500
2.24	11.18	.0039	.0017	61800
2.63	11.18	.0039	.0015	74000
2.95	13.12	.0049	.0017	113200

* Taken from *Machinery*, vol. III., No. 9, May, 1897.

TABLE XXXI.*
BRIEF SUMMARY OF GENERAL PRACTICE.

	Diameter, Inches	Total Allowance, Inches	Allowance per inch of Diam., in.	Pressure to enter pin, tons per inch of diameter.	Remarks
E. P. Allis Co., Milwaukee, Wis.	8	.008	.001	10 to 16	Allowance is more on smaller sizes, less [on larger.
Baldwin Locomotive Works, Phila., Pa.	5 to 9	.015 to .027	.003	10 to 15	Axles.
H. Bollinckx, Brussels, Belgium	3 to 6	.0015 to .009	.0015	5 to 10	Wrist-pins.
	2.2	.0039	.0018	9	Crank-pins.
	1.18	.0027	.0023	2 to 3	Lever-pins.
	11.4	.0098	.0009	1 to 2	Cylinder-cases.
	1.49	.0059	.04	12	Piston-rods into piston.
Buckeye Engine Co., Salem, O.	3 to 10	.01 to .015	.003 to .0015	9 to 10	
Buffalo Forge Co., Buffalo, N.Y.	4 to 10	.01 to .03125	.0025 to .003		
	15 to 40	.047 to .0625	.003 to .0015		
Dickson Mfg. Co., Scranton, Pa.	4.5	.01	.0022	8	Diameter over 5 inches, 7 tons.
Juniata Shops, P. R. R., Altoona, Pa.	5.375	.017	.0031	10	With 80 pins, pressure varied from 35 to 80 tons; average, 56.
	6	.023	.0038	10	With 20 pins, pressure varied from 35 to 90 tons; average, 71.5.
	4	.013	.0032	10	With 60 pins, pressure varied from 30 to 60 tons; average, 45.
Lane & Bodley Co., Cincinnati, O.	2.74	.0025 o .0068	.001 to .0025	8.35 to 9.1	See Table XXVIII for more details.
	3.36	.006 to .0068	.0018 to .0021	6.15 to 13.8	
	4.25	.0103	.0024	10.3	
	5.93	.0127	.0021	4.21	
	6.88		.0022		
Phila. Engineering Wks., Phila., Pa.	7.88	.0132 to .0142	.002 to .0021	6.12 to 9.5	Least allowance takes greater pressure.
	8.90	.0154 to .016	.0018 to .0018	8.27 to 10.02	Same allowance and pressure for sizes on each line.
	8 to 6	.019 to .0112	.0021 to .0013	10.35 to 7.65	
	6 to 9	.002	.0007 to .00085	40 tons total.	
	9 to 12	.001	.0007 to .0005	75 to 80 total.	
	12 to 16	.008	.0007 to .00045	100 total.	
Rogers Locomotive Co., Paterson, N.J.	4 to 9	.015 to .022	.004 to .0024	150 total.	See Table XXIX.
Russell & Co., Massillon, O.	4 to 7.5	.009 to .006	.0025 to .0012	12	
Schenectady Loco. Wks., Schenec., N.Y.	10 to 18	.006 to .008	.0005 to .00016	4	Axles.
	3 to 8			10 to 11	Crank-pins.
Straight-Line Engine Co., Syracuse, N.Y.	3.25 to 3.9375	.007 to .0075	.002 to .0019	5 to 8	Both crank-pins.
	4.0675 to 5.0675	.008 to .009	.0016 to .0015		Both shafts.
	3. to 3.4675	.006 to .007	.002 to .0022		
Union Iron Works, San Francisco, Cal.	4.375 to 4.75	.0075 to .008	.0017 to .0017		
	Generally	about	.001		Endeavor to obtain 300 to 500 lbs. pressure per square inch of contact surface.

*Taken from Machinery, vol. III., No. 9, May, 1897.

Table XXXII.

FORCED FITS. PRACTICE OF CHICAGO, MILWAUKEE, AND ST. PAUL RAILWAY.

	Diameter at large end, inches.	Approximate length, inches.	Taper = variation of diameter per foot of length.	Pressure to force parts together, tons.
Parallel-rod crank-pin *............	4¹³⁄₁₆	6¼	1/32	20
Main crank-pins *....	4	6	1/32	25
	4⅜	6	1/32	30
	5¼	6	1/32	35
Engine driving-axles	7¾	7½	none	60 to 80; †
	7⅞	8¼		80 preferred
Engine truck-axle...	5¹³⁄₁₆	7	none	30 to 40; 25 will hold.
Car-axles............	5⅜	7 to 9¼	1/16	40 ‡

* End riveted over after forcing into place.
† If only 60 tons is used, more care is taken in fitting key.
‡ Only slight variations from this amount allowed.

CHAPTER VIII.

AXLES, SHAFTING, AND POSITIVE SHAFT-COUPLINGS.

93. Notation.—The following notation is used in the formulas for the strength and deflection of shafting:

$D =$ diameter of round shaft, or width of side of square shaft, inches;

$E_s =$ shearing modulus of elasticity of material of shaft, used for torsion, pounds per square inch;

$E_t =$ tensile modulus of elasticity of material of shaft, pounds per square inch;

$I =$ moment of inertia of cross-section of shaft about a gravity axis in the plane of the section, biquadratic inches;

$J =$ polar moment of inertia of cross-section of shaft, biquadratic inches;

$L =$ length of shaft, inches;

$M_b =$ bending moment, inch-pounds;

$M_t =$ twisting moment, inch-pounds;

$(M_b)_i =$ ideal bending moment which would induce the same fibre-stress as the combined bending and twisting moments;

$(M_t)_i =$ ideal twisting moment which would induce the same fibre-stress as the combined bending and twisting moments;

$P =$ turning force, pounds;

$S_b =$ section modulus of shaft for bending $= I \div r$;

$S_t =$ section modulus of shaft for torsion $= J \div r$;

$W =$ bending force, pounds;

$c =$ distance from centre of shaft to most remote part of section, inches;

$d =$ diameter of bore in hollow shaft, inches;

$f =$ fibre-stress in material of shaft due to bending, pounds per square inch;

$l =$ length of lever-arm of turning force, inches;

AXLES, SHAFTING, AND POSITIVE SHAFT COUPLINGS. 217

$n =$ the ratio $d \div D$;
$s =$ shearing-stress in shaft due to turning force, pounds per square inch;
$w =$ weight of shaft per inch of length, pounds;
$\theta =$ angular deflection or twist of shaft in degrees $= 57.29578 \times$ (angular deflection in radians).

ROUND SHAFTING.

94. Torsional strength of round shafts.—When a shaft is used simply for transmitting power from one point to another, there being no intermediate devices, such as gears or pulleys and belts, for receiving or delivering power, the torsional strength and angular deflection of the shaft are all that ordinarily need consideration. This assumes that the bearings of a horizontal shaft are near enough together to prevent excessive sagging or bending between them, and that a vertical shaft is supported by thrust-bearings so as to prevent excessive local end thrust or buckling. The angular deflection or stiffness needs especial attention when steady running under a variable load is important.

The relation between the turning moment M_t and shearing-stress s in a shaft is expressed by the equation

$$M_t = Pl = s\frac{J}{c}. \quad \ldots \ldots \quad (94)$$

For a solid round shaft $J = \frac{\pi D^4}{32}$ and $c = \frac{D}{2}$.

Substituting these values in the last equation, it becomes:
For solid round shafts

$$M_t = Pl = .1963 s D^3, \quad \ldots \ldots \quad (95)$$

whence

$$D = \sqrt[3]{\frac{5.1 M_t}{s}} = \sqrt[3]{\frac{5.1 Pl}{s}}. \quad \ldots \ldots \quad (96)$$

For a hollow round shaft $J = \dfrac{\pi(D^4 - d^4)}{32}$, and $c = \dfrac{D}{2}$ as before. Equation (94) becomes, by substituting these values in it:
For hollow round shafts

$$M_t = Pl = .1963s(D^4 - d^4). \quad \ldots \ldots (97)$$

Putting $d = nD$ in equation (97), and transposing, puts the equation into a convenient form to solve for D when the ratio n of the bore to the diameter is assumed. The equation thus becomes:
For hollow round shafts

$$D = \sqrt[3]{\dfrac{5.1 M_t}{(1-n^4)s}} = \sqrt[3]{\dfrac{5.1 Pl}{(1-n^4)s}}. \quad \ldots \ldots (98)$$

The slight weakening effect of removing the centre of a shaft is worthy of note. A shaft having an axial hole of a diameter $d = .3D$ through its centre has $1 - (.3)^4 = .973$ of the torsional strength of a solid shaft of the same material; if $d = .4D$, the strengths are as .936 to 1; for $d = .5D$, the ratio is .875 to 1; and for $d = .6D$ the ratio is .784 to 1.

The torsional strength of a hollow shaft having a bore half as large in diameter as the shaft is 1.347 times that of a solid shaft of an equal sectional area (i.e., the same weight per linear foot) and of the same material.

The above statements regarding the relative strength of solid and hollow shafting are based upon the assumption that the material in each is of the same strength. In shafts of the same sectional area, however, the material of the hollow-forged shaft can be made stronger than that of the solid one, because it can be worked more thoroughly under the hammer or forging-press. The ordinary method of making a hollow forging is to cast the ingot solid and bore out the centre. This takes out the poorest material, which is always at the centre. Practically the same result is obtained when the piece is forged solid and the centre bored out afterwards. For forgings of very large diameter, the method of boring before forging is undoubtedly preferable.

In modern practice a working shearing-strength as high as

12000 pounds per square inch is used for hollow-forged steel shafts for the screw-propellers of large vessels, etc.

Example.—The turning force F, applied at a distance $l = 40$ inches from the centre of a shaft 15 inches in diameter when working at a maximum shearing fibre-stress $s = 9000$ pounds per square inch, may be found by equation (95). By substituting the given values in this equation,

$$P \times 40 = .1963 \times 9000 \times (15)^3,$$

whence

$$P = 149000 \text{ pounds.}$$

Sectional area of shaft $= 176.71$ square inches.

Example.—The diameter of a hollow shaft having a hole half as large as the shaft, to resist a turning force $P = 149000$ pounds, acting on a lever-arm of a length $l = 40$ inches, and working at a maximum shearing fibre-stress of 9000 pounds per square inch, can be found by substituting in equation (98), thus obtaining

$$D = \sqrt[3]{\frac{5.1 \times 149000 \times 40}{[1 - (.5)^3]9000}} = 15.68 \text{ inches.}$$

Sectional area of shaft $= 144.84$ square inches.

The area of a solid shaft for the same strength, as shown in the preceding example, is 22 per cent greater than this.

95. Twist of a shaft under torsional stress.—The relation between the angle of twist and the turning moment acting on a solid round shaft is expressed by the following equation, which gives θ in degrees:

$$\theta = 583.6 \frac{M_t L}{E_s D^4} = 583.6 \frac{PlL}{E_s D^4}. \quad \ldots \quad (99)$$

For hollow shafts the same relation is expressed by the following equation, giving θ in degrees:

$$\theta = 583.6 \frac{M_t L}{E_s(D^4 - d^4)} = 583.6 \frac{PlL}{E_s(D^4 - d^4)}. \quad \ldots \quad (100)$$

The following equation of the relation between the shearing-stress per unit area and the twist is applicable to both solid and hollow round shafts:

$$\theta = 114.6 \frac{sL}{E_s D} \text{ degrees.} \qquad (101)$$

Example.—The angular twist of a solid machine-steel shaft 15 inches in diameter and 100 feet = 1200 inches long, when subjected to a turning force $F = 149000$ pounds, applied at a distance $l = 40$ inches from the centre, assuming that the modulus of elasticity for shearing $E_s = 12000000$,* is by equation (99)

$$\theta = 583.6 \frac{149000 \times 40 \times 1200}{12000000 \times (15)^4} = 6.87°.$$

The data in this example, with the exception of the length of the shaft, are the same as used in the examples of the preceding section. The fibre-stress used in these examples is 9000 pounds per square inch. Substituting this value of s in equation (101), and solving, gives

$$\theta = 114.6 \frac{9000 \times 1200}{12000000 \times 15} = 6.87°,$$

which is, of course, the same as obtained by equation (99).

The shaft, if working at a shearing-stress of 12000 pounds per square inch, would have a twist of 9.17° per 100 feet of length; the value of the turning force for this stress would be $P = 198700$ pounds.

96. Bending-strength of round shafting.—A round shaft that is subjected to bending only, as when a horizontal rotating shaft supports a load that does not exert a twisting moment upon it, can be calculated for strength as if it were a beam, the working fibre-stress f being selected with due regard to the fact that the stress is repeated at every revolution. Each fibre is alternately put into tension and compression during every revolution of the shaft. The safe working stress cannot be taken so high, on this account, as is

---

* $E_s = 10000000$ is more nearly correct for the materials generally used for line shafting in factories, etc.

permissible for a stationary beam supporting a static load. In the majority of cases the fibre-stress, as determined by the required rigidity of the shaft, is lower than the safe working strength for repeated stress.

The values of the maximum bending moment M_b acting on a round shaft when resisting a bending force W, applied according to the more common methods of loading, and for the weight of the shaft itself, are as follows:

Load applied to projecting end at a distance L from supporting bearing (cantilever),

$$M_b = WL. \qquad \qquad (102)$$

Load applied midway between two end supports,

$$M_b = \frac{WL}{4}. \qquad \qquad (103)$$

Load applied between two supports at distances a and b respectively from them $(a + b = L)$,

$$M_b = \frac{Wab}{L}. \qquad \qquad (104)$$

Two equal loads applied, each at the same distance a from the end nearest it,

$$M_b = Wa. \qquad \qquad (105)$$

Bending moment due to weight of horizontal projecting shaft (cantilever), length of projection $= L$,

$$M_b = \frac{wL^2}{2}. \qquad \qquad (106)$$

Bending moment due to weight of shaft, end supports,

$$M_b = \frac{wL^2}{8}. \qquad \qquad (107)$$

The relation between the bending moment M_b, tensile or compressive fibre-stress f, and the diameter of the shaft is given by the following four equations:

For solid round shafts,

$$M_b = .0982 f D^3; \quad \ldots \ldots \ldots (108)$$

$$D = 2.168 \sqrt[3]{\frac{M_b}{f}}. \quad \ldots \ldots (109)$$

For hollow round shafts,

$$M_b = .0982 f (D^3 - d^3); \quad \ldots \ldots (110)$$

$$D = 2.168 \sqrt[3]{\frac{M_b}{f\left(1 - \frac{d^3}{D^3}\right)}}. \quad \ldots \ldots (111)$$

The actual point of support of a shaft is frequently difficult to decide upon. If the bearings are self-aligning, the point of support may be taken in the plane, perpendicular to the shaft, containing the point about which the bearing-sleeve swivels; this is generally at the middle of the sleeve. When a bearing is rigidly supported, however, the point of pressure against the shaft may be near one end of the box at the time the shaft is subjected to the greatest bending moment. This applies to a shaft working under a varying bending force, as the crank-shaft of a steam-engine.

A long, tight-fitting hub, as of a fly-wheel supported on a horizontal shaft, prevents the shaft from bending where it is encircled by the hub. If the shaft is of the same cross-section throughout, the maximum fibre-stress in it will be at the end of the hub. While this shortens the lever-arm of the bending force and reduces the maximum fibre-stress to a lower value than if the load were (theoretically) applied at the middle of the hub, it also localizes the strain at the ends of the hub, and thus increases the liability to rupture, on account of fatigue of the metal, by the repetition of stress due to the rotation of the parts. Probably the best way of reducing the liability to fracture at the end of the hub is to enlarge the shaft where the hub is placed upon it, joining the enlargement to the smaller part by a fillet of as large a radius as can be conveniently used.

It is assumed in the above discussion that the centre of the hub contains the centre of gravity of the fly-wheel.

In practice it has been found that shafts and axles which are subjected to repeated stress, and especially to shocks and vibrations, are more durable when proportioned so that the stress and strain are distributed over them with at least an approximate degree of uniformity, and all re-entrant angles well filleted, than when made of uniform section throughout, or with sharp re-entrant angles.

It is common practice in car-axles to make them small in the middle in order that they may spring more readily, and thus reduce the strain upon the parts near the wheels and journals. This is an application of the same principle that makes a hammer-handle more durable when reduced in section between the head and the part that is held in the hand.

97. The lateral deflection of shafting on account of its own weight seldom needs consideration, except, possibly, for the smaller sizes. The remedy for the excessive deflections is to decrease the distance between bearings, or to increase the diameter of the shaft. The same is generally true of deflections caused by a load or belt pull.

The amount of the deflection can be calculated by the formulas ordinarily given for beams, but such a calculation is seldom needed.

It should be remembered, however, that a small shaft having supports so far apart as to allow it to bend considerably may, when rotated at a high speed, run out of true to a dangerous extent on account of the centrifugal action. This is due, not to the deflection of the shaft, which may be considered only as a measure of the liability to excessive bending, but to the fact that such a shaft when rotating will not keep its axis in exactly the same position. As soon as the shaft moves laterally to the least extent, the centrifugal action increases its deflection.

98. Shaft subjected to both torsion and bending, general case. —While this is the most common case that is found in practice with shafting, the experiments that have been made with a view to establishing the formulas that have been developed to meet it are very few and meagre. Of all the formulas that have been presented by different writers the following two, (112) and (113), seem to be the most convenient and correct.

The method adopted in these formulas is to find a bending or torsional moment equivalent to the actual moments acting simultaneously, and design the shaft accordingly. The shaft must, of course, be designed to resist the moment which has the greatest tendency to break it.

The value of the ideal or equivalent bending moment $(M_b)_i$ which will produce the same tensile or compressive fibre-stress in the material as the combined bending and twisting moments, M_b and M_t, is expressed by the equation

$$(M_b)_i = \tfrac{3}{8}M_b + \tfrac{5}{8}\sqrt{M_b^2 + M_t^2}. \quad \ldots \quad (112)$$

And the value of the ideal torsional moment which will cause the same shearing-stress in the material as the combined bending and torsional moments is shown in the formula

$$(M_t)_i = \tfrac{3}{8}M_b + \sqrt{M_b^2 + M_t^2}. \quad \ldots \quad (113)$$

Solid Round Shafts.

99. Solid round shafts subjected to more than one force.—The value of the resisting moment required to withstand the ideal bending moment $(M_b)_i$ of equation (112) is expressed by the formula, for a solid round shaft,

$$(M_b)_i = \frac{\pi f D^3}{32} \quad \ldots \quad (114)$$

By substituting this value of $(M_b)_i$ in equation (112) it reduces to equation (115).

For a solid round shaft:

$$fD^3 = 6.366[.6M_b + \sqrt{M_b^2 + M_t^2}], \quad \ldots \quad (115)$$

whence

$$D^3 = \frac{6.366}{f}[.6M_b + \sqrt{M_b^2 + M_t^2}], \quad \ldots \quad (116)$$

and

$$f = \frac{6.366}{D^3}[.6M_b + \sqrt{M_b^2 + M_t^2}]. \quad \ldots \quad (117)$$

The value of the resisting moment required to withstand the ideal twisting moment $(M_t)_i$ of equation (113) is, for a solid round shaft,

$$(M_t)_i = \frac{\pi s D^3}{16}. \qquad \ldots \quad (118)$$

Substituting this value of $(M_t)_i$ in equation (113), it becomes, by a slight reduction,
For a solid round shaft:

$$sD^3 = 5.093[.6M_b + \sqrt{M_b^2 + M_t^2}], \quad \ldots \quad (119)$$

whence

$$D^3 = \frac{5.093}{s}[.6M_b + \sqrt{M_b^2 + M_t^2}] \quad \ldots \quad (120)$$

and

$$s = \frac{5.093}{D^3}[.6M_b + \sqrt{M_b^2 + M_t^2}]. \quad \ldots \quad (121)$$

In equations (115) to (121) the bracketed quantities are identical. In (119) the exponent 5.093 is 0.8 that of (115); therefore, if the shearing-strength of the material is taken as 0.8 of the tensile strength, which is very commonly done for iron and steel, the same value will be obtained for D^3 by both equation (116) and (120.) Hence either one of these equations may be used for determining D when $s = .8f$.

If the shearing-strength is taken greater than .8 of the tensile (i.e., $s > .8f$), then equation (116) should be used for finding D; but if the shearing-strength is taken as less than .8 of the tensile, (i.e., $s < .8f$), equation (120) should be used.

Example.—Suppose that the belt whose sectional area is determined in the example in §41 is to run on a pulley 5 feet in diameter, whose shaft is supported by bearings on both sides of the pulley, one at a distance of 5 feet and the other 3 feet from the centre of the pulley-hub; also that all the power is to be transmitted in one direction through the shafting leading from the pulley. What should be the diameter of the shaft for working strengths of 12000 pounds per square inch tensile, and 10000 pounds per square inch shearing?

226   FORM, STRENGTH, AND PROPORTIONS OF PARTS.

The values of the turning force P and the belt-tensions T_n and T_o, as determined in the example in § 41, are:

$$P = 1467 \text{ pounds};$$
$$T_n = 2960 \text{ pounds};$$
$$T_o = 1493 \text{ pounds}.$$

The torsional moment M_t that the shaft must resist is

$$M_t = Pl = 1467 \times (2.5 \times 12) = 44010 \text{ inch-pounds};$$

and the bending force W that the shaft must resist is the resultant of T_n and T_o, which act at an angle of $(180° - 160°) = 20°$ with each other. This resultant is 4360 pounds. (The difference between this resultant and the sum of the two belt-tensions $(T_n + T_o) = 4453$ is so small comparatively as not to need consideration in general practice. The sum of the belt-tensions would ordinarily be used for so small an angle between the two stretches of belt. The actual resultant will be used, however, in the following calculations.)

The bending moment M_b which the shaft must resist is, according to equation (104),

$$M_b = \frac{4360(3 \times 12)(5 \times 12)}{8 \times 12} = 98100 \text{ inch-pounds}.$$

Since the ratio (10000 : 12000) is greater than .8, equation (116), involving the tensile strength of the material, must be used, whence

$$D^3 = \frac{6.366}{12000}[.6 \times 98100 + \sqrt{(98100)^2 + (44010)^2}];$$
$$D = 4.45 \text{ inches}.$$

100. Overhanging solid round crank-shafts and other overhanging shafts acted on by a single rotative force.—When a single rotative force P is applied to an overhanging shaft, the equations of turning and resisting moments take a simpler form than those in § 99. This is due to the fact that both the bending moment M_b and the twisting moment M_t are due to the same torsional force P, instead of different forces. In this case $M_b = PL$ and $M_t = Pl$.

Substituting these values of M_b and M_t in equation (116), it reduces to—

For a solid round shaft:

$$D^3 = 6.366\frac{P}{f}[.6L + \sqrt{L^2 + l^2}] ; \quad \ldots \quad (122)$$

$$f = 6.366\frac{P}{D^3}[.6L + \sqrt{L^2 + l^2}] ; \quad \ldots \quad (123)$$

$$P = \frac{fD^3}{6.366[.6L + \sqrt{L^2 + l^2}]} ; \quad \ldots \quad (124)$$

and similarly for the shearing-stress equation (120) reduces to—
For a solid round shaft:

$$D^3 = 5.093\frac{P}{s}[.6L + \sqrt{L^2 + l^2}] ; \quad \ldots \quad (125)$$

$$s = 5.093\frac{P}{D^3}[.6L + \sqrt{L^2 + l^2}] ; \quad \ldots \quad (126)$$

$$P = \frac{sD^3}{5.093[.6L + \sqrt{L^2 + l^2}]}. \quad \ldots \quad (127)$$

Example.—An overhanging crank-shaft is to be designed to drive a piston 35 inches in diameter against a constant pressure of 60 lbs. per sq. in. The following dimensions are also fixed: length of stroke, 40 inches; length of connecting-rod 6 feet; "overhang" of crank 15 inches, measured from middle of crank-pin to a point 3 inches back from the front of the crank-shaft bearing; tensile stress in material of shaft not to exceed 12000 lbs. per sq. in.; shearing-stress not to exceed 10000 lbs. per sq. in. The speed is to be so slow that the inertia of the moving parts need not be considered.

Since the limiting shearing-stress is greater than .8 of the tensile stress, the diameter of the shaft can be determined by equation

(122). The values of the quantities entering into this equation for this particular case are:

$$P = \frac{\pi(35)^2}{4} 60 = 57726.6 \text{ pounds};$$

$$L = 15 \text{ inches};$$

$$l = 20 \text{ inches};$$

$$f = 12000 \text{ lbs. per sq. in.}$$

Therefore

$$D^3 = 6.366\tfrac{57726.6}{12000}[.6 \times 15 + \sqrt{(15)^2 + (20)^2}]$$
$$= 1041;$$

whence

$$D = 10.13 \text{ inches.}$$

Hollow Round Shafts.

101. The equations for hollow round shafts are similar to those for solid ones and may be written directly from them. The more important ones are given below for convenience of reference.

102. Hollow round shafts acted on by more than one force.
For tensile fibre-stress:

$$D^3 = \frac{6.366}{f(1-n^4)}[.6M_b + \sqrt{M_b^2 + M_t^2}]; \quad \ldots \quad (128)$$

$$f = \frac{6.366}{D^3 - d^3}[.6M_b + \sqrt{M_b^2 + M_t^2}]. \quad \ldots \quad (129)$$

For shearing fibre-stress:

$$D^3 = \frac{5.093}{s(1-n^4)}[.6M_b + \sqrt{M_b^2 + M_t^2}]; \quad \ldots \quad (130)$$

$$s = \frac{5.093}{D^3 - d^3}[.6M_b + \sqrt{M_b^2 + M_t^2}]. \quad \ldots \quad (131)$$

103. Overhanging hollow round shafts acted on by a single turning force. Crank-shafts, etc.
For tensile fibre-stress:

$$D^3 = \frac{6.366P}{f(1 - n^4)}[.6L + \sqrt{L^2 + l^2}]; \quad \ldots \quad (132)$$

$$f = \frac{6.366P}{D^3 - d^3}[.6L + \sqrt{L^2 + l^2}]; \quad \ldots \quad (133)$$

$$P = \frac{f(D^3 - d^3)}{6.366[.6L + \sqrt{L^2 + l^2}]}. \quad \ldots \quad (134)$$

For shearing fibre-stress:

$$D^3 = \frac{5.093P}{s(1 - n^4)}[.6L + \sqrt{L^2 + l^2}]; \quad \ldots \quad (135)$$

$$s = \frac{5.093P}{D^3 - d^3}[.6L + \sqrt{L^2 + l^2}]; \quad \ldots \quad (136)$$

$$P = \frac{s(D^3 - d^3)}{5.093[.6L + \sqrt{L^2 + l^2}]}. \quad \ldots \quad (137)$$

104. Experimentally determined values of the breaking tensile stress of round shafting subjected to combined bending and torsion. —Professor Gaetano Lanza made a series of experiments on shafting subjected to both bending and torsion while rotating.[*] Some of the values obtained by him are given in Table XXXIII. They represent the breaking stress by fatigue of the metal, but not the static strength.

---

[*] Trans. Amer. Soc. Mech. Eng., .ol. VIII., 1887, p. 130.

TABLE XXXIII.*

BREAKING-STRENGTH OF SHAFTING SUBJECTED TO COMBINED BENDING AND TWISTING.

Diameter of Shaft, inches.	Total Revolutions.	Horse-power Transmitted.	M_b Max. Bending Moment. Inch-lbs.	M_t Max. Twisting Moment. Inch-lbs.	f Fibre-stress Causing Fracture. Lbs. per sq. in.
1.25	7,040	11.717	11,514.1	3,926.4	62,168
1.25	38,839	8.181	10,507.8	2,656.8	55,876
1.25	31,041	5.291	9,891.0	1,714.6	52,062
1.25	108,002	4.331	9,241.7	1,399.2	48,539
1.25	80,694	6.276	9,241.7	2,027.6	48,911
1.25	19,333	6.342	8,917.1	2,028.2	47,245
1.25	82,741	6.283	8,917.1	2,029.7	47,246
1.25	108,739	6.192	8,592.5	2,031.6	45,582
1.25	88,208	6.338	8,267.8	2,026.8	43,914
1	185,233	6.283	3,781.5	2,029.7	41,768

* Taken from Transactions Am. Society Mechanical Engineers, 1887, vol. VIII., p. 136.

Samples of the shafting tested for combined stress were also tested for static tensile strength. The results are given in Table XXXIV.

TABLE XXXIV.

STATIC TENSILE STRENGTH OF SHAFTING.

Refers to shafting represented in Table XXXIII.

```
                                        Breaking Tensile Strength,
                                              lbs. per sq. in.
   1.25″ diam. { No. 1...................... 46,800
               { No. 2...................... 49,865
                                              ──────
                    Average................. 48,333

   1″ diam.    { No. 1...................... 58,687
               { No. 2...................... 61,812
                                              ──────
                    Average................. 60,250
```

105. Practically determined formulas for round shafting.— There are many cases in practice where the relations between the torsional and bending moments, or even the value of either, cannot be estimated with much accuracy. To meet such conditions, formulas have been devised to accord with transmission-machinery that has given satisfactory service. These formulas are given in many hand-books and treatises on power transmission. They may be found in Kent's Mechanical Engineers' Pocket-book, together with tables for different speeds, etc., calculated from the formulas.

SHAFTS OF SYMMETRICAL SECTIONS OTHER THAN ROUND.

106. Shafts of any section other than round are seldom used in machine construction. Since they are sometimes used, however, the method of determining their strength for combined stress will be pointed out.

In order to obtain the equation of strength of any form of section, it is only necessary to substitute the resisting moment of the shaft against bending for the ideal bending moment $(M_b)_i$ in equation (112), or its resisting moment against torsion for the ideal torsional moment $(M_t)_i$ in equation (113), and solve as for round shafts after making the substitution.

Calling f the working strength of the material for tensile or compression stress, and S_b the section modulus of the shaft for resisting bending, the following equation may be written:

$$(M_b)_i = fS_b ; \quad \ldots \ldots \quad (138)$$

and, similarly, calling s the working shearing-strength of the material, and S_t the section modulus of the shaft for resisting torsion,

$$(M_t)_i = sS_t. \quad \ldots \ldots \quad (139)$$

The values given in these two equations are the ones to be substituted in equation (112) or (113) in order to obtain equations of the same general nature as (115), (116), and (117).

POSITIVE COUPLINGS FOR SHAFTS.

107. Rigid shaft-couplings.—When no allowance is to be made for lack of alignment of the connected shafts, a rigid coupling is commonly used.

The most common form of rigid coupling for large shafts is a pair of flanged hubs or collars. The hub of each is solid and bored to a diameter to fit the shaft, to which it is generally keyed. The flanges are bolted together after putting in place. The shafts are generally brought into line either by allowing one shaft to enter into the bore of the half coupling on the other, by a shoulder on one flange fitting into a corresponding recess in the other, or by making an accurate fit of the bolts in the flanges. Other less common methods are also used.

For the medium and smaller sizes of shafting used for power transmission, couplings bored to a slightly smaller diameter than that of the shaft, and split longitudinally, either upon one side or both, are very commonly used.

A very simple form of split-sleeve coupling is shown in Fig. 119. It consists of a sleeve bored to the same diameter from end to end,

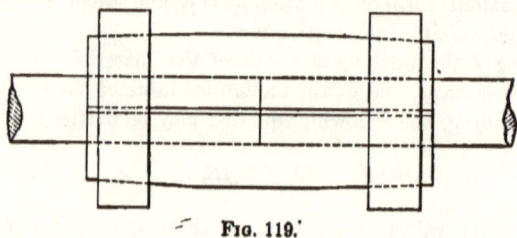

Fig. 119.

and given a slight conical taper toward each end on the outside. It is either split longitudinally down one side, or cut completely in two longitudinally, so as to form two halves. When used, it is placed upon the two sections of shafting, which are brought together, and then a ring, bored to the same taper as the outside of the sleeve, is either driven or shrunk upon each end of the coupling, thus causing it to grip the shafts tightly. When the torsional force to be transmitted is small, the coupling may be used without a key, but for heavy work a key is generally used after the manner common to solid couplings.

It can readily be seen that a split-sleeve coupling of much the same nature as the above one can be made by using bolts, instead of the rings, to clamp the two halves together.

A largely used form of coupling, commonly known as Seller's

AXLES, SHAFTING, AND POSITIVE SHAFT-COUPLINGS. 233

coupling, consists of an outer sleeve, bored at each end with a conical hole to receive a bushing with a corresponding taper on the outside. The bushing has a bore slightly smaller than the shaft. Bolts, extending from end to end through the sleeve, serve to draw the two conical bushings towards each other, and thus cause them to grip the shaft. These bolts also serve as keys to prevent the rotation of the bushings in the outer sleeve. Keys may be used or not between the shaft and bushing, according to the nature of the work.

108. Flexible shaft-couplings.—When there is a probability that there will be a slight relative movement of the shafts coupled together, as in the case of the shafts of a dynamo and engine resting on separate foundations and direct-connected together, it is necessary to have a coupling which will adjust itself to the slight throwing out of alignment of the shafts that occurs under such conditions.

The coupling Fig. 120 has been successfully applied for mod-

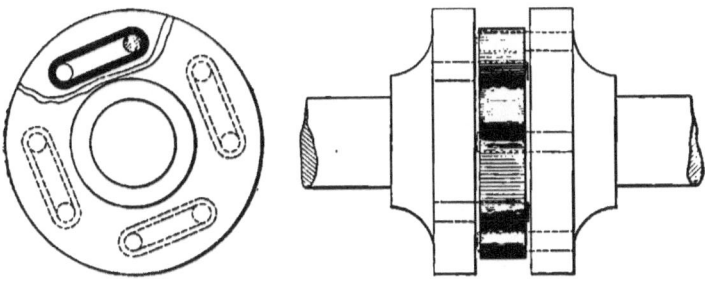

FIG. 120.

erate amounts of power transmitted. It consists of a pair of disks very similar to those of the rigid coupling, but instead of being held firmly together by bolts they are separated as shown, and each has a number of strong pins projecting from it so as to almost touch the other. The other disk has the same number of pins. The pins are spaced uniformly around both flanges near their peripheries. Each pin of one flange is connected to one of the other by a short link of some elastic material, as rubber or leather. As one flange rotates it draws the other after it by means of the connecting-links, whose elasticity allows them to adjust themselves to the lack of alignment of the shafts. The links can be made longer or shorter than shown in the figure if desired.

Another flexible coupling, which allows the shafts to have bending motion relatively to each other, but does not allow lateral motion, is shown in Fig. 121.* It consists of a large flange upon one of the shaft-ends to be coupled, and a small one upon the other. The two flanges are connected together by a washer-shaped ring, which may

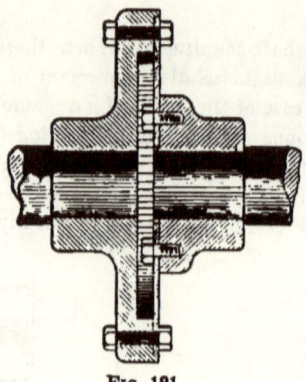

Fig. 121.

be made by cutting it from sheet metal, the outer diameter being made much larger than the inner. This flat ring, being of an elastic material, allows a slight bending motion of the shaft as stated without injury to the parts.†

109. Positive clutch-couplings are often used for connecting parts of machinery which are to be disengaged at times. The most common form consists of two parts resembling, in a manner, the halves of the coupling shown in Fig. 120, but having projecting teeth or jaws instead of pins. These jaws are made so that they will engage with each other for transmitting rotative motion. The coupling is generally disengaged by slipping one of its halves along the shaft so as to separate the jaws.

---

* Trans. Amer. Soc. Mech. Eng., vol. VII., p. 526.

† For Oldham's and Hooke's universal couplings see Part I. of Machine Design.

CHAPTER IX.

FRICTION-COUPLINGS AND BRAKES.

110. In the operation of machinery it is often desirable to start and stop some part which receives its motion from a constantly running driver, or to bring two parts into engagement without changing their relative positions. For this purpose "**friction-couplings**" are used almost universally. Such a coupling transmits power by means of two smooth surfaces, held together with sufficient pressure to make their frictional resistance to slipping great enough to produce the required transmission of power.

The friction-brake, for retarding the motion of a part, is practically the reverse of the friction-coupling, so far as its general principle is concerned, but differs from it in that there is generally much more rubbing between the surfaces, and consequently the materials must be selected with more attention to their qualities of not abrading and cutting. This applies to brakes such as are used on hoisting-machinery, but hardly to those for railway-wheels, etc., where the conditions are such that it is impossible to prevent abrasion on account of foreign matter getting between the rubbing surfaces.

FRICTION-COUPLINGS

111. Cone friction-couplings, consisting of two parts having conical surfaces, one external and the other internal, fitting together as in Fig. 122, are often used in machine construction, and especially in machine tools, for connecting the ends of two shafts, or for transmitting power from a shaft by means of a gear or other device attached to the part of the coupling which is free to rotate upon the shaft when the clutch is open, but is driven by the shaft when the two parts of the clutch are forced together.

The notation in the equations for a cone friction-clutch is:

M = torsional moment transmitted through clutch;
N = total normal pressure between the conical surfaces of the clutch;
P = turning force acting at a distance R from the centre of the shaft;

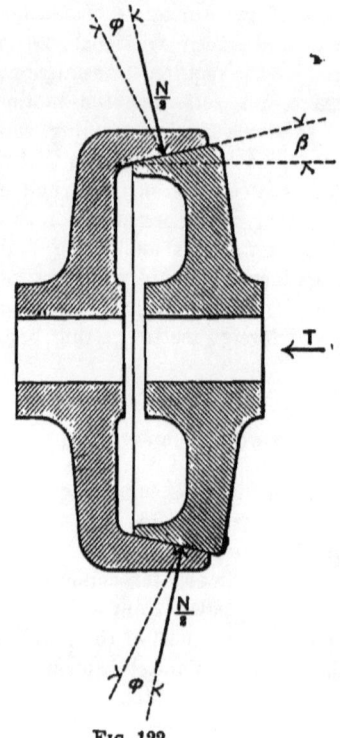

Fig. 122.

R = mean radius of conical surfaces;
T = axial force acting to close the clutch;
β = angle between the conical surfaces and the axis of shaft; the apex angle of the conical surfaces is therefore 2β;
$\mu = \tan \phi$ = coefficient of friction between the conical surfaces;

ϕ = angle of friction between the conical engaging surfaces of the clutch.

If the two parts of the coupling are forced together when they have no relative rotation, the pressure between the conical surfaces would make an angle ϕ with a normal to the surfaces, as indicated by the dotted lines in Fig. 122. This frictional resistance would have to be taken into consideration if it were desired to set the clutch so that there would be no slip whatever between the parts at the time of closing or afterwards. But if the clutch is closed while one part is rotating relatively to the other, or if a slight slip at the time of throwing on the load is not objectionable, it is not necessary to consider the angle of friction when determining the axial force T required for closing it. A slight slip at the time of closing is almost invariably, probably always, allowable. Of the following equations, the ones not including ϕ are therefore applicable to cone friction-clutches in nearly all cases.

For cone friction-clutches closed while having a rotative slip between the engaging surfaces:

$$N = T \csc \beta; \qquad (140)$$

$$P = \mu N = \mu T \csc \beta; \qquad (141)$$

$$M = PR = \mu TR \csc \beta. \qquad (142)$$

For a cone friction-clutch closed without rotative slipping between the engaging surfaces:

$$N = T \csc (\beta + \phi); \qquad (143)$$

$$P = \mu N = \mu T \csc (\beta + \phi); \qquad (144)$$

$$M = PR = \mu TR \csc (\beta + \phi). \qquad (145)$$

112. Multiple-ring friction-coupling.—When it is desirable to keep the size of the coupling small, especially in diameter, the form shown in Fig. 123 is applicable.* As illustrated, the outer

---

* Weston friction-coupling used by Yale & Towne Mfg. Co. on hoisting-machinery.

cylindrical sleeve A is fitted on the shaft with a feather key, so that it may slip along it. The hub B has a running fit on the shaft, but is prevented from moving along it by a shoulder. A

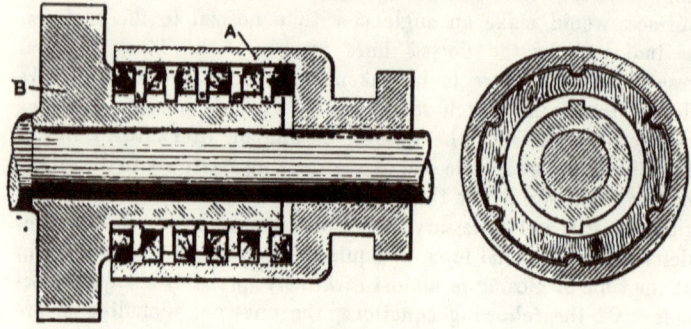

Fig. 123.

number of rings are placed between the sleeve and hub, each alternate one engaging with the sleeve by means of several projections on the latter corresponding to feather keys. The remaining rings engage with the hub in a similar manner. By applying an axial force to press the sleeve toward the hub, a pressure, equal to the axial force, is induced upon each side of every ring.

Putting, for multiple-ring friction-couplings:

$M =$ torsional moment transmitted through coupling;
$P =$ turning force exerted;
$R =$ mean radius of rubbing surfaces of rings;
$T =$ axial force exerted to press the sleeve toward the ring;
$n =$ number of pairs of rubbing surfaces in contact;
$\mu =$ coefficient of friction of the rubbing surfaces;

then:

$$P = \mu n T; \qquad \qquad (146)$$

$$M = PR = \mu n TR. \qquad \qquad (147)$$

As represented in the figure, there are as many pairs of rubbing surfaces as there are rings. This is not always true of such couplings, however.

113. Materials and coefficient of friction for friction-couplings.
—In machine tools and the counter-shafts for driving them, unless they are exceptionally large, the material of a cone friction-coupling is generally cast iron for both rubbing surfaces. When accurately ground together with a fine abrasive, as flour-emery, the cast iron gives excellent satisfaction for this use. The coefficient of friction μ must be taken with regard to whether the surfaces are oily or dry. Since there is nearly always a certainty of oil getting on them at some time, the value $\mu = 0.15$ is probably as high as it is safe to assume for cast iron on cast iron.

When a friction-coupling is to perform heavy service with considerable slipping at the time of setting the machinery into motion, cast iron on wood or leather gives good service. The wood may be either set on end or with the grain parallel to the direction of rubbing. Any of the metals that run well together for journal-bearings may be used if the coupling is kept clean and lubricated. Otherwise wood or leather rubbing against metal is safer. For dry wood on cast iron μ may be taken from 0.15 to 0.18; for oily wood μ falls as low as 0.10. The value of μ for leather on cast iron does not seem to be well determined for such service as is required for friction-clutches, but it can probably be taken as high as 0.20 for oily leather; it is much higher for dry leather. Leather should not be used where there is sufficient slipping to burn it.

The best service for heavy work and hard usage seems to be obtained with wood rubbing on metal. In practice the latter is generally cast iron.

FRICTION-BRAKES.

114. Strap brake.—A form of strap brake much used for hoisting-machinery is shown in Fig. 124. The brake-drum, whose centre is at O, is partly encircled by a strap whose ends are attached to a lever for tightening it upon the drum. The strap may be made of any material, according to the requirements to be met. For heavy service on mine-hoists, a strap of Swedish wrought iron, lined with blocks of basswood placed with the grain parallel to the length of the strap, is very commonly used. Experience has shown that a steel strap is very liable to fracture under such service, apparently on account of fatigue of the metal, caused by the repeated application of stress due to tightening and loosening the

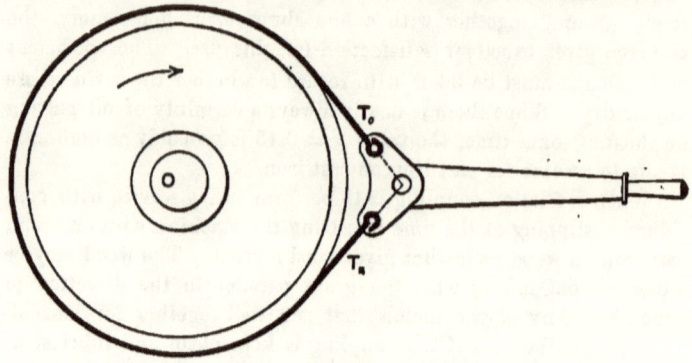

Fig. 124.

drum, and to the almost incessant vibrations common to such service. Basswood is selected because in it are combined both durability and uniform gripping power. For lighter service, a wrought-iron strap is often used directly against the cast-iron brake-drum. This can be done successfully only where the rubbing surfaces can be kept lubricated.

The tensions T_a and T_o in the two free portions of the brake-strap between the drum and brake-lever, for a given force P resisting the rotation of the drum, and applied at the surface of the latter, may be found by the equations for belting. The value of z in these equations is zero for a brake-strap, since no centrifugal force due to its own weight acts upon it.

Equations (39) to (43) are applicable to the solution. In these equations the coefficient of friction μ may be taken as 0.1 for wet or oily surfaces of wood on cast iron, and from 0.15 to 0.18 for dry rubbing surfaces of the same materials.

115. The Prony brake, Fig. 125, is a special form of strap brake. Its almost universal application is as an absorbent dynamometer for tests of the power developed by motors. As shown in the figure, it consists of an iron strap A encircling a brake-drum whose centre is at O. The strap is lined with wooden blocks which bear against the surface of the drum. The ends of the straps are held together by a bolt which affords a means of adjustment of

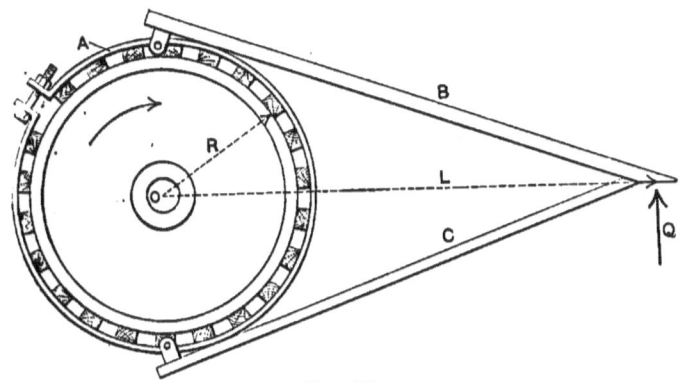

Fig. 125.

tension in the strap. The two members B and C each have one end attached to the brake-strap, and are joined together so as to form a rigid arm. As the brake-drum rotates in the direction of the arrow, the tendency of this arm to rotate with it is resisted by a force Q acting against the end of the arm.

If the two members of the brake-arm are attached to the strap at points diametrically opposite, so that each portion of the strap covers 180° of the drum, and if Q acts vertically upward and is of a value equal to the portion of the weight of the brake that is supported by the drum, then the following equations are applicable for determining the relation between the maximum tension T_n in the brake-strap and the force Q. It can be seen that these equations are of the same nature as those for belting. The brake-strap corresponds to two portions of a belt having an arc of contact of 180° upon the pulley.

The notation for the Prony brake is:

L = distance from centre of drum to line of action of Q (not necessarily the distance from the centre of the drum to the point of application of Q);

242   FORM, STRENGTH, AND PROPORTIONS OF PARTS.

P = total frictional resistance at surface of brake-drum;
Q = force applied at end of brake-arm to resist its rotation;
R = radius of brake-drum;
T = greatest tension in brake-band;
$\epsilon = 2.71828$;
$\pi = 3.1416$.

By taking moments about Q

$$PR = QL \quad \text{or} \quad P = \frac{L}{R}Q. \quad \ldots \quad (148)$$

In accordance with the equations for belting, taking into account the fact that there is no centrifugal force acting on the brake-strap,

$$T = \frac{P}{2} \frac{\epsilon^{\mu\pi}}{\epsilon^{\mu\pi} - 1} = \frac{P}{2} \frac{10^{.00758\mu180}}{10^{.00758\mu180} - 1}. \quad \ldots \quad (149)$$

Whence, by substituting the value of P given in equation (148),

$$T = \frac{QL}{2R} \frac{\epsilon^{\mu\pi}}{\epsilon^{\mu\pi} - 1} = \frac{QL}{R} \frac{10^{.0075\mu180}}{10^{.00758\mu180} - 1}. \quad \ldots \quad (150)$$

If the points of attachment of the two members of the brake-arm to the brake-strap are not diametrically opposite, it is hardly possible to get equations for the maximum tension in the strap that are of any practical value. If the material of the strap were totally inelastic, such equations might be deduced, but on account of its elasticity they cannot be. The equations just given may be applied with a fair degree of accuracy, however, when the arc between points of attachment does not differ greatly from 180°, and if Q does not differ greatly from the portion of the weight of the brake supported by the drum.

The coefficient of friction μ, and the materials for the Prony brake, may be taken the same as for the strap brake of the preceding section. Since the service required of a Prony brake is seldom so trying as that of the strap brake for hoisting-machinery, it is generally not necessary to exercise so much care in selecting the materials.

CHAPTER X.

FLY-WHEELS AND PULLEYS.

116. A fly-wheel, as commonly used in machine construction, is a rotating wheel which serves as a storehouse for energy. At certain periods during the operation of the machine the rotative speed of the fly-wheel is increased, thus increasing its store of kinetic energy, and at other periods the speed is decreased, the wheel thus returning or supplying energy to the machine.

Examples of the application of fly-wheels are very common. Three cases may be cited, however, to point out different classes of service.

In the ordinary double-acting steam-engine with a single cylinder the fly-wheel has its speed slightly increased twice during each revolution, and also slightly decreased twice during the same revolution, provided the average speed of rotation does not change during a series of successive revolutions. The function of the fly-wheel in this case is to prevent great fluctuation of speed during a single revolution, as well as to maintain an approximately uniform speed when the load against which the engine works varies suddenly.

In a punching and shearing machine the fly-wheel receives a store of energy while being brought up to its normal speed. As the punch or shears are driven through the metal operated upon, the speed of the fly-wheel is decreased and a portion of its energy given up to perform the work upon the metal and to overcome frictional resistances. The time interval of working on the material punched or sheared, which corresponds to the period of reduction of speed, is small in comparison with the time the machine is running idly. After punching the material the fly-wheel again receives energy while being brought back to its normal speed.

In the Howell torpedo a fly-wheel is brought up to an exceedingly high speed. When the torpedo is launched into the

244 FORM, STRENGTH, AND PROPORTIONS OF PARTS.

water, the kinetic energy stored in the wheel is utilized to drive a screw propeller for forcing the torpedo through the water. The fly-wheel gradually stops as it gives up its energy.

117. Determination of moment of inertia and kinetic energy of a webbed wheel.—The total kinetic energy, K.E., stored in a rotating fly-wheel is expressed by the equation

$$K.E. = \tfrac{1}{2}(\text{mass}) \times (\text{angular velocity})^2 \times (\text{mean radius of rotation})^2,$$

or

$$K.E. = \tfrac{1}{2}(\text{moment of inertia}) \times (\text{angular velocity})^2.$$

The form of fly-wheel that is most generally adopted has a rim of rectangular cross-section with slightly rounded corners, a hub that is approximately cylindrical, and arms tapering uniformly from the hub to the rim, growing smaller toward the rim; for small sizes of wheels a solid web is often used, the web generally being thinner at the rim than at the hub.

The moment of inertia of a wheel with uniformly tapering arms can be found by reducing the arms to an equivalent web. This reduction can be made with an accuracy far within the limits required for ordinary practice, and the results obtained by this method of substituting a web for the arms are ordinarily more accurate than those obtained by dealing with the arms directly. (See § 120.) The method of finding the moment of inertia of a webbed wheel of the form shown in Fig. 126 will therefore be given

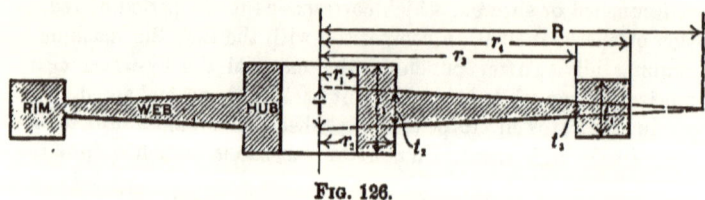

Fig. 126.

as being more generally applicable than that for any other form of wheel.

The notation for the equations applying to a webbed wheel is:

$E =$ energy given out by wheel for a given change of speed of rotation, foot-pounds;

FLY-WHEELS AND PULLEYS. 245

I = moment of inertia of entire wheel about its axis, foot-pound-seconds;

I_{hub}, I_{web}, and I_{rim} = moment of inertia of the hub, web, and rim, respectively, foot-pound-seconds;

K.E. = total kinetic energy stored in the rotating wheel, foot-pounds;

$N = \omega \div 2\pi$ = revolutions per second;

N_1 = initial speed of rotation, revolutions per second;

N_2 = speed of rotation after slowing down, revolutions per second;

$$R = \frac{r_2 t_3 - r_3 t_2}{t_3 - t_2} = \text{radial distance at which sides of web would}$$

intersect if extended, used for convenience only, feet;

$$T = \frac{t_2 R}{R - r_2} = \frac{t_3 R}{R - r_3} = \text{length of axis intercepted between}$$

sides of web if extended to axis of wheel, used for convenience only, feet;

V = velocity, feet per second;

g = acceleration due to force of gravity = 32.2 feet per second per second;

p = tensile stress in rim due to rotation, pounds per square inch;

r = mean radius of rim of wheel, feet;

r_1 = radius of bore of hub, feet;

r_2 = outer radius of hub = inner radius of web, feet;

r_3 = outer radius of web = inner radius of rim, feet;

r_4 = outer radius of rim, feet;

t_1 = thickness (or length) of hub, feet;

t_2 = thickness of web at inner edge, feet;

t_3 = thickness of web at outer edge, feet;

t_4 = thickness of rim measured parallel to axis of wheel, feet;

w = weight of material, pounds per cubic foot;

λ = increase in radius of ring due to centrifugal action, feet;

$\pi = 3.1416$;

$\omega = 2\pi N$ = velocity of rotation, radians per second;

ω_1 = initial speed of rotation, radians per second;

ω_2 = speed of rotation after slowing down, radians per second.

246 FORM, STRENGTH, AND PROPORTIONS OF PARTS.

The moments of inertia of the three parts, the hub, web, and rim, are:

$$I_{hub} = \frac{w\pi t_1}{2g}(r_2^4 - r_1^4); \quad \ldots \ldots \quad (151)$$

$$I_{web} = \frac{2\pi w T}{g}\left[\frac{r_3^4 - r_2^4}{4} - \frac{r_3^5 - r_2^5}{5R}\right]; \quad \ldots \quad (152)$$

$$I_{rim} = \frac{w\pi t_4}{2g}(r_4^4 - r_3^4). \quad \ldots \ldots \quad (153)$$

The moment of inertia of the entire wheel is the sum of the I's for all the parts; whence

$$I = I_{hub} + I_{web} + I_{rim}. \quad \ldots \ldots \quad (154)$$

The total kinetic energy K.E. stored in the wheel when rotating at an angular velocity of ω radians, or N revolutions, per second, is

$$\text{K.E.} = \tfrac{1}{2}\omega^2 I = 2\pi^2 N^2 I. \quad \ldots \ldots \quad (155)$$

The energy E transformed from kinetic energy into mechanical energy or heat, or both, when the speed of rotation of the fly-wheel drops from ω_1 to ω_2 radians, or from N_1 to N_2 revolutions, per second, is shown by the expression

$$E = \tfrac{1}{2}(\omega_1^2 - \omega_2^2)I = 2\pi^2(N_1^2 - N_2^2)I. \quad \ldots \quad (156)$$

The same amount of mechanical energy E must, of course, be applied to the wheel to bring it back from N_2 to N_1 revolutions per second, friction neglected.

Example.—It is required to find the kinetic energy K.E. at a speed of 300 revolutions per minute of a cast-iron fly-wheel of the form shown in Fig. 126 and having the following dimensions:

$t_1 = 6$ in. $= .5$ ft.; $\quad r_1 = 1\tfrac{1}{4}$ in. $= \tfrac{5}{48}$ ft.;
$t_2 = 1\tfrac{1}{2}$ in. $= .125$ ft.; $\quad r_2 = 3$ in. $= .25$ ft.;
$t_3 = 1$ in. $= \tfrac{1}{12}$ ft.; $\quad r_3 = 18$ in. $= 1.5$ ft.;
$t_4 = 5$ in. $= \tfrac{5}{12}$ ft.; $\quad r_4 = 24$ in. $= 2$ ft.

The weight of cast iron may be taken as 450 pounds per cubic foot.

If the acceleration due to force of gravity is taken as 32.2 feet per second per second, then the units of measurement must be the foot, pound, and second.

The moment of inertia of the three parts may first be determined.

The moment of inertia of the hub is, by equation (151),

$$I_{hub} = \frac{450\pi \times .5}{2 \times 32.2}\left[(.25)^4 - \left(\frac{5}{48}\right)^4\right]$$
$$= 10.97(.0039062 - .0001174)$$
$$= .0416.$$

For the web equation (152) can be applied. In this equation the quantities T and R enter. T may be measured on the drawing with considerable accuracy, but the intersection of the lines determining R is generally difficult to determine. The values of both will be calculated for this problem according to the equations given in the notation; whence

$$R = \frac{18 \times 1.5 - 3 \times 1}{1.5 - 1} = 48 \text{ inches} = 4 \text{ feet},$$

and

$$T = \frac{1.5 \times 48}{48 - 3} = 1.6 \text{ inches} = \tfrac{2}{15} \text{ foot.}$$

By substituting in equation (152)

$$I_{web} = \frac{2\pi 450 \times 2}{32.2 \times 15}\left[\frac{(1.5)^4 - (.25)^4}{4} - \frac{(1.5)^5 - (.25)^5}{5 \times 4}\right]$$
$$= 11.7 \times .885$$
$$= 10.4.$$

And for the rim, by equation (153),

$$I_{rim} = \frac{450\pi 5}{2 \times 32.2 \times 12}[(2)^4 - (1.5)^4]$$
$$= 100.3.$$

248 FORM, STRENGTH, AND PROPORTIONS OF PARTS.

For the entire wheel the moment of inertia is therefore

$$I = .0416 + 10.4 + 100.3$$
$$= 110.74.$$

It can be seen that the effect of the hub on the total moment of inertia is practically inappreciable, and may therefore be neglected in a wheel whose rim diameter and weight of rim are as great in proportion to the similar quantities for the hub as for the wheel just considered.

The kinetic energy of the entire wheel when rotating at 300 revolutions per minute is, according to equation (155),

$$\text{K.E.} = 2\pi^2 N^2 I$$
$$= 2\pi^2 (\tfrac{300}{60})^2 \, 110.74$$
$$= 54648 \text{ foot-pounds.}$$

118. Problem. To design a fly-wheel for a given moment of inertia and according to a given form.—Let the required moment of inertia $I = 300$, the wheel to be similar to Fig. 126.

Since the required wheel is to be similar to the one considered in the preceding problem, it is only necessary to change the dimensions of the wheel, all in the same proportion, to such an extent as will give the required I—in other words, to apply such a scale to the drawing as will give the required I.

The proportionate change of the dimensions can readily be determined by making use of the fact that the I of similar wheels is proportional to the fifth power of their linear dimensions. Therefore the linear dimensions of the wheel in the preceding example, which has an $I = 110.74$, must each be multiplied by

$$\sqrt[5]{\frac{300}{110.74}} = 1.22$$

in order to obtain the required $I = 300$.

This scale gives the outer radius of the rim

$$r_4 = 1.22 \times 24 = 29.28 \text{ inches,}$$

and the other dimensions must be increased in the same proportion.

The above solution shows that when it is desired to design a wheel for a required I it can be done by making a drawing according to the form desired, considering it full size, and finding its I accordingly. The scale which must be applied to the drawing to obtain the required I can then be determined by dividing the required I by the I of the drawing, considered full size, and extracting the fifth root of the quotient.

119. Problem. To design a fly-wheel which will furnish a given amount of energy for a given variation of speed.—Let it be required that the wheel shall furnish 40000 foot-pounds of energy for a 5% speed reduction, and that the full speed shall give the outer circumference of the rim a velocity of 4800 feet per minute.

This problem can most readily be solved by making use of the fact that in similar wheels having the same circumferential linear velocity the kinetic energy is proportional to the cubes of similar linear dimensions.

For convenience it may be assumed that the form of wheel selected is that of Fig. 126, and that the proportions selected for the first trial are those of § 117.

For the wheel considered in § 117, $I = 110.74$. When this wheel runs at a circumferential velocity of 4800 feet per minute, its angular velocity is

$$\omega_1 = \frac{4800}{60 r_1} = \frac{4800}{60 \times 2} = 40 \text{ radians per second.}$$

When the speed has dropped 5%, the angular velocity is

$$\omega_2 = 40 - .05 \times 40 = 38 \text{ radians per second.}$$

The energy given out by the wheel while slowing down 5% is, by equation (156),

$$E = \tfrac{1}{2}[(40)^2 - (38)^2]110.74$$
$$= \tfrac{1}{2} \times 156 \times 110.74$$
$$= 8638 \text{ foot-pounds.}$$

250 FORM, STRENGTH, AND PROPORTIONS OF PARTS.

In order to secure a fly-wheel, similar to the one just considered, which, when running at 4800 feet per minute circumferential speed, will furnish 40000 foot-pounds of energy for a 5% reduction of speed, the proportions given in § 117 must all be multiplied by the same factor, whose value is

$$\sqrt[3]{\frac{40000}{8638}} = 1.6668 = 1\tfrac{2}{3} \text{ about.}$$

The outer radius of the rim must therefore be $1.6668 \times 2 = 3.3336$ feet, and the other dimensions must be increased in the same proportion.

In the fly-wheel considered above, part of the energy given out by it during its reduction of speed is converted into heat on account of journal friction, etc. If the frictional resisting forces remain constant, the mechanical energy converted into heat while the speed is being checked is proportional to the number of revolutions made during the change of speed. If the speed of the wheel is checked by frictional resistance only, all of the energy given up by the change of speed is converted into heat.

120. Moment of inertia of a fly-wheel with arms.—The I of the hub and rim can be found by equations (151) and (153), the numerical solution being similar to that in § 117. Only the method of dealing with the arms will therefore be considered. It is assumed that the hub and rim have the form shown in Fig. 126.

Suppose that the arms are sheared off from the hub and rim so as to leave the surfaces of the latter two parts smooth, and that the arms are flattened out to form a web which will just fit in between the hub and rim. The moment of inertia of the web will be practically the same as that of the arms from which it was made, and the area of its outer edge will be the same as the total area of all the sheared outer ends of the arms; the area of the inner edge will be equal to the total of all the sheared surfaces at the inner ends of the arms.

If, in a pulley with arms:

A_1 = total sheared area of the inner ends of the arms;
A_2 = total sheared area of the outer ends of the arms;
t_1 = thickness of inner edge of equivalent web having the same area as the sheared area of all the inner ends of the arms;
t_2 = thickness of outer edge of equivalent web;

and the remainder of the notation is the same as for the webbed wheel, Fig. 126; then

$$t_1 = \frac{A_1}{2\pi r_1} \quad \text{and} \quad t_2 = \frac{A_2}{2\pi r_2}. \quad \ldots \ldots \quad (157)$$

As an example, let it be required to find the moment of inertia of six pulley-arms according to the following data:

$A_1 = 80$ square inches;
$A_2 = 48$ square inches;
$r_1 = 11$ inches (see Fig. 126);
$r_2 = 40$ inches (see Fig. 126).

By equation (157)

$$t_1 = \frac{80}{2\pi 11} = 1.157 \text{ inches};$$

$$t_2 = \frac{48}{2\pi 40} = .1909 \text{ inch}.$$

These are the values of t_1 and t_2, according to Fig. 126, for the web whose moment of inertia is the same as that of all the arms. The corresponding values of R and T are:

$$R = \frac{40 \times 1.157 - 11 \times .1909}{1.157 - .1909} = 45.73 \text{ inches};$$

$$T = \frac{1.157 \times 45.73}{45.73 - 11} = 1.524 \text{ inches}.$$

The moment of inertia of all the arms is, therefore, by equation (152), using the values for the equivalent web,

$$\begin{aligned}
I_{arms} = I_{web} &= \frac{2\pi 450 \times 1.524}{32.2 \times 12}\left[\frac{(40)^4 - (11)^4}{4(12)^4} - \frac{(40)^5 - (11)^5}{5 \times 45.73(12)^4}\right] \\
&= 11.151\left[\frac{2545359}{82944} - \frac{102238949}{4741300}\right] \\
&= 11.151(30.688 - 21.566) \\
&= 11.151 \times 9.122 \\
&= 101.7.
\end{aligned}$$

It is not generally convenient to obtain the areas of the sheared ends of the arms, since these ends are curved surfaces which are portions of cylindrical surfaces of radii equal to the distances from the centre of the pulley to the outside of the hub and inside of the rim. In any ordinary design the error is inappreciable when the area taken is that of a plane section of the arm at right angles to its length.

Even if the arms taper uniformly from end to end, and the areas of the curved sheared surfaces are taken, there is a slight error in this method; for if the arms were flattened out so that each particle remains at its original distance from the centre of the wheel, the web thus formed will not have straight lines for its sides, as shown in Fig. 126, but the sides will be slightly concave.

The greater the width of the arm in the direction of the wheel's rotation, the greater the curvature of the side of the web. In ordinary wheel designs, however, this curvature is so slight as to be negligible.

The two errors, one due to taking plane cross-sections of the arm, and the other to assuming that the equivalent web has straight lines bounding its axial section, are compensative, instead of accumulative.

121. Stresses in fly-wheels with arms.—The stresses which occur in a fly-wheel of the ordinary design when in operation are so complicated that it is not believed they can be computed with even a practical degree of accuracy. The general nature of the stresses may be shown in such a manner, however, as to be a guide to the designer when considering the methods of reducing them with a view to decreasing the liability of the wheel to rupture on account of excessive speed or sudden variation of speed.

Fig. 127 is a portion of a built-up pulley. The hub is complete, but the ends of sections of the rim are not fastened. The pulley is represented as being under stress applied by a weight W hanging from the end of an arm attached to the hub or shaft which supports the wheel. The wheel is prevented from rotating by cords or tension-bars attached to pins at A, B, and C, extending out from both sides of the rim and having a cord attached to each end, the cords all having the same tension and pulling at right angles to the arms.

FLY-WHEELS AND PULLEYS.   253

Each arm acts as a cantilever to resist the bending action of the cords pulling on the pin near its end, and is accordingly bent or deflected as a cantilever. This bending makes the centre line of

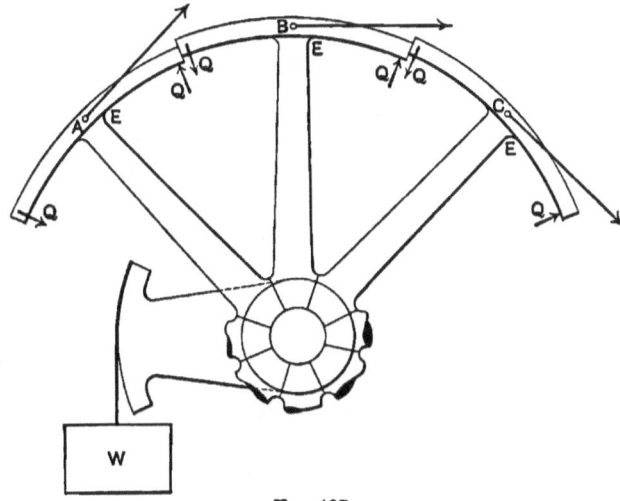

Fig. 127

each arm convex on the side opposite that toward which the cords are pulling, assuming that the centre lines were straight before stress was applied, and the sections of the rims are consequently tipped forward so that their ends are slipped over each other as shown.

Now suppose that, while the arms are still under stress due to the hanging weight W, the ends of the rim-sections are brought into line by radial forces Q applied near the ends of each section as shown. This will cause stress in the rim and arms. In the neighborhood of the angles E where the arms join the rim, the material will be in tension, whose maximum value will occur near E. The maximum value of this stress may occur in the arm or rim, according to their relative strengths.

The material of each arm is in tension near the hub on the side opposite E on account of the pull of the cords.

The bending of the arm shortens the radial distance from the centre of the pulley to the outer end of the arm; hence, when all the sections of the entire pulley have their ends brought into line, the rim will be drawn in somewhat at the end of each arm, and, instead of being circular, will be slightly wavy around its circumference. The effort of the rim to assume its circular form, assuming that this is the form when not under stress, will cause tension in the arm; there will also be bending tensile stress in the rim near E on account of its being bent inward.

If, instead of having cords attached to the pins in the rim of the stationary fly-wheel, it were started or stopped from rotating by power applied through its shaft, the inertia of the parts would induce stresses similar to those caused by the pull of the cords. A belt running on the rim and transmitting power to or from the shaft has much the same effect.

In addition to the stresses corresponding to those caused by the pull of the cords on a stationary wheel, there are in a rotating wheel others due to the centrifugal tendency of the rotating parts. These stresses may be considered by dealing with the rim and arms separately.

When a ring, thin radially, of a mean radius r, rotates about its geometrical axis with a linear velocity of V feet per second, circumferential tension is induced in the material. The value of this stress p in pounds per square inch is expressed by the following equation, in which $(w \div 144)$ equals the weight of a piece of the material 1 inch square and 1 foot long:

$$p = \frac{w}{144g}V^2 = \frac{w}{144g}(\omega r)^2 = \frac{w}{144g}(2\pi N r)^2. \quad . \quad . \quad (158)$$

Equation (158) shows that the tensile stress in a thin rotating ring is proportional to the square of the linear velocity V, and that, if V remains constant, the tensile stress is constant whatever the radius of the ring.

The radius of the ring is increased on account of this stress by an amount λ in feet given by equation (159), in which $E_t =$ tensile modulus of elasticity of the material:

$$\lambda = \frac{pr}{E_t}. \quad . \quad . \quad . \quad . \quad . \quad (159)$$

If the ring is thick radially, say 1 foot thick for 20 feet of diameter, the mean radius $(r_i + r_o) \div 2$ can be used without an error great enough to require practical consideration.

For the arms equations of the same nature as (158) and (159) can be written. They are probably of no practical value, however, so they will be omitted. The general nature of the effect of rotation upon the arms may be expressed, however, by the statement that, if a thin, prismatic bar is rotated about an axis at one end and normal to the length of the bar, the maximum tensile stress in the bar is one half as great as in a ring having a radius equal to the length of the bar and rotating at the same angular velocity; the increase of the length of the bar is one third that of the increase of the radius of the ring.

Suppose that a ring of metal, corresponding to the rim of a fly-wheel, is placed on a horizontal table, and a "spider," corresponding to the hub and radial arms, is placed in the ring, the arms being of such a length as to just fit in the ring without pressure against it. If the table is rotated about a vertical axis coincident with that of the ring and spider, carrying these parts with it, the ring, on account of the centrifugal action, will increase in diameter more rapidly than the arms will increase in length, and will therefore separate from them, leaving a space between the end of each arm and the ring. If, on the contrary, the arms are attached to the ring so that no separation can occur, the arms will be somewhat elongated on account of the pulling-out action of the ring, and the ring will be bent inward where the arms are attached. Tensile stress will therefore be induced in the arms and in the inner side of the ring at and near the arms. The relative intensity of the stress in the arms and ring will depend on their relative strength.

The pulling-in action of the arms, when there are several, reduces the tension in the rim caused by centrifugal action.

As a summary of the above, it may be stated that, if a fly-wheel which has no stress in its parts when at rest is rotated uniformly about its axis, tensile stress will be induced in the arms and rim, that in the rim having the greatest intensity midway between the arms and at or near the arms; also, if the speed of rotation is varied by a driving or resisting force applied through the shaft or hub, there will be additional tensile stress induced about one of the angles between the end of each arm and the rim. A pulley carrying a belt that is transmitting power has similar stresses.

Only tensile stresses have been considered in the discussion of the fly-wheel, because they are greater than those of compression, and also because most fly-wheels, and a considerable proportion of large pulleys, are made of cast iron, which has a much lower tensile than compressive strength.

When a fly-wheel is supported on a horizontal shaft, there is an additional tension in the lower arms. The tension in an arm extending vertically downward from the hub is much greater than that due to the portion of the rim belonging to the arm, i.e., greater than that due to a weight equal to that of the whole rim divided by the number of arms.

122. Numerical example of stress in, and enlargement of, a rotating ring due to centrifugal action.—The tensile stress in a cast-iron ring 20 feet in diameter, rotating about its geometrical axis at a circumferential velocity of 6000 feet per minute, is, by equation (158), taking $w = 450$,

$$p = \frac{w}{144g} V^2 = \frac{450}{144 \times 32.2} \times \left(\frac{6000}{60}\right)^2 = 970 \text{ lbs. per sq. in.}$$

The increase of the radius caused by the rotation is, by equation (159), taking $E_t = 14000000$ lbs. per sq. in.,

$$\lambda = \frac{970 \times 10}{14000000} = .000693 \text{ ft.} = .0083 \text{ in.}$$

The corresponding increase in the circumference is

$$2\pi \times .0083 = .052 \text{ in.}$$

The arms in a fly-wheel of ordinary design are considerably smaller near the rim than at the hub. The elongation is therefore not uniformly distributed along an arm, but is greatest at or near its smallest cross-section. For a rotating wheel having a heavy, stiff rim, it seems fair to assume that the greatest elongation per unit length in the arm is at least as great as would be obtained per unit length in a prismatic bar of a total length equal to the diameter of the rim, by subjecting it to a tensile stress sufficient to elongate it to the same extent as the diameter of the free rim would be increased by the rotation.

Upon this assumption the tension in each arm supporting the ring just considered, taking the arm as 83 inches long and allowing for no distortion of the hub, would be

$$\text{Tension in arm} = \frac{.0083}{83} \times 14000000 = 1400 \text{ lbs. per sq. in.}$$

The rim-tension of 970 lbs. per sq. in., and the tension of 1400 lbs. per sq. in. in the arms are of comparatively small value when considered separate from stresses due to other causes. But, since the rim- and arm-tension increase as the square of the velocity, they may attain considerable values for very high speeds, such as sometimes occur when an engine "races."

123. Sectional-rim fly-wheels and pulleys.—When the rim of a fly-wheel or pulley is divided into sections by being cut through midway between each pair of arms, as in Fig. 127, the ends of the sections must be held together, of course, by some kind of fastening. As has already been stated, centrifugal action tends to bend or bow the rim out between the arms. If the wheel is rotating uniformly and has no belt upon it, each part of the rim between two arms acts much like a beam supported at the ends and uniformly loaded from end to end. The centrifugal action in the wheel corresponds to a uniformly distributed load acting radially outward. This bending outward of the rim has a tendency to open up the joint more at the outer than the inner side of the rim. For this reason the fastenings, which are tension-members, should be placed as near as possible to the face or outer circumference of the rim, and should have a considerable thickness of metal, measured radially, between them and the inner circumference of the rim.

If the rim is thin, as for band-wheels, lugs or flanges must generally be added for the fastenings. These lugs should be of sufficient radial height to make the distance from the fastenings to the inner side of the lug large in comparison with the distance from the centre of the fastenings to the face of the pulley, and they should be braced well with circumferential ribs running back towards the arms. In order to give the rim rigidity to resist bending, these ribs should also have a considerable radial depth.

In recent years a number of large pulleys have been designed with sectional rims divided at the end of each arm, there being no joint in the rim between the arms.

124. Bursting tests of small cast-iron fly-wheels by centrifugal action.—A number of tests were made upon small cast-iron fly-wheels by Prof. C. H. Benjamin in order to determine the speed at which they would burst when rotating with practical uniformity.* The wheels tested were all of cast iron, and most of them were miniatures of large fly-wheels designed by leading manufacturers of such machinery. Some of the miniatures were 15 inches, and the others 24 inches, in diameter. Part had solid rims, others had rims made in two sections and fastened together with bolts, while still others had rims in two sections fastened together with links. The wheels were direct-connected to a steam-turbine, which drove them at a gradually increasing speed until they flew to pieces. No brake or pulley was applied to cause tangential resistance to rotation, and no belt was run on the wheel.

The tests showed that, as would naturally be expected, the wheels having sectional rims fastened together midway between two arms were much weaker than those having solid rims; and that, although the tension-members of the fastenings at the ends of the sections were of less tensile strength than the rim itself, they gave way in but one instance, in which case the bolts broke; the rupture occurred in the jointed rims of all the other wheels.

It was also shown that a wheel cast from the same pattern as another, but having its rim turned down so as to be much thinner radially, broke at a much lower speed than the one having the rim left thick, as it had come from the mould.

Tests upon wheels similar in every respect except the number of arms, some having three and some six, showed that the three-arm pulleys were weaker than those having six. The pattern for the three-arm wheels was made by removing the alternate arms from the patterns for the six-arm wheels.

Tables XXXV to XLI show the dimensions of the fly-wheels tested and the results of the tests. When bolts were used to fasten the joints, flanges, extending across the inner surface of the rim,

---

* The Bursting of Small Cast-iron Fly-wheels. Trans. Amer. Soc. Mech. Eng., vol. xx., 1899.

furnished holes for the bolts. When two links were used, they were imbedded in the side of the rim, after the method of Fig. 128. The third link, when used, gripped a pair of lugs extending out from the inner surface of the rim.

TABLE XXXV.

FIFTEEN-INCH WHEELS.

No.		Rim.				Arms.		Weight of Wheel, pounds.
	Style.	Diameter, inches.	Breadth, inches.	Depth, inches.	Area, sq. in.	No.	Area, sq. in.	
1	Solid	15½	2	.70	1.4	6	.46	20.87
2	"	15⅛	2	.65	1.3	6	.46	20.44
3	"	15	2	.615	1.23	6	.46	19.12
4	"	14⅝	2	.52	1.04	6	.46	16.62
5	Sectional	15$\frac{3}{16}$	....	....	....	6	.46	20.87
6	Solid	15$\frac{3}{16}$	2	.69	1.38	3	.46	19.25
7	"	15	2	.615	1.23	3	.46	16.56
8	"	14¾	2	.475	.95	3	.46	13.68
9	"	14⅛	1⅞	.400	.75	6	.46	12.68
10	"	14½	1⅞	.347	.65	6	.46	13.00

TABLE XXXVI.

FIFTEEN-INCH WHEELS.

No.	Bursting Speed.		Centrifugal Tension $= \frac{V^2}{10}$.	Remarks.
	Revolutions per Minute.	Feet per Second $= V$.		
1	6,525	430	18,500	
2	6,525	430	18,500	
3	6,035	395	15,600	Thin rim.
4	5,872	380	14,400	" "
5	2,925	192	3,700	Joint.
6	5,600*	368	13,600	Three arms.
7	6,198	406	16,500	" "
8	5,709	368	13,600	" "
9	5,709	365	13,300	Thin rim.
10	5,709	361	13,000	" "

* Doubtful.

Table XXXVII.

TWENTY-FOUR-INCH WHEELS.

No.	Shape and Size of Rim.					Weight of Wheel, pounds.
	Diameter, inches.	Breadth, Inches.	Depth, Inches.	Area, sq. in.	Style of Joint.	
11	24	2½	1.5	3.18	Solid rim.	75.25
12	24	4 1/16	.75	3.85	Internal flanges, bolted.	93
13	24	4	.75	3.85	" " "	91.75
14	24	4	.75	3.85	" " "	95
15	24	4 1/16	.75	3.85	" " "	94.75
16	24	1.2	2.1	2.45	Three lugs and links.	65.1
17	24	1.2	2.1	2.45	Two " " "	65

Table XXXVIII.

FLANGES AND BOLTS.

No.	Flanges.			Bolts.		
	Thickness, inches.	Effective Breadth, inches.	Effective Area, inches.	No. to Each Joint.	Diameter, inches.	Total Tensile Strength, pounds.
12	11/16	2.8	1.92	4	5/16	16,000
13	11/16	2.75	1.89	4	5/16	16,000
14	15/16	2.75	2.58	4	5/16	16,000
15	15/16	2.5	2.34	4	3/8	20,000

BY TESTING-MACHINE.

Tensile strength of cast iron = 19,600 pounds per square inch.
Transverse strength of cast iron = 46,600 pounds per square inch.
Tensile strength of 5/16 bolts = 4,000 pounds.
Tensile strength of 3/8 bolts = 5,000 pounds.

FLY-WHEELS AND PULLEYS.

TABLE XXXIX.
FAILURE OF FLANGED JOINTS.

No.	Area of Rim, sq. in.	Effective Area Flanges, sq. in.	Total Strength Bolts, pounds.	Bursting Speed.		Centrifugal Tension		Remarks.
				Rev. per Minute.	Ft. per Second $= V$.	Pounds per Sq. In. $= \frac{V^2}{10}$.	Total Pounds.	
11	3.18	....	......	3,672	385	14,800	47,000	Solid rim.
12	3.85	1.92	16,000	....	....	......	......	Flange broke.
13	3.85	1.89	16,000	1,760	184	3.400	13,100	" "
14	3.85	2.58	16,000	1,875	196	3,850	14,800	Bolts broke.
15	3.85	2.34	20,000	1,810	190	3,610	13,900	Flange broke.

TABLE XL.
LINKED JOINTS.

No.	Lugs.			Links.			Rim.	
	Breadth, in.	Length, in.	Area, sq. in.	Number Used.	Effective Breadth, in.	Thickness, in.	Effective Area, sq. in.	Max. Area, sq. in.	Net Area, sq. in.
16	.45	1.0	.45	3	.57	.327	.186	2.45	1.98
17	.44	.98	.43	2	.54	.380	.205	2.45	1.98

BY TESTING-MACHINE.

Tensile strength of cast iron = 19,600.
Transverse strength of cast iron = 40,400.
Av. tensile strength of each link = 10,180.

TABLE XLI.
FAILURE OF LINKED JOINTS.

No.	Strength of Links, pounds.	Strength of Rim, pounds.	Bursting Speed.		Centrifugal Tension.		Remarks.
			Rev. per Minute.	Ft. per Second $= V$.	Pounds per Sq. In. $= \frac{V^2}{10}$.	Total, Pounds.	
16	30,540	38,800	3,060	320	10,240	25,100	Rim broke.
17	20,360	38,800	2,750	290	8,410	20,600	Lugs and rim broke.

Special Forms of Fly-wheels and Pulleys.

125. Hollow cast-iron arms with wrought-iron or steel tension-rods for fly-wheels and pulleys.—On account of the comparatively low tensile strength of cast iron and the predominance of tensile stress in the arms of pulleys and fly-wheels, it is obvious that the liability of the arms to fracture will be reduced if, by some means, the cast iron can be relieved partly or wholly of tensile stress, although by doing so it may be subjected to high compressive stress. A simple and comparatively inexpensive way of doing this is to cast the arms hollow and put a steel or wrought-iron tension-rod through the centre of each from rim to hub. The rod is put in tension by means of its end fastenings when the pulley is constructed. The surrounding cast-iron walls act as a column to resist this tension and are thereby subjected to compression. A pulley so constructed is unquestionably much safer under heavy service than one having only cast-iron arms. This device has been adopted by at least one concern which has placed many large pulleys in commercial service.

126. "Tangent" arms for pulleys and fly-wheels.—Pulleys of this description have arms which, instead of being radial, are tangent to a circle of fixed diameter, as in the construction common to bicycle-wheels. Tangent-arms have been adopted for wheels of considerable size performing service where the demands upon the fly-wheel for energy are sudden and severe. Such fly-wheels have been applied to mine hoisting-machinery. The arms, instead of being round, are generally of rectangular cross-section. Being purely tension-members, they are of course made of some such material as wrought iron or steel, cast iron being totally unfit for this purpose. The tangential direction of the arms prevents any necessity of their resisting the torsional moment of the shaft or rim by a bending moment in the arm. It is not probable that even a very approximate calculation of the stresses in the arms can be arrived at.

127. "Built-up" plate fly-wheel.—A fly-wheel made of rolled structural-steel plates is in use in the power-station of the Union Railroad Co. at Providence, R. I. The web of the wheel is made of a number of segmental plates placed so as to break joints, and held together by rivets passing through from side to side. The rim is built up of several pieces of plate metal cut so as to form a ring when placed together. These sections are placed together so as to break joints and are riveted through from side to side. Steel plate 1 inch thick is used for the web, and $1\frac{3}{4}$ and $1\frac{1}{2}$ inches thick for the rim. The hub is of cast iron 72 inches in diameter, made of two disks. It can be seen that the pulley has what might be called a built-up web. The pulley is 18 feet in diameter and has a rim $15\frac{1}{4} \times 16$ inches. The web is made of two thicknesses of one-inch plate. The factor of safety is about forty.

128. Wire-wound fly-wheel.—A fly-wheel having a rim composed of wire wound circumferentially under tension around a web was constructed for use in a rolling-mill at Ladore, Wales, using the Mannesmann process of rolling tubing from the solid bar. In this class of rolling a very great amount of power must be supplied to the machinery in a short time. The strain upon the fly-wheel is accordingly very great on account of the sudden reduction of speed which it must undergo in order to deliver its energy, as well as on account of the high speed at which it runs in order to have sufficient energy stored in it.

The construction adopted is the strongest and safest that can be devised. Two steel disks, 20 feet in diameter, are bolted to a cast-iron hub. The outer edges of the disks form a groove into which wire is wound to form the heavy rim. The groove is filled with 70 tons of No. 5 steel wire wound on under a tension of 50 pounds. The wheel is run at 240 revolutions per minute, which corresponds to a circumferential velocity of 15080 feet = 2.85 miles per minute.

129. Other special forms of pulleys.—Many large pulleys are now constructed with built-up rims and cast-iron arms. The rims are frequently made heavy enough to serve as fly-wheels.

Smaller pulleys are often made completely of wood built up in sections and having but a few arms, generally two or four.

Medium-size pulleys are often made with wrought-iron or sheet-steel rims and cast-iron arms. Small-size pulleys are also frequently constructed in this way.

An "all-steel" pulley has recently been placed on the market. It is made up completely of sheet metal, the arms and hub being pressed into form.

Two sets of arms are very commonly used for wide pulleys, and in some cases as many as three sets are used.

TABLE XLII.*

DIMENSIONS OF STANDARD ENGINE-PULLEY ARMS OF ELLIPTICAL CROSS-SECTION. (SIX ARMS IN EACH PULLEY.)

For very wide faces two sets of arms should be used. The diameter of hub is taken as twice that of the shaft on which it fits.

Diameter of Pulley, inches.	Light Pattern, Size of Arms.				Medium Pattern, Size of Arms.				Heavy Pattern of Fly-wheels, Pulleys with Thick Rims, and Pulleys with Faces up to 26 inches Wide—Size of Arms.			
	At Rim.		At Hub.		At Rim.		At Hub.		At Rim.		At Hub.	
	Dimens., inches.	Area, sq. in.	Dimens., inches.	Area, sq. in.	Dimens., inches.	Area, sq. in.	Dimens., inches.	Area, sq. in.	Dimens., inches.	Area, sq. in.	Dimens., inches.	Area, sq. in.
10	2 × 1	1.57	2¼ × 1¼	2.2	2 × 1	1.57	2¼ × 1¼	2.2				
12	2 × 1	1.57	2¼ × 1¼	2.2	2 × 1	1.57	2¼ × 1¼	2.2				
14	2 × 1	1.57	2¼ × 1¼	2.2	2 × 1	1.57	2¼ × 1¼	2.2				
16	2 × 1	1.57	2¼ × 1¼	2.2	2 × 1	1.57	2¼ × 1¼	2.2				
18	2 × 1	1.57	2¼ × 1¼	2.2	2 × 1	1.57	2¼ × 1¼	2.2				
20	1¼ × ⅞	.858	2 × 1	1.57	2⅛ × 1¼	2.08	2⅝ × 1⅜	2.56				
22	1¼ × ⅞	.858	2 × 1	1.57	2⅛ × 1¼	2.08	2⅝ × 1⅜	2.56				
24	1¼ × ⅞	.858	2 × 1	1.57	2⅛ × 1¼	2.08	2⅝ × 1⅜	2.56				
26	1¼ × ⅞	.858	2 × 1	1.57	2¼ × 1¼	2.20	2½ × 1½	2.94				
28	1¼ × ⅞	.858	2 × 1	1.57	2⅜ × 1¼	2.32	2⅞ × 1½	3.38				
30	2 × 1⅛	1.767	2⅛ × 1⅝	2.69	2⅜ × 1¼	2.32	2⅞ × 1½	3.38				
32	2 × 1⅛	1.767	2⅛ × 1⅝	2.69	2½ × 1⅜	2.91	3 × 1½	3.53				
34	2 × 1⅛	1.767	2⅛ × 1⅝	2.69	2⅞ × 1⅜	3.10	3¼ × 1½	3.82	3¼ × 1½	3.82	3¾ × 2	5.89
36	2¼ × 1⅛	1.98	2¾ × 1⅜	2.96	3 × 1½	3.53	3¼ × 1⅝	4.14	3¼ × 1⅝	4.14	3¾ × 2	5.89
38	2¼ × 1⅛	1.98	2¾ × 1⅜	2.96	3 × 1½	3.53	3¼ × 1⅝	4.14	3¼ × 1⅝	4.14	3¾ × 2	5.89
40	2¼ × 1⅛	1.98	2¾ × 1⅜	2.96	3 × 1½	3.53	3¼ × 1⅝	4.14	3¼ × 1⅝	4.14	3¾ × 2	5.89
42	2⅜ × 1¼	2.32	2⅞ × 1½	3.38	3¼ × 1½	3.82	3¾ × 1¾	5.15	3½ × 1¾	4.81	4 × 2⅛	6.67
44	2⅜ × 1¼	2.32	2⅞ × 1½	3.38	3¼ × 1½	3.82	3¾ × 1¾	5.15	3½ × 1¾	4.81	4 × 2⅛	6.67
46	2⅜ × 1¼	2.32	2⅞ × 1½	3.38	3¼ × 1½	3.82	3¾ × 1¾	5.15	3½ × 1¾	4.81	4 × 2⅛	6.67
48	2¾ × 1⅜	2.96	3¼ × 1⅝	4.14	3⅝ × 1½	4.27	4¼ × 1¾	5.84	3¾ × 2	5.89	4½ × 2¼	7.94
50	2¾ × 1⅜	2.96	3¼ × 1⅝	4.14	3⅝ × 1½	4.27	4¼ × 1¾	5.84	3¾ × 2	5.89	4½ × 2¼	7.94
54	3⅜ × 1¾	4.97	4¼ × 2	6.67	4 × 1⅝	5.10	4¾ × 1⅞	6.99	4½ × 2⅛	7.09	5 × 2½	9.79
58	3⅜ × 1¾	4.97	4¼ × 2	6.67	4 × 1⅝	5.10	4¾ × 1⅞	6.99	4½ × 2⅛	7.09	5 × 2½	9.79
60	3¾ × 2	5.89	4¾ × 2⅜	8.85	4¼ × 2	6.67	5 × 2⅝	10.3	4⅝ × 2¼	7.73	5⅜ × 2¾	11.60
66	3⅞ × 2⅛	6.46	4⅞ × 2½	9.56	4¼ × 2	6.67	5 × 2⅝	10.3	4⅝ × 2¼	7.73	5⅜ × 2¾	11.60
68	3⅞ × 2⅛	6.46	4⅞ × 2½	9.56	4¾ × 2⅜	8.85	5⅛ × 2⅞	13.4	4⅞ × 2½	9.50	5⅝ × 2⅞	12.68
72	4 × 2¼	7.06	5 × 2½	9.79	4¾ × 2⅜	8.85	5⅛ × 2⅞	13.4	5 × 2⅝	10.3	5¾ × 3	13.54
78	4 × 2¼	7.06	5 × 2½	9.79	4¾ × 2⅜	8.85	5⅛ × 2⅞	13.4	5 × 2⅝	10.3	5¾ × 3	13.54
84	4¼ × 2⅜	7.92	5¼ × 2⅝	10.82	4⅞ × 2½	9.56	5¾ × 3	13.83	5 × 2¾	10.79	6 × 3¼	15.31
96	5 × 2⅝	10.3	6 × 3	14.13	5⅛ × 2⅞	13.4	6½ × 3¼	16.58	7 × 3½	19.24	8⅛ × 4⅛	29.71

* Pulley and Fly-wheel Construction, by Theo. F. Scheffler, Jr., *Power*, April, 1898, p. 8.

*Designs and Proportions of Fly-wheels and Pulleys Taken from Practice.**

130. Since it is evidently impossible to make calculations for fly-wheels and pulleys with regard to their strength, it seems advisable to present some of the larger sizes, as made by leading builders, which have shown themselves to be satisfactory in use; also to give some of the proportions of the smaller sizes of pulleys.

Table XLII gives the sizes of arms for pulleys from 10 to 96 inches in diameter and having rims with faces up to 26 inches wide. These are the proportions adopted by Struthers, Wells & Co. for the pulleys used upon their engines. Table XLIII gives the thicknesses of rims adopted by the same company.

TABLE XLIII.

AVERAGE THICKNESS OF ENGINE-PULLEY RIMS.

Wide Pulleys.			Narrow Pulleys.		
Diameter, inches.	Face Width, inches.	Thickness of Rim, inches.	Diameter, inches.	Face Width, inches.	Thickness of Rim, inches.
12	3½	5/16	6	3½	1/4
18	4½	3/8	8	4½	5/16
20	6½	7/16	10	5½	3/8
28	8½	1/2	14	7½	7/16
36	9½	9/16	16	10½	7/16
44	10½	1/2	20	10½	7/16
44	12½	9/16	32	10½	1/2
48	12½	9/16	32	10½	5/8
54	12½	11/16	36	10½	5/8
60	14½	3/4	36	12½	7/8
60	14½	15/16	42	12½	7/8
72	16½	7/8	48	12½	1
72	16½	15/16	48	14½	1⅛
72	16½	1⅛	54	14½	1⅛
78	16½	1¼	54	16½	1⅛
78	18½	1⅜	54	16½	1⅛
78	20½	1½	60	16½	1¼
78	22½	1½	72	18½	1¼
84	24½	1⅝	72	20½	1¼
84	26½	1¾	78	22½	2
96	31	1⅞	78	24½	2
96	31	2			
108	34	2¼			
108	37	2½			
120	42	2¾			
120	48	3			

* The following illustrations of fly-wheels and pulleys are taken from drawings and blue-prints kindly supplied by the establishments whose names are mentioned in connection with them.

266 FORM, STRENGTH, AND PROPORTIONS OF PARTS.

Fig. 128 is a 20-foot band-wheel with a 62-inch face and having a rim divided into eight sections as made by the Edward P. Allis Co. The rim-sections meet at the ends of the arm. The weight is 65000 pounds.

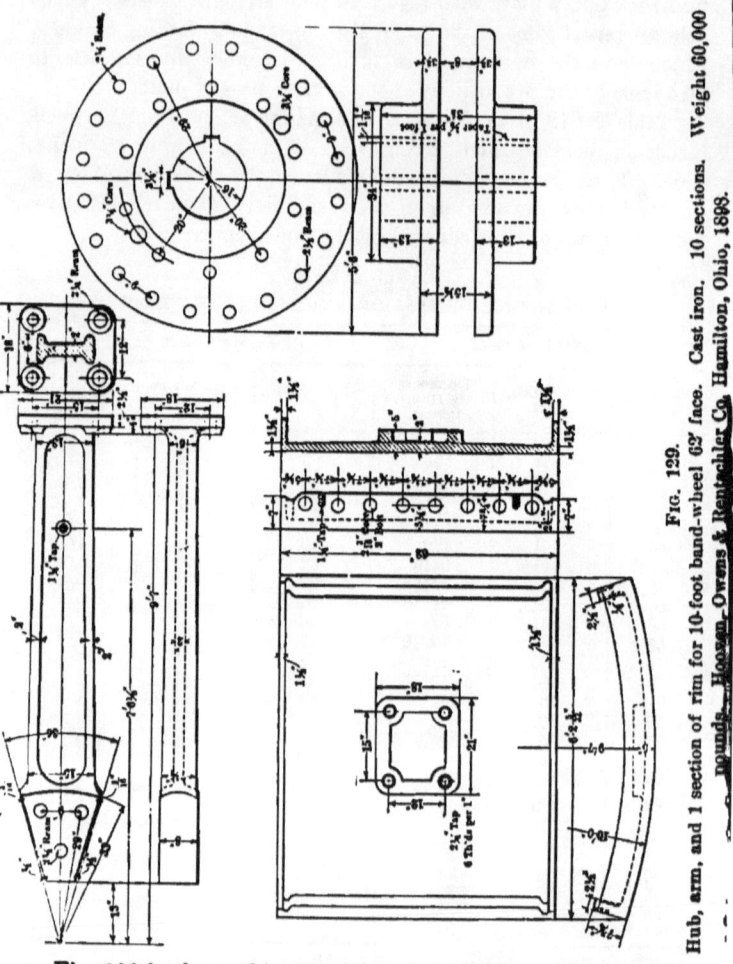

Fig. 129.—Hub, arm, and 1 section of rim for 10-foot band-wheel 63' face. Cast iron. 10 sections. Weight 60,000 pounds. Hooven, Owens & Rentschler Co., Hamilton, Ohio, 1898.

Fig. 129 is the working drawing of the hub, one arm, and one

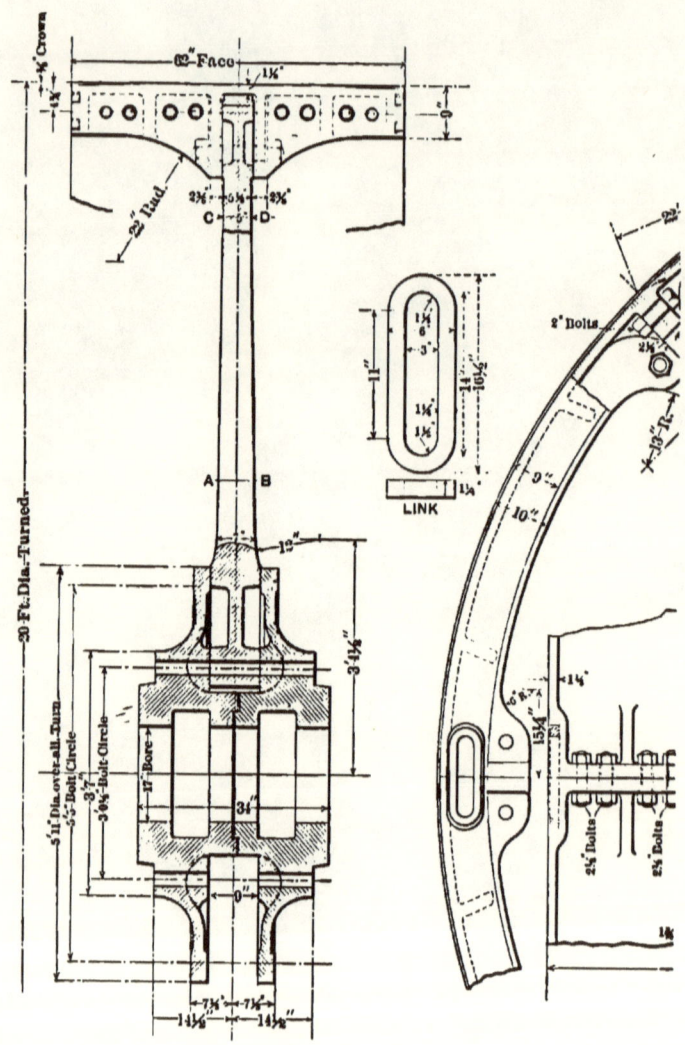

FIG. 128

20-foot band-wheel, 62-inch face. Cast iron. 8 rim sections. Weigh

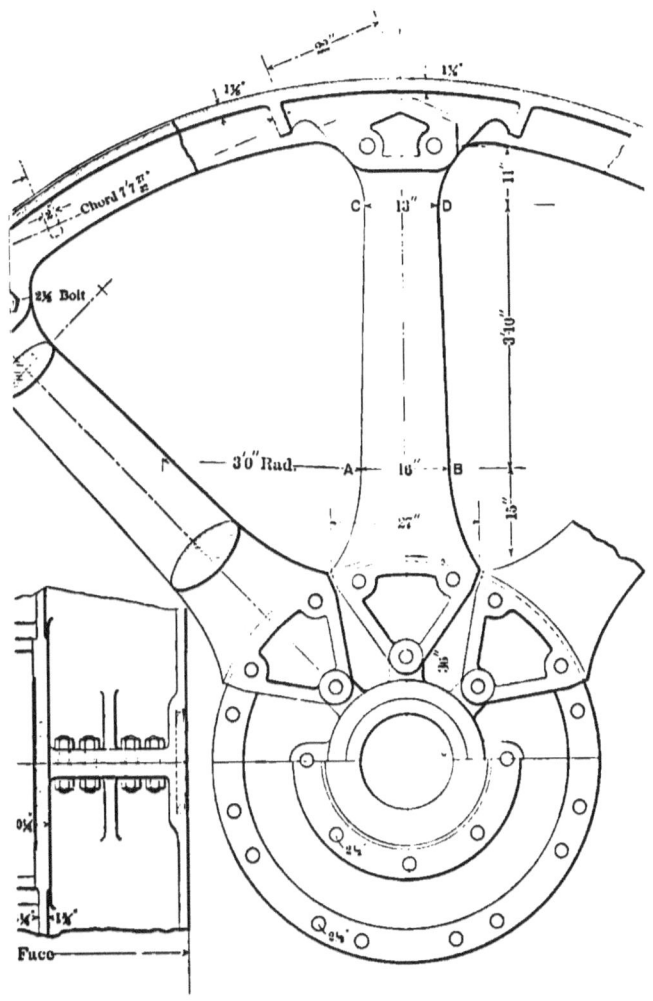

To face page 256.

,000 pounds. Edward P. Allis Co., Milwaukee, Wis., 1895.

FLY-WHEELS AND PULLEYS.

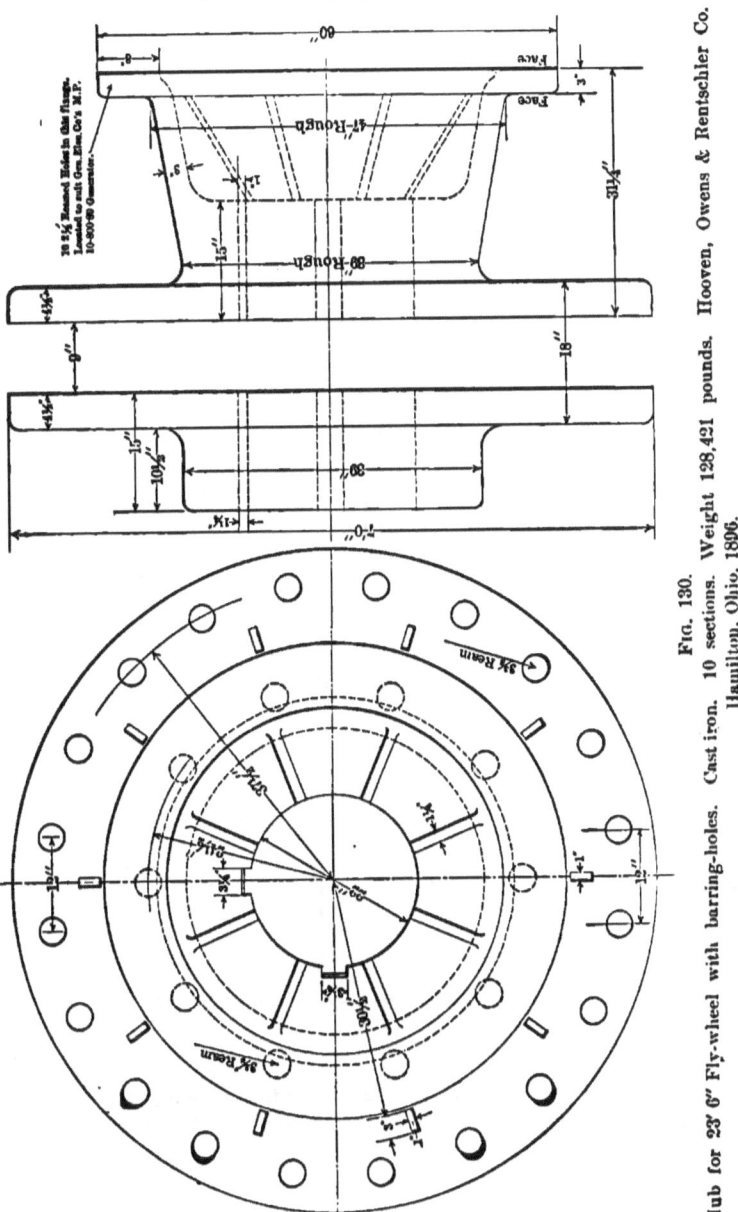

FIG. 130. Hub for 23′ 6″ Fly-wheel with barring-holes. Cast iron. 10 sections. Weight 128,491 pounds. Hooven, Owens & Rentschler Co. Hamilton, Ohio, 1896.

268 FORM, STRENGTH, AND PROPORTIONS OF PARTS.

section of the rim of a 10-foot band-wheel having a 62-inch face and weighing 60000 pounds as made by the Hooven, Owens &

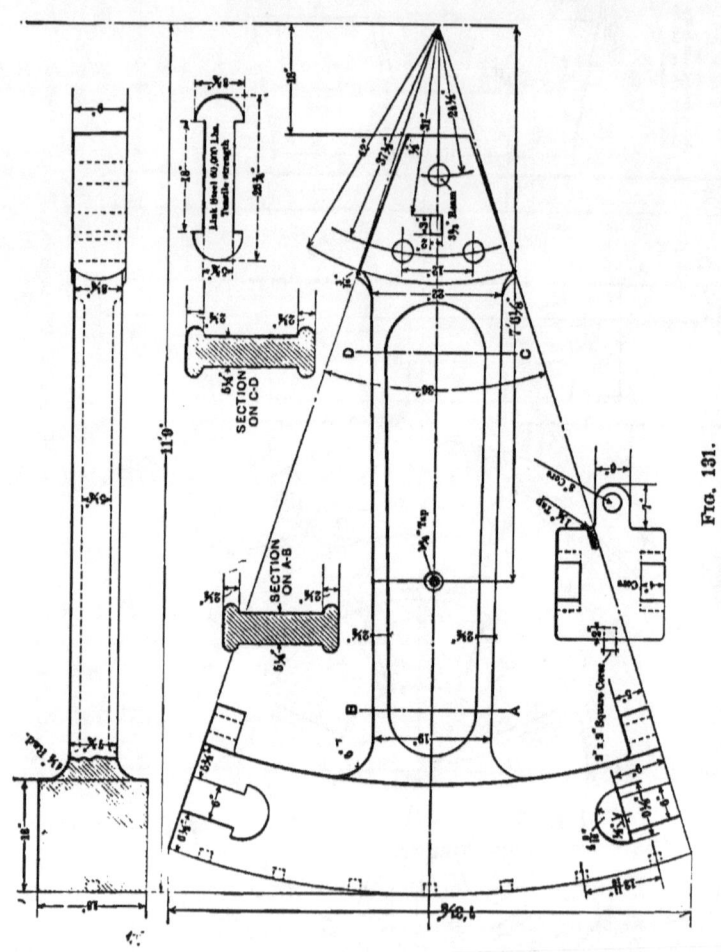

FIG. 131.—Arm and 1 section of rim for 23' 6" fly-wheel with barring-holes. Cast iron. 10 sections. Weight 128,421 pounds. Hooven, Owens & Rentschler Co., Hamilton, Ohio, 1896.

Rentschler Co. The rim is divided into ten sections, each section having one arm. The arm in this design is different from those

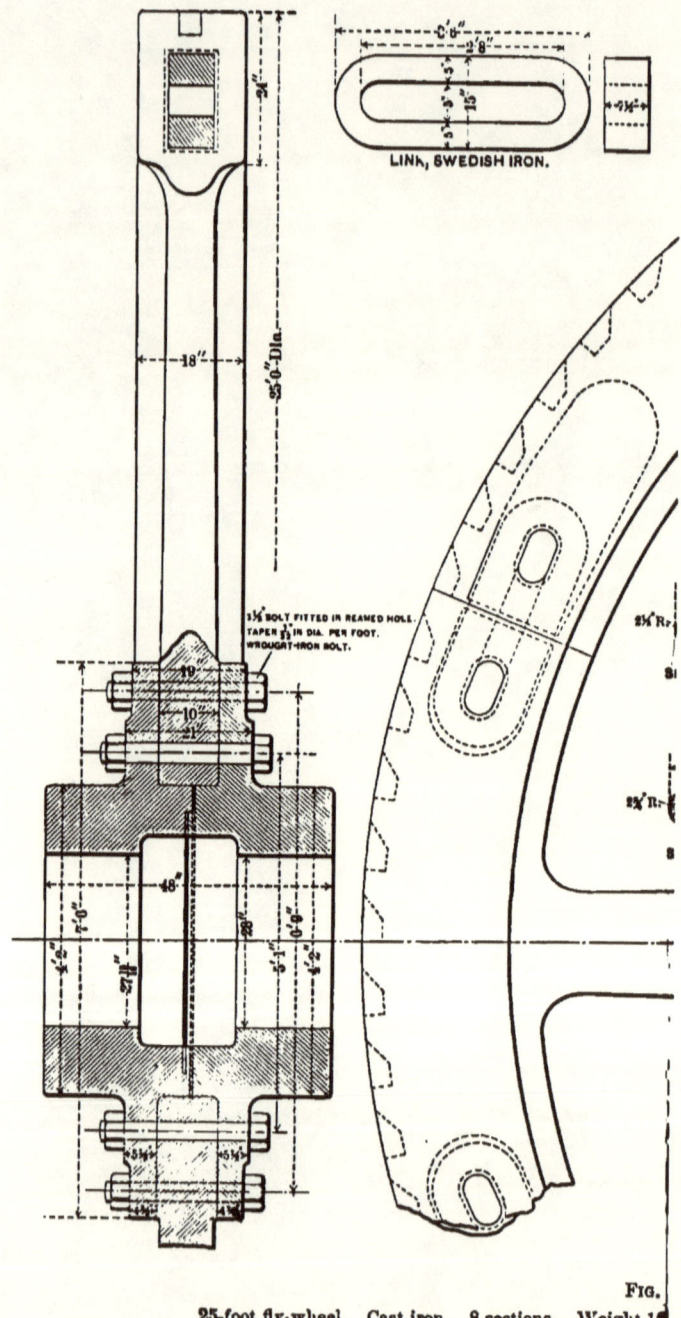

Fig.

25-foot fly-wheel. Cast iron. 8 sections. Weight 16

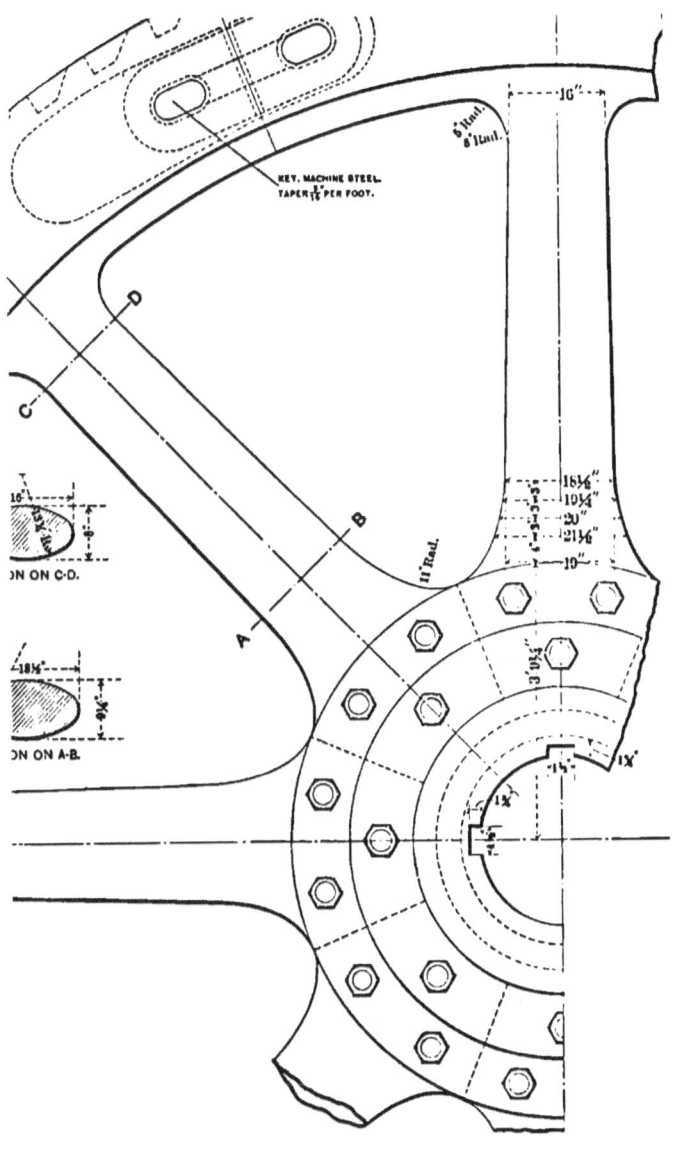

pounds. Fraser & Chalmers, Chicago, Ill. 1894.

To face page 268.

FLY-WHEELS AND PULLEYS.

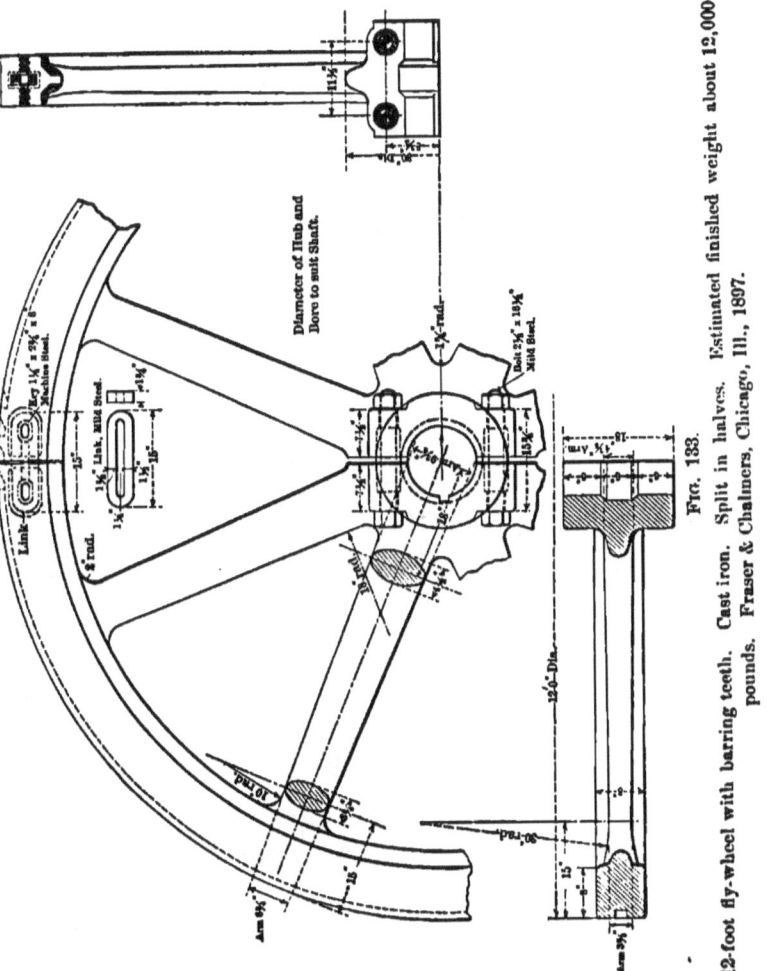

Fig. 133.—12-foot fly-wheel with barring teeth. Cast iron. Split in halves. Estimated finished weight about 12,000 pounds. Fraser & Chalmers, Chicago, Ill., 1897.

usually adopted. It is made in the form of a cast-iron I beam. This form presents advantages in the way of getting a sound casting and in making an arm which is strong enough to resist a turning force acting upon the rim or hub. A wheel of this design is operating successfully at 75 revolutions per minute.

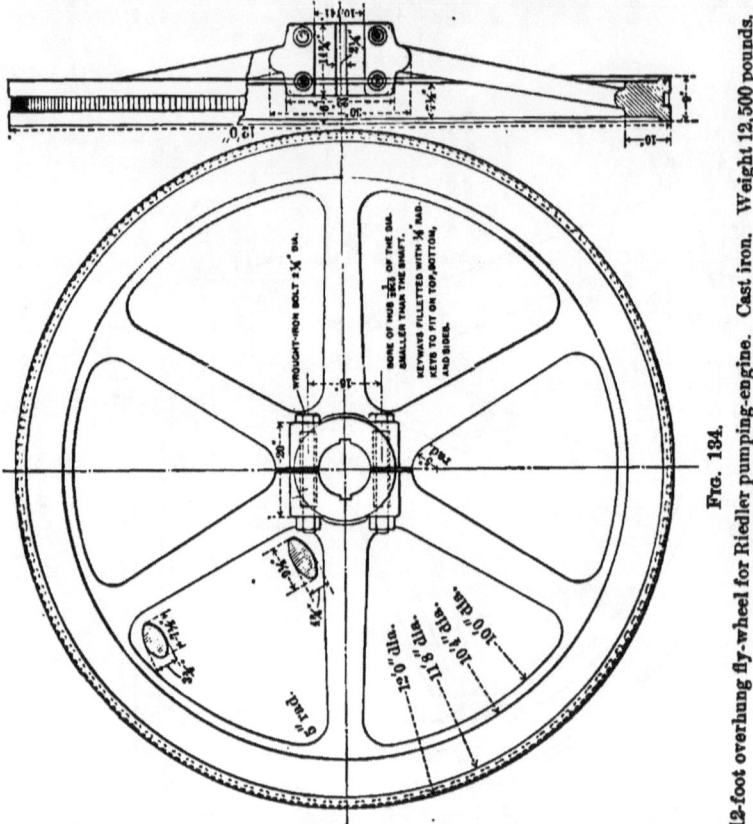

Fig. 134.— 12-foot overhung fly-wheel for Riedler pumping-engine. Cast iron. Weight 12,500 pounds. Fraser & Chalmers, Chicago, Ill., 1895.

Figs. 130 and 131 are from the working prints of a fly-wheel 23 feet 6 inches in diameter as made by the same company. This wheel is also divided into ten sections. The net weight is 128421 pounds. One of the wheels made after this design is direct-con-

FLY-WHEELS AND PULLEYS. 271

nected to an 800-kilowatt generator speeded at 80 revolutions per minute.

Fig. 132 is a 25-foot fly-wheel weighing 160000 pounds and divided into eight sections. Four of these wheels have been made

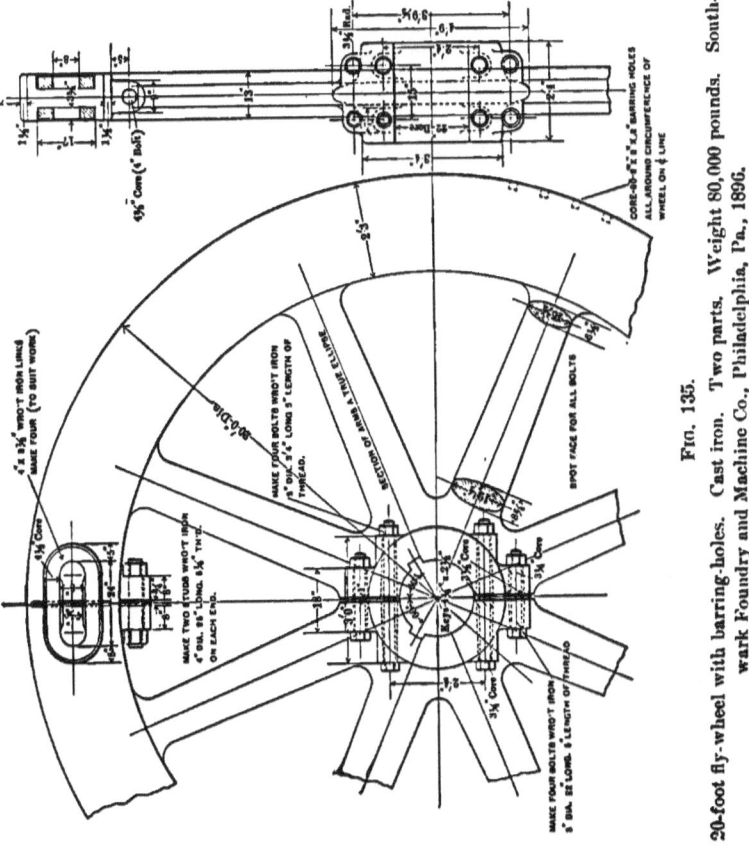

FIG. 135.—20-foot fly-wheel with barring-holes. Cast iron. Two parts. Weight 80,000 pounds. Southwark Foundry and Machine Co., Philadelphia, Pa., 1896.

by Fraser & Chalmers, and put into operation in the power-station of the West Chicago Street Railway Co. They are on 34 and 54 × 60 inches compound condensing Corliss engines which are direct-connected to electric generators.

272   FORM, STRENGTH, AND PROPORTIONS OF PARTS.

Fig. 133 represents another fly-wheel as built by the same company. It is 12 feet in diameter and made in two parts. The weight is about 12000 pounds.

Fig. 134 is another 12-foot fly-wheel made by Fraser & Chalmers. The design is peculiar in that the rim is overhung in order to bring its weight over the bearing which supports it. The weight of this pulley is 12500 pounds. Two of these wheels are in use on each of the pumping-engines for delivering water at high pressure to the hydraulic accumulators in the works of the Pope Tube Co. The water pumped is utilized for tube-drawing benches, etc.

Fig. 135 represents a 20-foot fly-wheel weighing 80000 pounds as constructed by the Southwark Foundry & Machine Co. The wheel is made in two parts, held together by links at the rim and bolts at the hub.

Fig. 136 shows a 28-foot fly-wheel as proposed by the Edward P. Allis Co. for the Third Avenue Railway Co. of Milwaukee, Wis. Only the principal dimensions are given.

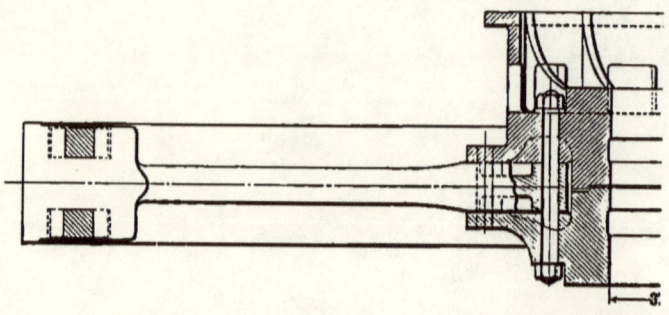

FIG. 1
Proposed 28-foot fly-wheel for General Electric Co. Third Aven

The Edward P. Allis Co., Milwaukee, Wis.

CHAPTER XI.

CYLINDERS, TUBING, PIPES, AND PIPE-COUPLINGS.

131. In the following discussion of cylinders, tubing, and pipes, they are separated into two classes, according to whether the walls are thin or thick, on account of the difference in the nature of the stress occurring in thin and thick walls. It is assumed that all the cross-sections discussed are circular.

The notation for cylinders, tubing, and pipes is as follows:

D = external diameter, inches;
F'' = cross-sectional area of wall of cylinder, square inches;
T = total circumferential tension per inch of length in the wall of the cylinder, pounds;
T'' = total longitudinal tension in the wall of the cylinder, pounds;
a = thickness of wall, inches;
d = internal diameter, inches;
p = internal pressure, pounds per square inch;
t = circumferential tension, pounds per square inch;
t' = longitudinal tension, pounds per square inch;
π = 3.1416.

132. Tension in a thin circular cylinder due to internal pressure.—When a pipe or tube is closed at the end and contains a liquid or gas under pressure, there is a tendency to burst the pipe which produces tensile stress in its walls.

If the pipe is long and the walls thin in comparison with its diameter, the relation between the internal bursting-pressure and the total circumferential tension in an inch of length, taken at some distance from the end so that there will be no strengthening effect of the cap or head, is expressed by the equation

$$T = \tfrac{1}{2}pd. \quad \ldots \ldots \ldots \quad (160)$$

This expression is obtained in the same manner as that for the tension in a shrunk-on ring (see § 90 and Fig. 118).

The circumferential tension per square inch in the walls when they are thin in comparison with the diameter is

$$t = \frac{T'}{a} = \frac{1}{2}\frac{pd}{a} \text{ (approximately).} \quad . \quad (161)$$

The total longitudinal force tending to separate the ends of the pipe is

$$T' = \frac{p\pi d^2}{4}. \quad . \quad . \quad . \quad . \quad . \quad (162)$$

The wall area resisting this force is that of the cross-section cut by a plane normal to the length of the pipe, and is

$$F' = \pi\frac{D^2 - d^2}{4}. \quad . \quad . \quad . \quad . \quad (163)$$

The walls being very thin in comparison with the diameter of the cylinder, their cross-sectional area is

$$F'' = \pi da \text{ (approximately)} \quad . \quad . \quad (164)$$

The longitudinal tensile stress per square inch, as obtained by combining equations (162) and (163), is

$$t' = \frac{pd^2}{D^2 - d^2}. \quad . \quad . \quad . \quad . \quad (165)$$

And the approximate longitudinal stress per square inch, as obtained by combining equations (162) and (164), is

$$t' = \frac{1}{4}\frac{pd}{a} \text{ (approximately)} \quad . \quad . \quad (166)$$

Equations (161) and (166) show that for pipes with thin walls the circumferential tensile stress per unit area is twice as great as the longitudinal. In other words, the tendency to burst the pipe longitudinally is twice as great as circumferentially.

The cylinder-head, flange, or cap upon the end of the cylinder or pipe aids the material near the end of the pipe in resisting the circumferential tensile stress due to internal pressure. If, therefore, the cylinder is short in comparison with its diameter, its capacity to resist pressure will be greater than that of a long one.

Example.—The circumferential tensile stress in a pipe 7 inches inside diameter and $7\frac{5}{8}$ inches outside, when withstanding an internal pressure of 2500 lbs. per square inch, is, for thin walls, by equation (161),

$$t = \tfrac{1}{2}(2500 \times 7) \div \tfrac{5}{16} = 2735 \text{ lbs. per square inch.}$$

The total end pressure on the pipe may be obtained by equation (162), whence

$$T' = \frac{2500\pi(7)^2}{4} = 96212 \text{ lbs.}$$

And the longitudinal tension per unit area is, by equation (166),

$$t' = \tfrac{1}{4}(2500 \times 7) \div \tfrac{5}{16} = 1367 \text{ lbs. per square inch.}$$

133. Cylinder with thick walls. Stress in material due to internal pressure.—When the walls of the cylinder are thick in comparison with its diameter, the circumferential tensile stress in them due to the internal pressure is not uniformly distributed throughout the material, but, on account of the elasticity of the walls, is greatest at the inner layer of metal and least at the outer, gradually diminishing from the inside toward the outside.

Equations (167) and (168) agree with those given in several books on the mechanics of materials, etc. They are based upon the assumption that the volume of the material forming the cylinder does not change when put under stress (i.e., the material is assumed to be incompressible). Under this assumption the circumferential stress in any fibre of the material is proportional to the square of its distance from the axis of the cylinder.

The equations for cylinders with thick walls are:

$$t = \frac{p(d + 2a)}{2a}; \quad \ldots \ldots (167)$$

$$a = \frac{1}{2}\frac{pd}{t - p}. \quad \ldots \ldots (168)$$

The influence of flanges and closed ends is doubtless greater in thick cylinders than in thin ones. For such short cylinders as are commonly used for hydraulic presses, etc., the strengthening effect of the closed end is very great.

134. Bursting tests of cylinders and pipes. Lap-welded wrought-iron pipe.*—The results of a series of tests made by R. W. Hildreth & Co. of New York upon a large number of pieces of lap-welded wrought-iron pipe illustrate so well the manner in which piping and couplings may give way under pressure that a very brief abstract of the log of the tests will be given.

The pipes tested were all 7 inches inside diameter and $7\frac{5}{8}$ inches outside. The sections were held together by cast-iron flanges of the form shown in Fig. 137. Three 1-inch bolts were used in each pair of flanges. The pipes and flanges were threaded with 8 threads per inch, having a taper of $\frac{1}{16}$ of an inch in $2\frac{1}{2}$ inches. When screwed together, the pipe protruded beyond the flange $\frac{1}{16}$ of an inch.

The gaskets were of gutta-percha 8.7 inches external diameter, so as to fit into the recess of the flange.

In the first series of experiments six lengths of pipe, each about 20 feet long, selected at random, were coupled together and tested under high pressure. The longitudinal tension due to the water-pressure against the heads of the pipe was resisted by the pipe and flanges, no auxiliary devices being used to resist this end thrust. One flange broke at a pressure of 1700 pounds per square inch, and after this was repaired another broke at an internal pressure of 2000 pounds per square inch.

---

* *Engineering News*, March 21, 1895.

In the second series of tests the ends of the pipes were tied together from end to end with two 3-inch rods having sleeve-nuts for tightening; the pipe was also secured against buckling. At the

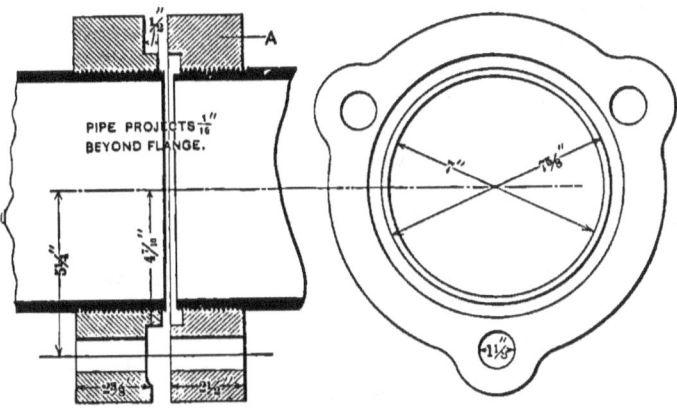

Fig. 137.

first trial one pipe burst six feet from the end at an internal pressure of 2400 pounds per square inch. The fracture showed a blister about 12 inches long. At the second trial a flange cracked at 800 pounds pressure. It was supposed that the bolts had not been tightened properly. At the third trial another flange broke with violence at 2300 pounds pressure. And at the fourth trial still another flange broke at 2500 pounds pressure per square inch.

All the flanges that broke were of the form A, Fig. 137. Each broke about midway between the lugs through which the bolts passed.

As a result of the tests it was recommended that a round flange with five or six $\frac{3}{4}$-inch bolts be specified instead of that with three lugs and bolts. It was stated that "with such flanges perfect pipe of the same quality and dimensions as those ordered should withstand a pressure of from 2500 to 3500 pounds per square inch."

Tests of drawn tubing. — A number of so-called "seamless" steel tubes were tested at Sibley College.* The tubes were made by welding or brazing the edges of steel skelp after the manner of making seamed tubing, and then drawing down in pipe-drawing dies, as is common for seamless-pipe manufacture. During the tests on the finished tubes no indication of a seam could be found, and in no case was there rupture along a seam.

The only tube which is recorded as having burst was 10 inches long, 1.254 inches external diameter and 1.202 inches internal, the thickness corresponding to a gauge of between 22 and 23. The ends were closed by clamping the tube endwise between two metallic plates, one of which was perforated and connected to the pressure-pump by tubing. The tube burst at an internal pressure of 4700 pounds per square inch, corresponding to a circumferential tensile stress of 108642 pounds per square inch. It gave way along one side about 3 inches from one end. The external diameter was increased about $\frac{1}{74}$ of an inch by the internal pressure.

Other similar tests were made upon pieces of tubing 30 inches long, but in each case the tube deflected so that the pressure could not be carried to the bursting-point. These tests showed that a tearing-stress of 80000 pounds per square inch could be applied without rupturing the material. It was also shown that a tearing-stress of 70000 pounds per square inch did not sensibly increase the diameter of the tube.

Cast-iron cylinders.—A series of experiments upon short cast-iron cylinders to determine their bursting-strength was conducted by Professor C. H. Benjamin.†

The form of the cylinders tested is shown in Fig. 138, and the proportions are given in Table XLIV. The lengths are approximately twice the diameters. The flanges and heads were made extra heavy in order that rupture might occur in the cylindrical shell. Each cylinder was examined for flaws, and if any small blowholes were found they were filled with lead or tin, hammered in, and the surface was then covered with a coating of paraffin.

The gaskets gave much difficulty on account of leaking and

---

* Trans. Amer. Soc. Mech. Eng., vol. XIX., 1898. Endurance, bending, crushing, and other tests were also made.

† Trans. Amer. Soc. Mech. Eng., vol. XIX., 1898.

TABLE XLIV.

DIMENSIONS OF CAST-IRON CYLINDERS TESTED FOR BURSTING-STRENGTH.

(Refers to Fig. 138.)

No.	A	B	C	D	E	Depth Counter-bore.	G	H	I	K	No. of Bolts in Each Head.
a	12.16	26.05	.70	16.25	......	....	1.07	1.12	1.0	....	24
b	9.16	17.95	.60	13.06	......	...	1.09	.70	1.0	....	16
c	6.09	12.19	.50	10.05	......	....	1.12	.70	1.0	....	8
d	12.45	26.5	.56	16.21	13.25	.12	1.75	1.35	1.5	....	24
e	9.12	19.0	.61	12.96	10.08	.11	1.5	1.25	1.25	....	16
f	6.12	13.0	.65	10.02	7.08	.11	1.25	1.00	1.25	....	8
1	9.58	18 7/8	.402	13.33	10.83	1/8	1 3/16	1 1/8	1 3/8	11 3/4	16
2	9.375	18 7/8	.573	13.13	10.63	1/8	1 3/16	1 1/8	1 3/8	11 3/4	16
3	9.13	18 7/8	.596	12.88	10.88	1/8	1 3/16	1 1/8	1 3/4	11 3/4	16
4	12.53	25 3/4	.571	16.4	13.34	1/8	1.34	1 1/2	15/16	14 3/4	24
5	12.56	25 5/8	.531	16.56	13 56	1/8	1.34	1 1/16	15/16	14 3/4	24
6	12.16	25 5/8	.93	16.22	13.41	1/8	1.18	1 1/8	15/16	14 3/4	24

NOTE.—The rough dimensions in this table are averages from a number of measurements. Cylinders bored to make walls of approximately uniform thickness, and the ends faced.

Cylinders *a*, *b*, and *c* made of ordinary foundry-iron showing a tensile strength of about 18000 pounds per square inch in the test-piece.

Cylinders *d* to 6 made of a special foundry mixture which showed a tensile strength of about 24000 pounds per square inch in the test-piece.

blowing out when used upon the flat-faced end of the cylinder. By counterboring, as shown in the figure, and packing with a

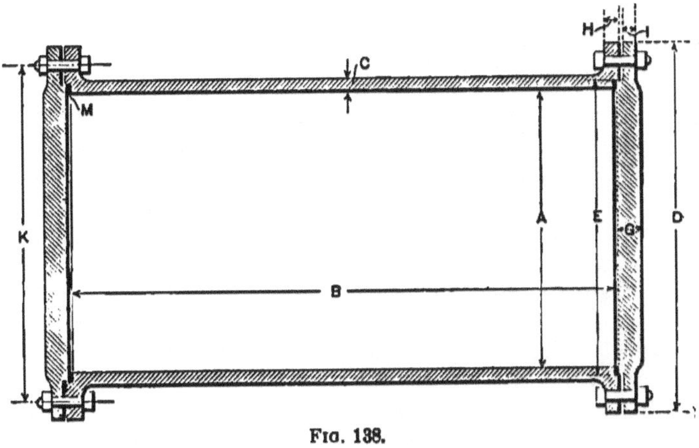

FIG. 138.

gasket of strawboard saturated with oil, this difficulty was largely obviated. This packing held except when the projection on the head was smaller in diameter than the counterbore in the cylinder.

One of the serious difficulties met during the test was leakage through the cylinder walls on account of minute blowholes. Some of them were almost invisible to the naked eye, but at high pressure the water would spurt out several feet in a slender stream. Water oozed through the walls at all points.

The log of the tests is as follows:*

"*Cylinder a.*—Wire-gauze packing; leaked at 400 pounds. Substituted copper wire No. 22, A. W. G.; this leaked at 600 pounds. Substituted soft-rubber gasket; pressure carried to 800 pounds several times. Leak at blowhole stopped by peening. On raising pressure to 775 pounds cylinder failed on a circumference just below the upper flange, the crack starting at blowhole and running each way about 90 degrees.

"*Cylinder b.*—Gasket of lead fuse wire $\frac{3}{32}$-inch diameter with ends fused together. Leakage at pressure of 450 pounds, and the flange cracked. Substituted rubber and graphite packing; leak at crack with pressure of 600 pounds; no further rupture.

"*Cylinder c.*—Rubber and graphite packing inserted, heated to 250 degrees Fahr. by live steam; bolts screwed down and packing left one day to harden. Leaked badly at 600 pounds; renewed packing, but it leaked again at 550 pounds. Flanges showed signs of failure and experiment was abandoned.

"*Cylinder d.* — Counterbored joint, with gasket of strawboard soaked in linseed-oil. Leakage at blowholes with 700 pounds pressure. Blowholes peened and coated with paraffin, when pressure was raised to 800 pounds several times. One blowhole calked on outside; on applying pressure of 700 pounds rupture occurred on longitudinal line through blowhole. Several small, blowholes found in line of fracture.

"*Cylinder e.* — (On this and all subsequent cylinders the counterbore and straw-board gasket were used.) Pressure raised gradually to 1325 pounds, when rupture occurred on circumference under flange. The crack began at several small blowholes.

---

* Illustrations and references to them are omitted.

"*Cylinder f.*—Pressure raised gradually to about 2500 pounds (above graduation of gauge), when cylinder failed in same manner as preceding one; cylinder leaking badly at time of rupture.

"*Cylinder No. 1.*—Broke at 600 pounds on a longitudinal line along a row of blowholes.

"*Cylinder No. 2.*—Broke at 1050 pounds around a circumference just under flange. Fracture very clean.

"*Cylinder No. 3.*—Broke at 975 pounds in the same manner as No. 2, the crack beginning where there was a slight flaw. Fracture clean.

"*Cylinder No. 4.*—A number of small blowholes near the centre of shell caused considerable trouble by leakage, and had to be calked inside and out. Rupture finally occurred at 700 pounds pressure along a longitudinal line.

"*Cylinder No. 5.*—Rupture occurred at 875 pounds, a crack starting under the flange running part way around and then up through flange and head.

"*Cylinder No. 6.*—At 475 pounds pressure the bottom head broke. On renewing this and raising pressure to 900 pounds the top head failed in the same manner. These heads had been used for several cylinders, and were probably weakened. The test was abandoned at this point for lack of time.

Great pains were taken in casting these cylinders, and they may be considered good examples of cast-iron cylinders as made for engine- or pump-work. The blowholes mentioned were most of them very minute, and under ordinary circumstances would have remained unnoticed."

The percolation of liquids through the walls of cast-iron cylinders at high pressure is of common occurrence. Such cylinders may be lined with copper, brass, etc., to prevent this leakage.

135. Special forms of pipes.—The high steam-pressures common to modern practice, as well as the high pressures used for hydraulic and other purposes, have necessitated the use of piping made of a stronger material than cast iron. For such purposes pipe made of steel plates riveted together, as is ordinarily done in the construction of shell boilers, has been adopted to a considerable extent. The bursting-strength of such a pipe is measured by the strength of the riveted joints.

"Spiral-riveted" pipe is used for nearly all the sizes of pipes common to practice, and especially for those which are thin in comparison with their diameters. It is made by wrapping a long strip of flat material, as a thin strip of paper would be wound around a lead-pencil, so that the edges overlap. In the pipe the overlap is made sufficient for a row of rivets used to fasten the successive convolutions together. The edges of the plate must be prepared before wrapping, by thinning or offsetting them, so that the diameter of the pipe will not increase with each turn of the strip from which it is made. On account of the necessity of such offsetting it is difficult to make this kind of pipe of any great thickness in comparison with its diameter. The helical joint is sometimes welded instead of riveted.

136. Pipe-couplings and -flanges.*—The most essential features of a pipe-coupling are that it shall be strong enough to prevent fracture when under pressure, and that it shall not leak either between the faces of the two flanges or between the pipe and flange.

The stresses which a flange must withstand are those due to internal pressure in the pipe, together with those caused by bolting the flanges together, expansion and contraction of the pipe, and its tendency to bend under its weight or to buckle on account of the elongation of the pipe by expansion when heated by steam, etc. Steel flanges are used for the high steam-pressures of modern practice, as well as for pipes for high hydraulic pressures.

The packing must be so held in position that it will not blow out under pressure. If it is held between two perfectly plane flange-faces, the friction between it and the flanges, together with its own strength, is all that holds it in position. The end thrust due to internal pressure has a tendency to separate the flanges, thus reducing their clamping force on the packing. This is apt to cause leakage and possibly to tear the packing to pieces and blow it out. This was plainly shown in the experiments on cast-iron cylinders cited in § 134. By recessing or counterboring one flange and forming its mate to fit into the counterbore, the packing, if of the

---

* The dimensions of the standard flanges adopted by the Amer. Soc. Mech. Engrs. are given in Kent's "Mechanical Engineers' Pocket-book," together with numerous other designs of flanges.

same external diameter as the counterbore and fitted into it, will be held in place much more effectively.

Fig. 139 represents a method of attaching pipe-flanges that is probably the best for high pressures that has ever been used. The

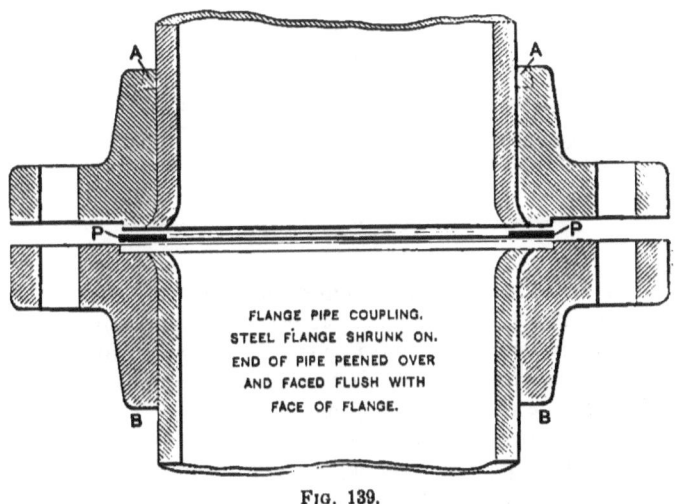

FIG. 139.

proportions shown in the figure are those for rolled-steel flanges. In order to attach the flange it is expanded by heat and then shrunk on the pipe so that the end of the latter protrudes beyond the face of the flange. The pipe is then peened or spun out so as to expand it over the rounded corner of the flange, and is then turned off so that it is just flush with the face of the flange. The packing P fits in the counterbore and covers the joint between the flange and pipe, thus serving the double purpose of preventing leakage between the flange-face and also between the pipe and bore of the flange. As an extra precaution a dovetail counterbore is sometimes made as shown by the dotted lines at A, into which a piece of some malleable material, such as copper, can be calked to prevent leakage between the pipe and flange.

For exceedingly high pressures and correspondingly thick pipe, the pipe and flange are both threaded and screwed together, the rest of the details remaining the same.

Fig. 140 and Table XLV show the form and proportions of

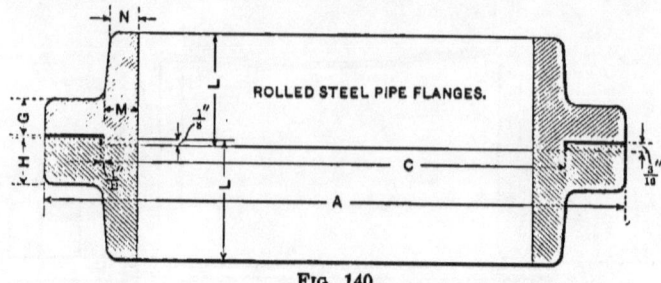

FIG. 140.

weldless rolled-steel flanges for pipes from 8 to 72 inches outside diameter and for pressures up to 150 pounds per square inch.

A pipe is sometimes riveted to the flange by a circumferential row of rivets extending radially through a lip running back over the pipe on the side of the flange opposite its face. This lip corresponds to the ones extending out toward A and B in Fig. 139. In such a joint the pipe extends only partly through the flange and is calked at the end to prevent leakage.

Flanges forged or cast as an integral part of the pipe are used to some extent, chiefly for thick pipes and high pressures.

137. Expansion-couplings for pipes.—In order to allow for the change of length which always occurs in pipes with change of temperature, some form of expansion-coupling must be introduced at intervals in order to prevent excessive stresses in the material. Numerous devices have been adopted for this purpose. Those most used may be divided into two general classes, however: 1st. Those which operate by telescopic action, generally having an accurately turned tube which slides back and forth through a stuffing-box packed in a manner suitable to allow such motion after the method common for piston-rod glands of steam-engines. 2d. Those in which the elasticity of one of the parts, called the expansion-piece, is depended upon to allow a slight end motion of the pipes. One form of this coupling consists of a short length of pipe made of

TABLE XLV.

DIMENSIONS OF WELDLESS ROLLED-STEEL FLANGES FOR PIPE CARRYING PRESSURE UP TO 150 POUNDS PER SQUARE INCH.

(Dimensions in Inches.) (Refers to Fig. 140.)

Diameter, inside.	8	9	10	11	12	13	14	15	16	17	18
A	14½	15½	16¾	17½	18½	20¼	21¼	22½	23½	24½	25¾
C	9½	10½	11½	12½	14	15	16	17	18	19	20
G	1 3/16	1¼	1 5/16	1⅜	1 7/16	1½	1 9/16	1⅝	1 11/16	1¾	1 13/16
H	1⅜	1 7/16	1½	1 9/16	1⅝	1 11/16	1¾	1 13/16	1⅞	1 15/16	2
L	3½	3⅝	3¾	3⅞	4	4⅛	4⅜	4½	4¾	4⅞	5
M	1	1	1⅛	1⅛	1 3/16	1 3/16	1¼	1¼	1 5/16	1 5/16	1⅜
N	⅝	⅝	¾	¾	13/16	13/16	⅞	⅞	15/16	15/16	1
Approx. Rough Wt.	67	77	91	97	113	186	154	172	192	209	235
Approx. Fin. Wt. Male	54	62	76	81	95	115	131	149	166	180	204
Finished Wt. Female	57	65	79	85	100	120	137	156	174	188	213

Diameter, inside.	19	20	24	30	36	42	48	54	60	66	72
A	27¼	28¼	32¾	39	45½	51½	57¾	64¼	70¾	76¾	83½
C	21	22½	26½	32½	39	45	51	57½	63½	69½	76
G	1⅞	1 15/16	2 3/16	2¼	2¼	2¼	2¼	2¼	2¼	2¼	2¼
H	2 1/16	2⅛	2⅜	2 7/16	2 7/16	2 7/16	2 7/16	2 7/16	2 7/16	2 7/16	2 7/16
L	5⅛	5⅜	6	6	6¼	6¼	6½	6½	6¾	7½	7½
M	1⅜	1 7/16	1⅝	1⅝	1 13/16	1 13/16	1¾	1¾	1 13/16	1 13/16	1⅞
N	1	1 1/16	1¼	1¼	1 5/16	1 5/16	1⅜	1⅜	1 7/16	1 7/16	1½
Approx. Rough Wt.	265	295	416	525	665	706	913	1053	1245	1444	1661
Approx. Fin. Wt. Male	232	257	371	466	595	686	820	949	1120	1301	1501
Finished Wt. Female	242	268	387	482	612	705	842	972	1148	1332	1534

rather thin copper, corrugated circumferentially. The corrugations allow the piece to change its length endwise under moderate end pressure. Another form consists of a pair of flanges, one much larger in diameter than the other, fastened together in a manner somewhat similar to that shown for the shaft-coupling in Fig. 121. Instead of having a plane disk, however, as shown in this figure, a corrugated copper disk is generally used, the corrugations running circumferentially around the disk. The corrugations allow a comparatively free end motion of the parts relatively to each other.

CHAPTER XII.

RIVETED JOINTS.

138. It is common practice in engineering to fasten together the edges of plates, angles, beams, etc., with rows of rivets. If the plates riveted together are to be used for the retaining-walls of a vessel to contain a fluid under pressure, the joint must be designed for tightness as well as strength. This means that the rivets must be put close enough together to prevent the plates from springing apart between the rivets to any considerable extent. In addition to placing the rivets near together it is generally necessary to calk the joints with a calking-tool, whose typical form is that shown in Fig. 141. This tool resembles a cold-chisel whose chipping edge has been ground off square. It is held in the posi-

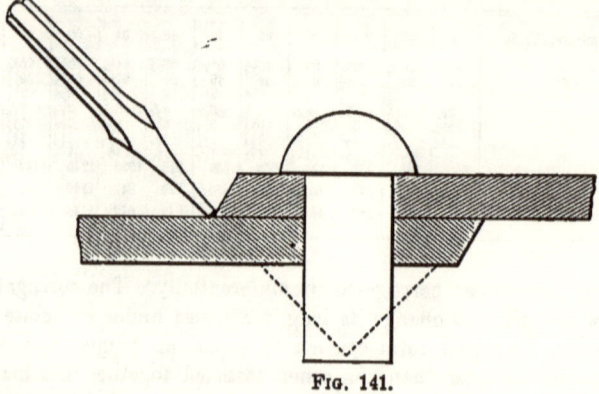

FIG. 141.

tion indicated in the figure and struck with a hammer to calk the edges of one plate down against the other.

If the joint is for vessels which are to hold solid material, or for the members of bridges and buildings, it is necessary to design it for strength only, there being no necessity for tightness.

In hand-riveting the rivet is heated to a full red heat and then passed through holes previously punched or drilled in the plates and brought into proper position to receive the rivet. The rivet is then held in place with a "dolly-bar" while the opposite end is first upset with hammers and then finished with a "set" or "snap." The "set" resembles a sledge-hammer with a depression cut in its face to conform with the kind of head that is desired, as button-head, cone-head, pan-head, etc. It is held against the rivet and struck with a sledge. Very small rivets are often put in place cold.

Machine-riveting is used much more in modern practice than hand-riveting. A machine-riveter has two dies, cupped to the form the rivet-head is to have, which press on the opposite ends of the rivet-blank after it has been put in place. The body of the rivet is swelled out in the hole and the head formed by the heavy pressure that is exerted. The dies of riveting-machines are commonly operated by either steam, air, or hydraulic pressure. In hydraulic riveting one method is to press the rivet down so as to form the heads while it is still very hot, then relieve it from the pressure of the die for a short time until it has cooled somewhat, and then put it under pressure again until it has become cold enough to prevent its stretching materially under the tendency which the plates may have to spring apart. In steam and pneumatic riveting the rivet-head is first formed by a steady pressure of the die, then allowed to cool somewhat, after which it is struck a few sharp blows with the die driven out by the action of the steam or air against it in order to bring the plates together so that they will be gripped tightly when the rivet has cooled completely. For very heavy riveting the riveter sometimes has a pair of closing- or gripping-dies, called a "plate-closer," which are used for pressing the plates together and holding them until the rivet has cooled.

If the work is well done, the rivets should fill the holes completely. Since the rivet blank is always smaller in diameter than the hole it is to fill, this necessitates its being upset from end to end so that it may swell to the size of the hole. It has been found that this can be accomplished better by having the blank hottest at the head end, so that it will swell first under the head, gradually filling the hole from the head toward the point, and finally forming the second head.

When the holes are made by punching, they are larger at one end than the other, being roughly conical on account of the punch being smaller than the die on which the plate rests during the operation of punching. The plates should be placed so that the small ends of the holes come together. When so placed, the end contraction of the rivet tends to draw it more tightly against the conical walls, thus eliminating, in a measure at least, the loosening effect of the decrease in diameter due to cooling, and giving the rivet a better fit in the hole; the rivet also grips the sides of the holes and draws the plates together, thus relieving the heads of a portion of the strain. If, on the contrary, the large ends of the holes are placed together, the swelling of the rivet when under the dies tends to force the plates apart, and its contraction to loosen it in the hole; also the end contraction is resisted by the heads only.

Numerous styles of riveted joints are in common use. Some of the simpler ones will be shown in order to explain the nature of the stresses that act upon the plate and rivets. The seams most com-

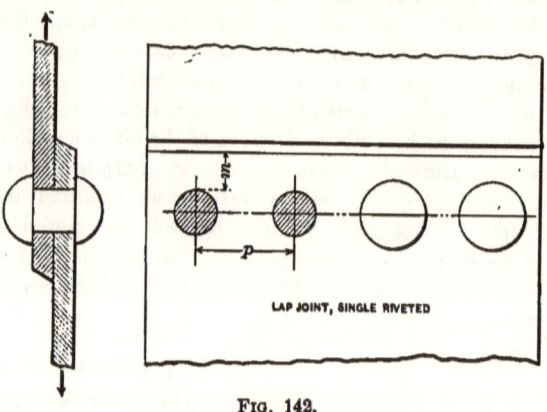

LAP JOINT, SINGLE RIVETED

FIG. 142.

monly used are of two general classes, namely, lap-joint and butt-joint. The names are derived from the manner in which the edges of the plates are placed relatively to each other. In the lap-joint the plates overlap each other, examples of this form of seam being shown in Figs. 142, 143, and 144; in the butt-joint the edges of

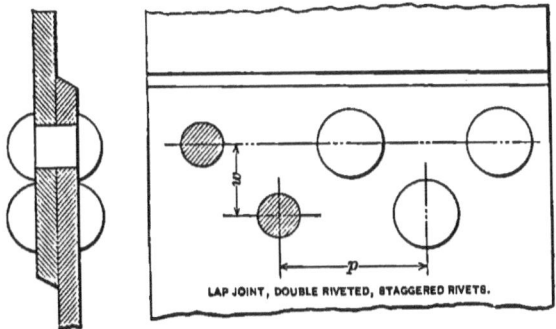

Fig. 143.

the plates are butted against each other, and one or two coverplates, straps, or welts placed over their junction, the rivets passing

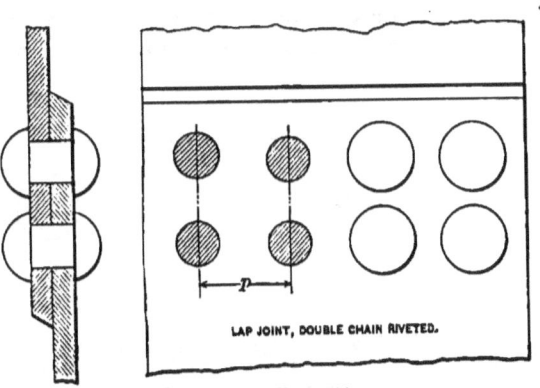

Fig. 144.

through one plate and one side of the strap or straps, as shown in Figs. 146 and 147. Fig. 145 is a lap-joint with cover-plate.

The seams are further classified according to the number of rows of rivets that are used, and the positions of the rivets of one row

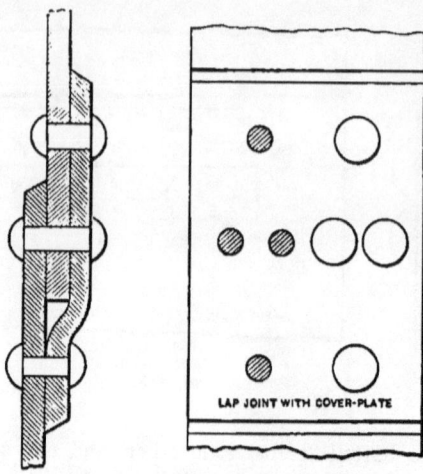

FIG. 145.

relatively to those of the other rows. The rows of rivets run parallel to the length of the seams. Single-riveted joints are shown in Figs.

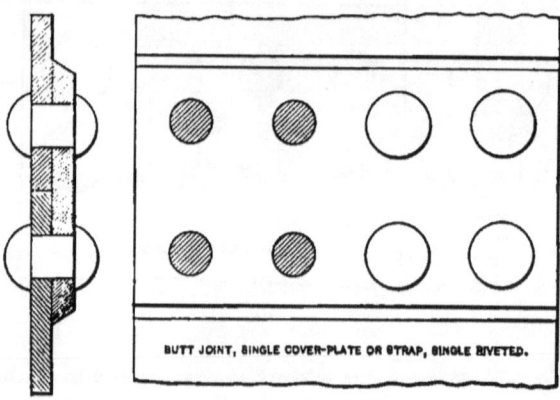

FIG. 146.

142, 146, and 147; in each of these joints the edges of the plates are pierced by but a single row of rivets, although two rows are

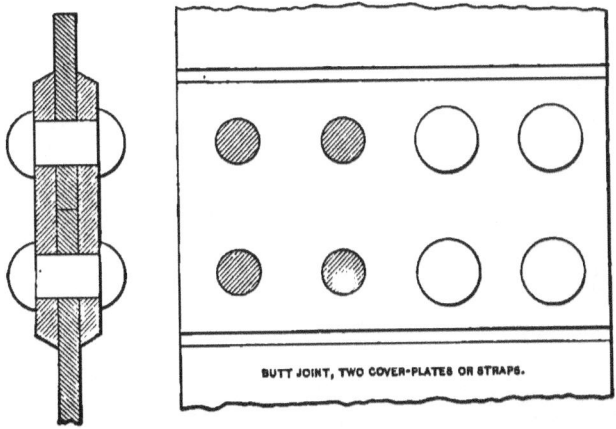

Fig. 147.

embodied in the seam in two of these three cases. Double riveting is shown in Figs. 143 and 144, the edges of the plates being pierced

Fig. 148.

with two rows of rivets. The rivets are said to be staggered when those of one row are opposite the spaces of the adjacent row, as in Figs. 143 and 145; when they are opposite each other, as in Figs. 144, 146, and 147, the seam is chain-riveted.

The **pitch** of rivets is the distance between the centres of adjacent rivets in the same row.

The **margin** is the distance from the edge of the plate to the edge of the rivet-hole. (See m, Fig. 142.)

The **overlap** is the distance one plate laps over the other.

All the rows of a seam do not need to have the same pitch, as can be seen by reference to Fig. 145, where the pitch of the outer rows is double that of the middle one.

PROPORTIONS OF RIVETED JOINTS.

139. A single-riveted joint may yield in one of the five following ways when forces are applied as indicated by the arrows in Fig. 142:

1st. Shearing the rivets in the plane of the plate surfaces that are in contact.

2d. Crushing the rivets, or the plate in front of them, by the "bearing-pressure" of the rivets against the plate.

3d. Tearing the plate between the rivets.

4th. Splitting or tearing the plate between the rivet and the edge of the plate.

5th. Shearing out the plate in front of the rivet, the piece sheared out having a width approximately equal to the diameter of the rivet.

Double-riveted joints may fail by either the first or third method, or by crushing the rivets. Any other manner of failure would be a combination of two or more of the five methods given above. The failure of joints with several rows of rivets may be still more complicated.

The frictional resistance to the slipping of the plates over each other, due to their being clamped together by the end contraction of the rivets, is, in a carefully made hydraulic machine-riveted joint, generally from one third to one half of the total strength of the joint. A hand-riveted seam generally offers less frictional resistance to the slipping of the plates over each other than one which is machine-riveted. In designing riveted seams it is not customary to take this frictional resistance into consideration.

The notation for riveted joints is as follows:

A = sectional area of rivet on a plane perpendicular to its axis, square inches. (It is assumed that the rivet fills the hole, and therefore its diameter is equal to that of the hole where the plates touch each other);

c = ultimate bearing- or crushing-strength of plate or rivet for single shear, pounds per square inch;
c' = ultimate crushing-strength of plate or rivets for double shear, pounds per square inch;
d = diameter of rivet-hole, inches;
f = ultimate tensile strength of plate between rivet-holes, pounds per square inch;
n = number of rivets passing through one plate along a length of edge equal to p;
p = pitch of rivets, inches;
s = ultimate shearing-strength of rivet for single shear, pounds per square inch;
s' = ultimate shearing-strength of rivet for double shear, pounds per square inch;
t = thickness of plate, inches;
w = distance between rows of rivets. (See Fig. 143.)

A rivet is in single shear when the shearing force acts only in one plane, as in a lap-joint, or a butt-joint with one cover-plate. It is in double shear when the tendency is to shear it off in two planes, as in the double-welt butt-joint, Fig. 147.

Table XLVI gives ultimate values that can be safely used in practice with good material. A suitable factor of safety must be introduced. The common practice in this country is to use steel plates with iron rivets for boiler construction. Steel plates with mild steel rivets are used to a considerable extent for other classes of work.

140. Rivets.—Since a rivet may fail either by crushing or shearing, it is first necessary, for a given thickness of plate, to find the diameter of rivet having equal crushing and shearing strengths by equating the strengths as follows:

For single shear,

$$cdt = \frac{\pi d^2}{4} s;$$

whence

$$d = \frac{ct}{.7854s} = 1.27\frac{ct}{s}. \quad \ldots \ldots \quad (169)$$

TABLE XLVI.

ULTIMATE STRENGTH OF RIVETS AND PLATES.

	Pounds per Square Inch.	
	Iron.	Steel.
Rivets, shearing-strength in single shear, s	40,000	50,000
Rivets, shearing-strength in double shear, s'	35,500	45,000
Bearing-pressure in single shear,* c	65,000	90,000
Bearing-pressure in double shear, c'	80,000	110,000
Plates, tensile strength between rivet-holes, t	40,000	55,000

* Bearing area taken as the projected area of the bearing surface.

For double shear,

$$c'dt = \frac{2\pi d^2}{4} s';$$

whence

$$d = \frac{c't}{1.5708 s'} = .635 \frac{c't}{s'} \quad . \quad . \quad . \quad . \quad (170)$$

Substituting in equation (169) the values given in Table XLVI gives for single shear:

Iron rivets,

$$d = 1.27 \tfrac{40000}{40000} t = 2.06 t. \quad . \quad . \quad . \quad (171)$$

Steel rivets,

$$d = 1.27 \tfrac{50000}{50000} t = 2.28 t. \quad . \quad . \quad . \quad (172)$$

Substituting in equation (170) gives for double shear:

Iron rivets,

$$d = .635 \tfrac{80000}{35500} t = 1.43 t. \quad . \quad . \quad . \quad (173)$$

Steel rivets,

$$d = .635 \tfrac{110000}{45000} t = 1.55 t. \quad . \quad . \quad . \quad (174)$$

The diameters given by equation (171) hold good in practice for plates from $\frac{3}{16}$ to $\frac{3}{8}$ of an inch thick, but beyond this point the diameter of the hole becomes smaller in proportion to the plate thickness. The difficulty of driving the larger sizes of rivets is largely accountable for this fact. Since the sectional area of a rivet varies as the square of its diameter, it is clear that any rivet smaller than the size giving equality of bearing and shearing strengths will fail by shearing; therefore in designing a joint whose rivet is not larger than the size having equal bearing and shearing strengths, it is not necessary to use the bearing-strength of the rivet and plate, the shearing-strength of the rivet being all that needs consideration; but if the rivet is larger than the size for equality of these two values the bearing-strength must be used.

Table XLVII gives average diameters of holes for plates as commonly used in practice.

Table XLVII.

DIAMETERS OF RIVET-HOLES COMMONLY FOUND IN PRACTICE FOR LAP-JOINTS AND SINGLE-STRAP BUTT-JOINTS.

Thickness of plate, inches	3/16	¼	5/16	⅜	½	⅝	¾	⅞	1	1⅛
Diameter of rivet, inches	⅜	½	⅝	¾	⅞	⅞	1	1	1⅛	1¼

The form and proportions of some of the rivet-heads most commonly used are shown in Figs. 149, 150, 151, and Table XLVIII.

141. Pitch of rivets.—For a given thickness of plate and diameter of rivet, the pitch required for equal strength of rivets and plates may be found by the following equations, in which it is assumed that the margin is large enough to prevent the rivets from breaking out toward the edge of the plate, and that rupture must occur either by shearing the rivets, tearing the plate along the line of rivet-holes, or crushing the rivets or plate.

When the diameter of the rivet equals or is smaller than the value for equality of shearing- and crushing-strength, then—

For single-riveted joints having the rivets in single shear,

$$\frac{\pi d^2}{4} s = (p - d) t f;$$

whence

$$p = \frac{\pi d^2}{4}\frac{s}{tf} + d = A\frac{s}{tf} + d. \quad . \quad . \quad . \quad (175)$$

TABLE XLVIII.

DIMENSIONS OF COUNTERSUNK RIVET-HEADS.*

(See Fig. 151.)

Dimensions in Inches.

Diam. of rivet, D	1/4	5/16	3/8	7/16	1/2	9/16	5/8	11/16	3/4	13/16	7/8	1	1 1/8	1 1/4
Diam. of head, A	1/2	19/32	11/16	25/32	7/8	31/32	1 1/16	1 5/32	1 1/4	1 11/32	1 7/16	1 5/8	1 13/16	2

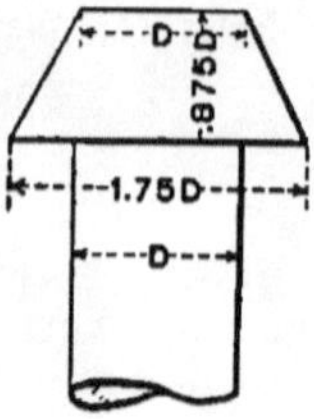

FIG. 149.

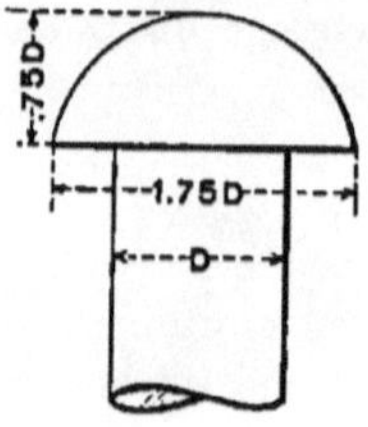

FIG. 150.

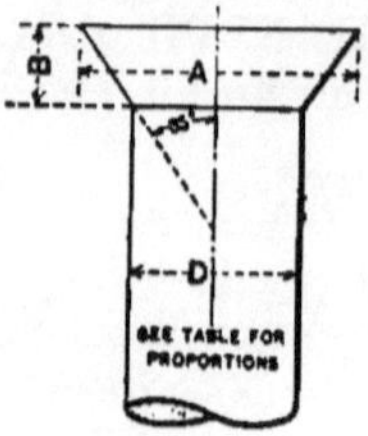

FIG. 151.

For double shear,

$$\frac{2\pi d^2}{4}s' = (p - d)tf;$$

whence

$$p = \frac{\pi d^2}{2}\frac{s'}{tf} + d = 2A\frac{s'}{tf} + d. \quad . \quad . \quad . \quad (176)$$

And in double-riveted joints, both rows of rivets being of the same pitch:

For single shear,

$$\frac{2\pi d^2}{4}s = (p - d)tf;$$

* Taken from drawings kindly furnished by J. H. Sternbergh & Co.

whence
$$p = \frac{\pi d^2}{2}\frac{s}{tf} + d = 2A\frac{s}{tf} + d. \quad . \quad . \quad . \quad (177)$$

For double shear,
$$\frac{4\pi d^2}{4}s' = (p - d)tf;$$

whence
$$p = \pi d^2\frac{s'}{tf} + d = 4A\frac{s'}{tf} + d. \quad . \quad . \quad . \quad (178)$$

If the rivet is larger than the theoretical value for equality of shearing and crushing strengths, failure will occur by crushing. In accordance with this—

For single-riveted joints having the rivets in single shear:
$$cdt = (p - d)tf;$$
whence
$$p = \frac{c + f}{f}d. \quad . \quad . \quad . \quad . \quad . \quad (179)$$

For double shear,
$$p = \frac{c' + f}{f}d. \quad . \quad . \quad . \quad . \quad . \quad (180)$$

For double-riveted joints whose rivets have the same pitch in both rows and are in single shear:
$$2cdt = (p - d)tf;$$
whence
$$p = \frac{2c + f}{f}d. \quad . \quad . \quad . \quad . \quad . \quad (181)$$

For double shear,
$$p = \frac{2c' + f}{f}d. \quad . \quad . \quad . \quad . \quad . \quad (182)$$

In general, in any joint of ordinary form having two or more rows of rivets passing through the edge of one plate, the pitch of

298 FORM, STRENGTH, AND PROPORTIONS OF PARTS.

the rivets can be determined, when p is taken equal to the pitch of the row farthest from the edge of the plate, by equation (183) or (184) when the rivets are small enough to fail by shearing, or by (185) or (186) when large enough for the joint to fail at the bearing surfaces.

For single shear,

$$p = \frac{n\pi d^2}{4}\frac{s}{tf} + d = nA\frac{s}{tf} + d. \quad \ldots \quad (183)$$

For double shear,

$$p = \frac{2n\pi d^2}{4}\frac{s'}{tf} + d = 2nA\frac{s'}{tf} + d. \quad \ldots \quad (184)$$

For failure by crushing when the rivets are in single shear,

$$p = \frac{nc + f}{f}d. \quad \ldots \ldots \ldots \quad (185)$$

For double shear,

$$p = \frac{nc' + f}{f}d. \quad \ldots \ldots \ldots \quad (186)$$

142. The efficiency of a riveted joint is the ratio of its strength to that of the solid plate. The theoretical efficiency can readily be calculated if it is assumed that the portion of the plate between the holes is of the same strength per unit area as at other places. Thus, for a single-riveted lap-joint of $\frac{3}{8}$-inch steel plates and $\frac{3}{4}$-inch iron rivets, the pitch, by equation (175), is

$$p = .442\frac{40000}{.375 \times 55000} + .75 = 1.6 \text{ inches (about).}$$

Taking $p = 1\frac{11}{16}$ inches,

$$\text{Efficiency} = \frac{p - d}{p} = (1\frac{11}{16} - \frac{3}{4}) \div 1\frac{11}{16} = .55 +.$$

The actual efficiency of a joint may differ considerably from the theoretical on account of various influencing causes, the principal ones being the effect of punching upon the strength, hardness, and ductility of the plate, and the pressure of the plates against each other due to the contraction of the rivets. It is probable that, with a well-made joint, working with a factor of safety of four, no slipping of the plates occurs so as to bring the rivets hard against the holes on the bearing side.

143. Shearing, punching, and drilling plates preparatory to riveting.—The plate metal, coming from the rolling-mill in irregular shapes, must be sheared to form and size suitable for its purpose. After this the rivet-holes are made in it, there being three methods of making them, as follows: 1st. Punching to the required size with a power-punch by a single stroke of the punch for each hole; 2d. Punching a hole from $\frac{1}{16}$ to $\frac{1}{8}$ of an inch smaller than required, and then enlarging it to full size with a reamer or similar cutting tool; 3d. Drilling. Of these three methods punching is by far the quickest and cheapest; the punching and reaming process is considerably more expensive; and drilling is the most costly of all.

When a hole is punched in a plate of metal, it is found that the "wad" or "plug" forced out is much smaller in volume than the hole formed by its removal (unless the plate is hard and brittle, and hence unsuitable for riveting up to withstand pressure). It is about the same diameter as the hole, but sometimes not more than half as long as the thickness of the plate. Since the metal in the wad is found to be little, if any, more dense than the plate, it is evident that there must be a flow of the metal as the punch presses upon it, of such a nature that part of the metal, instead of being punched out of the hole with the wad, is forced into the walls of the hole. The thickening of the plate around the hole is proof of this.

This flow of metal into the walls of the hole, which is really a cold working of the material, changes its physical qualities to a depth varying from $\frac{1}{16}$ to $\frac{1}{8}$ of an inch, according to the size of the hole and the quality of the metal, the result being that a thin ring or sleeve, harder and less ductile than the original plate, is formed around the hole. As to the effect of this hardened sleeve upon the strength of the plate along the line of holes, and consequently upon the efficiency of the joint, there has been much discussion with regard to whether it increases, decreases, or has any effect upon these qualities. The numerous experiments made upon riveted joints seem to indicate, however, that punching is injurious to hard plates and those made from a poor quality of iron or steel, but that soft plates may actually be benefited by the same treatment. Whether any benefit is ever derived or not, it is certainly true that when the operation is injurious its deleterious effects are more marked the poorer the quality and the greater the hardness of the plates.

By reaming out the holes after punching, the effects of punching are nearly or quite removed, and the joint shows practically the same efficiency as when the holes are drilled. The effects of punching can also be partly, sometimes completely, removed by annealing the plate after punching; this process is rather inconvenient and expensive for large plates, however.

The effects of shearing the plate to approximate size are removed by planing off a strip of metal along the edge, which at the same time reduces it to size and makes a smooth edge. The sheared edge always presents a rough surface when it comes from the shears, which in itself would not be acceptable for good work on heavy plates, especially where tight joints are to be made.

144. Faulty construction and grooving of riveted joints.— When the holes in the two plates do not coincide, it is evident that the rivet will be hard to drive into place, and an offset will be formed on it where the plates press together when it is headed; and if the heads are directly opposite each other they cannot both be concentric with the body of the rivet. The whole result is that the rivet is improperly formed and is not as efficient as a perfect one. By careful work very good seams with punched holes can be made.

By punching the holes small, clamping the plates together in the relative positions they will occupy when riveted, and reaming through both plates at the same time a practically perfect rivet-hole can be obtained; the reaming can be done by hand or power.

Drilled holes can be more accurately located than punched ones, consequently better coincidence of the holes can be obtained when the plates are drilled separately than when punched; and drilling has the further advantage that the plates can first be clamped together and then the holes drilled perfectly concentric.

In machine-riveting it is necessary for the dies to be held rigidly so that no side motion can occur, for if it does the head will be formed eccentrically with the body and the rivet weakened in consequence. In practice this fault is sometimes so great that the edge of the head is tangent to the body of the rivet. Fortunately such great faults are not found in boiler-work turned out by makers of any reliability.

The stress upon a lap-riveted seam, causing a tendency of the plates to bend, as in Fig. 148, localizes the tension to some extent at A and B. The localization does not generally need serious consideration, however, unless the plates are grooved by calking or corrosion along the edges at A and B. If deeply grooved, the tendency to bend becomes greater, and in badly grooved plates rupture may occur along the groove at the edge of the seam.

The slight bending of the plates at a seam also tends to localize their pressure against the rivets. This localization is at or near the plane of contact of the plates. It has the effect of causing the rivets to shear more readily than if there were no bending of the plates.

When lap-joints are tested to destruction, the head of the rivet is sometimes pulled off on account of the bending of the plate, as indicated in Fig. 148.

145. Examples of riveted joints taken from practice.—Figs. 152 to 163 represent the practice of the Baldwin Locomotive Works with regard to riveted joints.*

Figs. 164 to 171 illustrate the riveted seams adopted by the Continental Iron Works for cylindrical-shell marine-boilers.†

---

* Taken from blue-prints kindly furnished by Burnham, Williams & Co., Baldwin Locomotive Works.

† Taken from printed designs kindly furnished by the Continental Iron Works.

302 FORM, STRENGTH, AND PROPORTIONS OF PARTS.

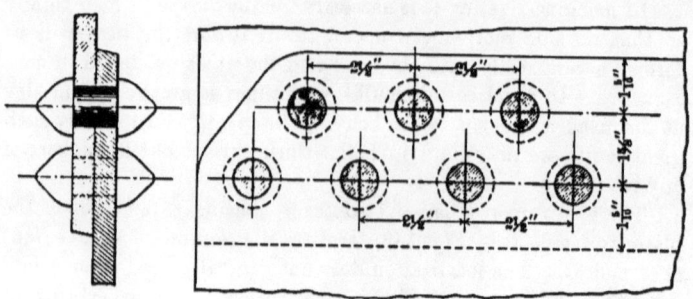

Fig. 152.
Double-riveted lap-joint.   1/2" plate, 13/16" rivets, 7/8" holes.
Baldwin Locomotive Works.

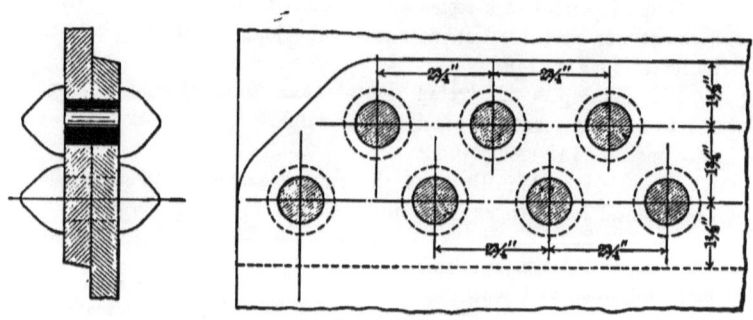

Fig. 153.
Double-riveted lap-joint.   5/8" plate, 15/16" rivets, 1" holes.
Baldwin Locomotive Works.

RIVETED JOINTS. 303

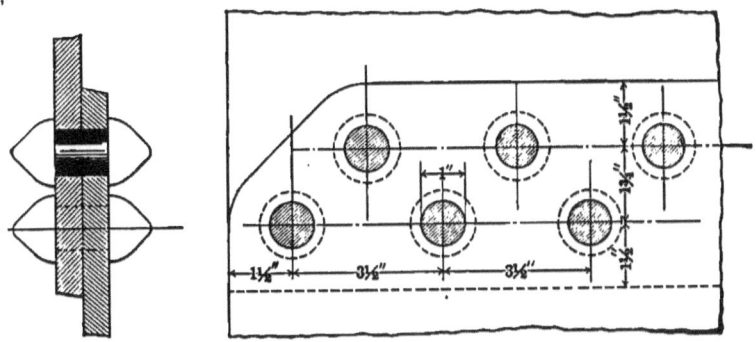

Fig. 154.
Double-riveted lap-joint. 5/8" plate, 15/16" rivets, 1" holes.
Baldwin Locomotive Works.

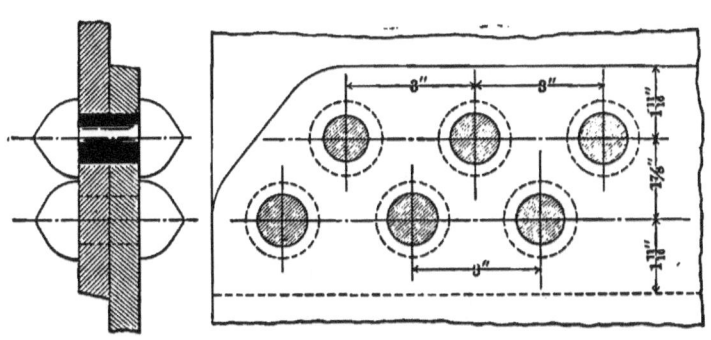

Fig. 155.
Double-riveted lap-joint. 11/16" plate, $1\tfrac{1}{16}$" rivets, $1\tfrac{1}{8}$" holes.
Baldwin Locomotive Works.

304 FORM, STRENGTH, AND PROPORTIONS OF PARTS.

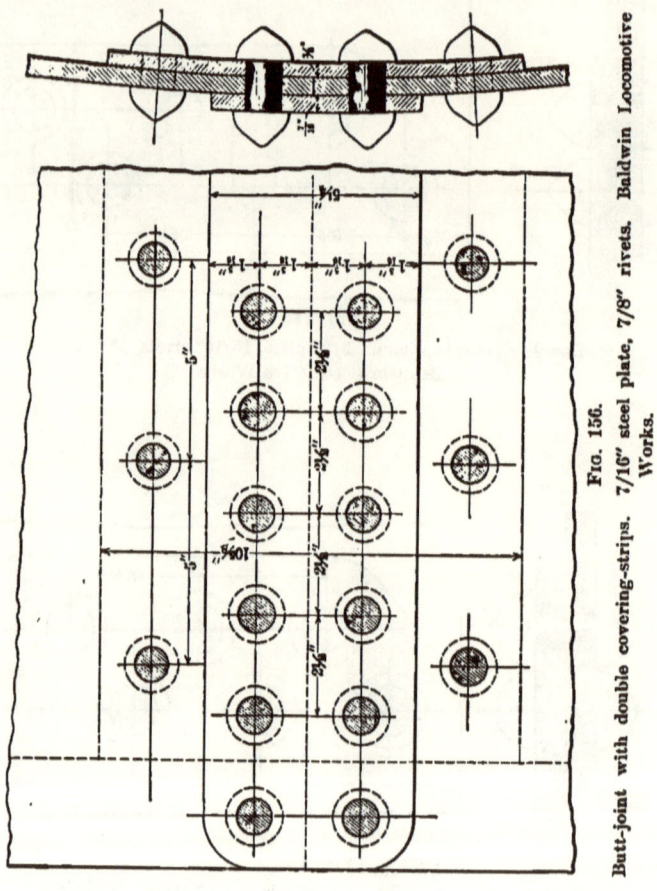

Fig. 156.

Butt-joint with double covering-strips. 7/16" steel plate, 7/8" rivets. Baldwin Locomotive Works.

RIVETED JOINTS.   305

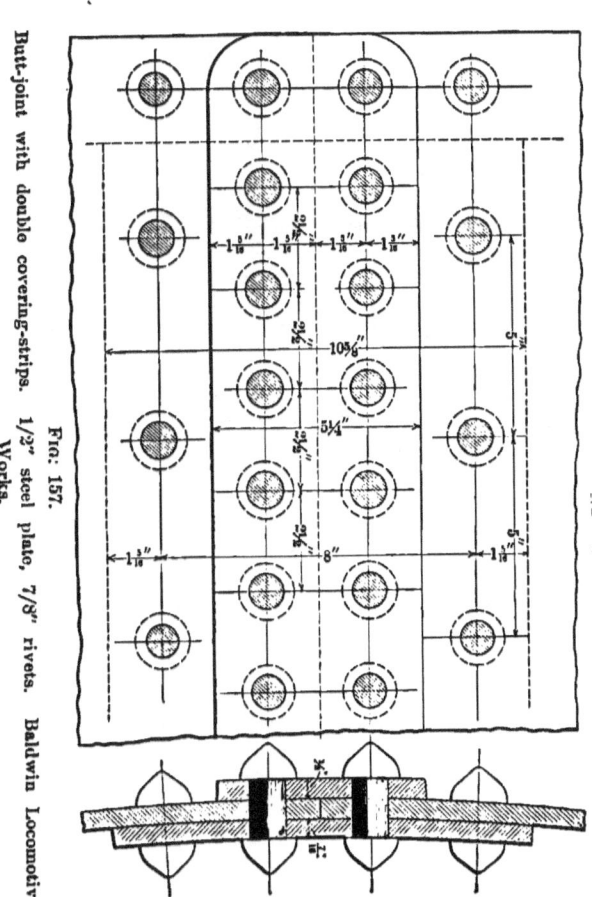

Fig. 157.
Butt-joint with double covering-strips. 1/2" steel plate, 7/8" rivets. Baldwin Locomotive Works.

306 FORM, STRENGTH, AND PROPORTIONS OF PARTS.

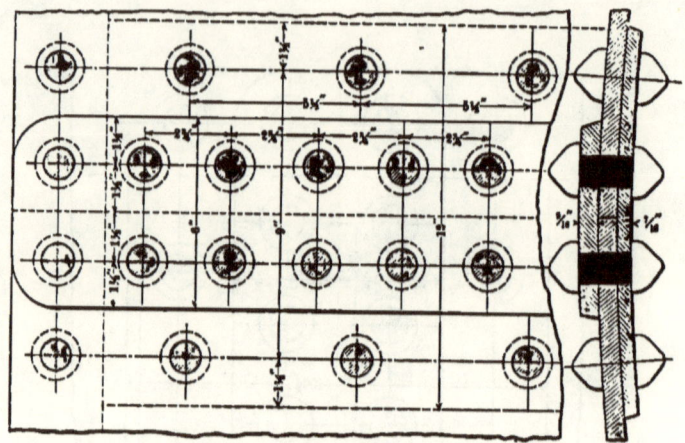

Fig. 158.
Butt-joint with double covering-strips. 9/16" steel plate, 1" rivets.
Baldwin Locomotive Works.

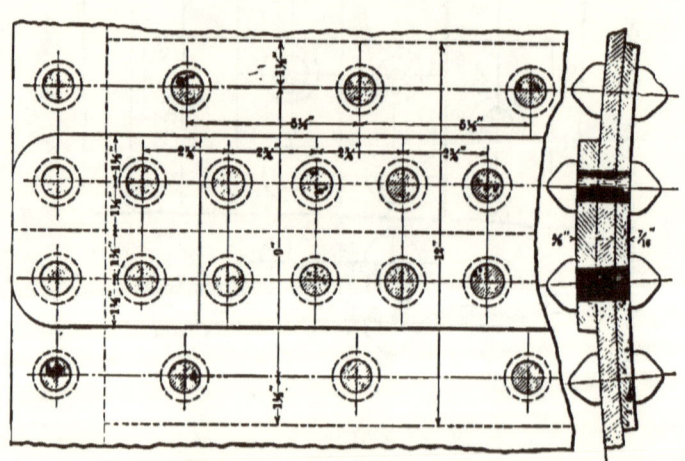

Fig. 159.
Butt-joint with double covering-strips. 5/8" steel plate, 1" rivets.
Baldwin Locomotive Works.

RIVETED JOINTS. 307

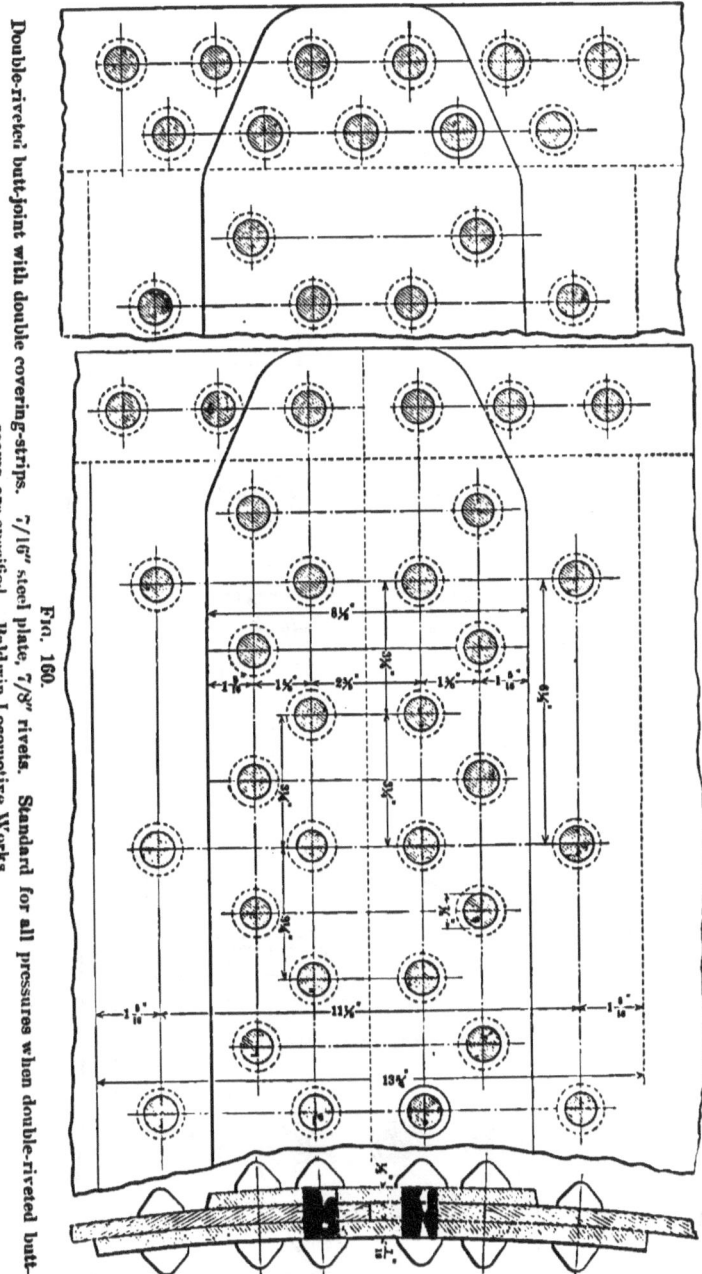

Fig. 160.
Double-riveted butt-joint with double covering-strips. 7/16" steel plate, 7/8" rivets. Standard for all pressures when double-riveted butt-seams are specified. Baldwin Locomotive Works.

308 FORM, STRENGTH, AND PROPORTIONS OF PARTS.

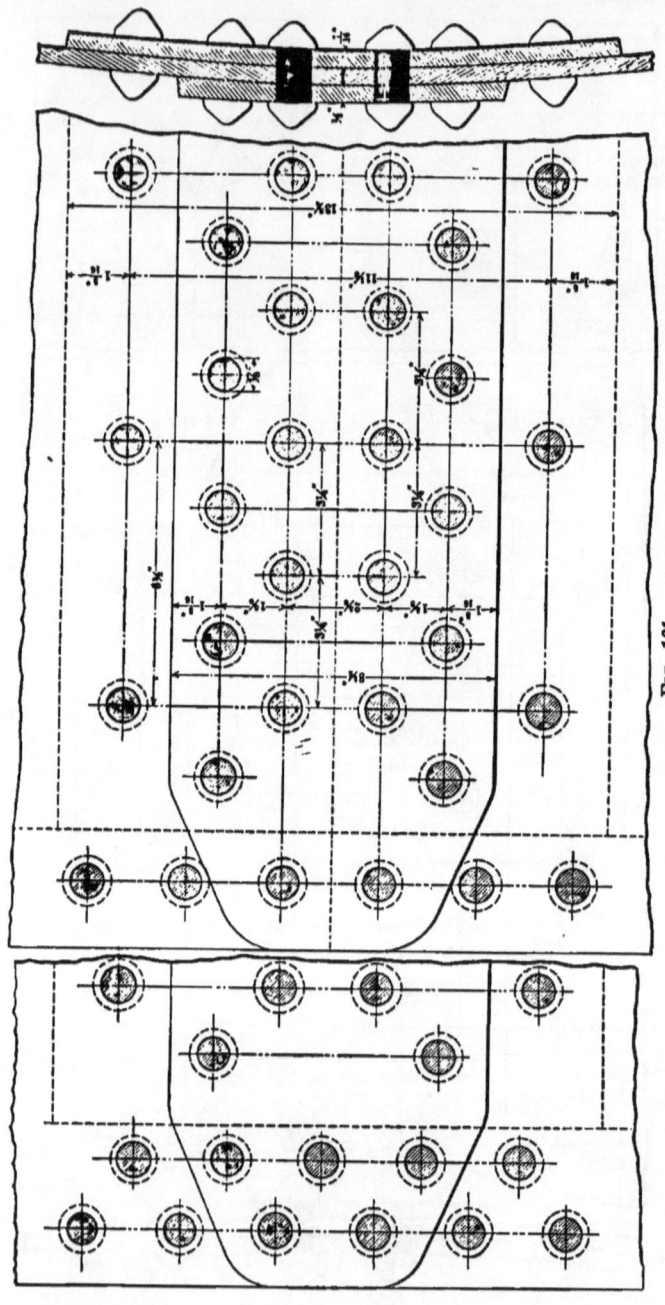

FIG. 161.

Double-riveted butt-joint with double covering-strips. 1/2" steel plate, 7/8" rivets. Standard for all pressures when double-riveted butt-seams are specified. Baldwin Locomotive Works.

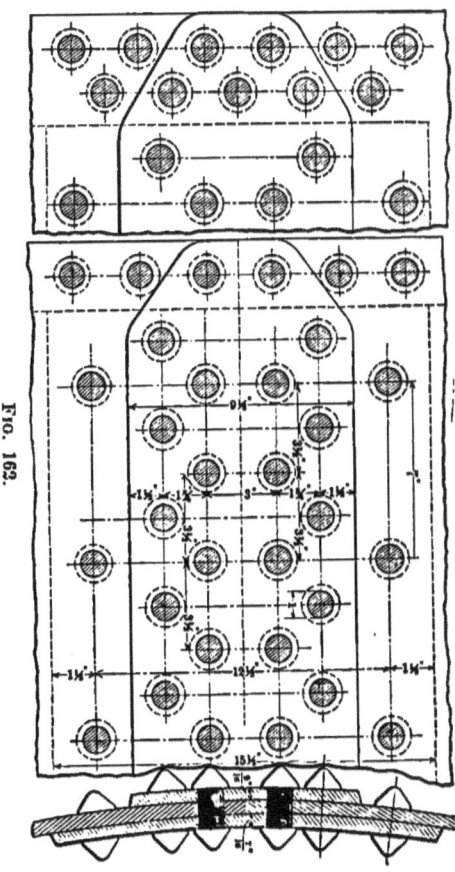

FIG. 162.

Double-riveted butt-joint with double covering-strips. 9/16″ steel plate, 1″ rivets. Standard for all pressures when double-riveted butt-seams are specified. Baldwin Locomotive Works.

310 FORM, STRENGTH, AND PROPORTIONS OF PARTS.

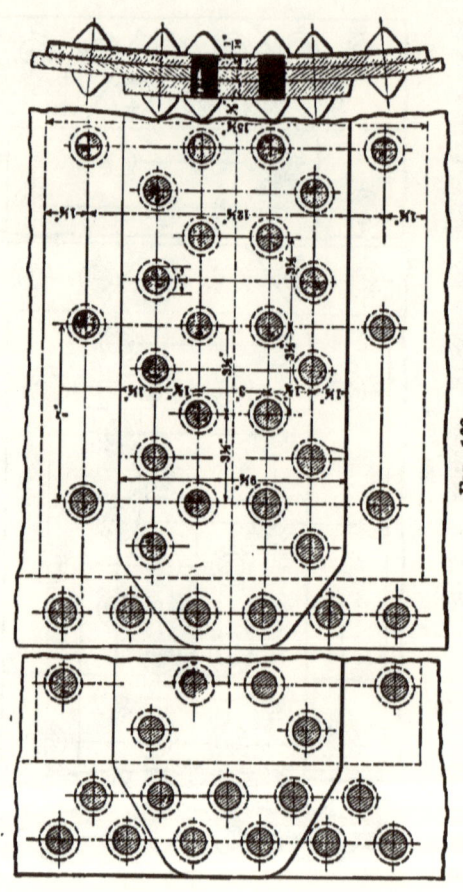

FIG. 168.—Double-riveted butt-joint with double covering-strips. 5/8" steel plate, 1" rivets. Standard for all pressures when double-riveted butt-seams are specified. Baldwin Locomotive Works.

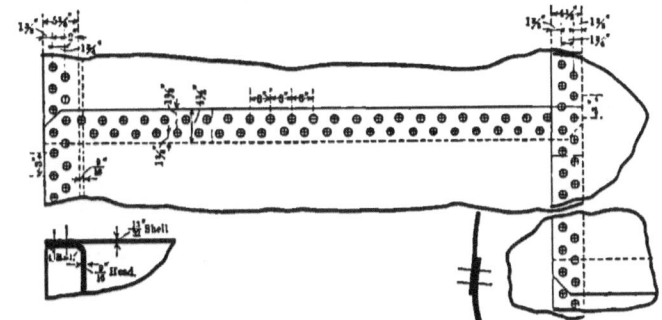

Fig. 164.

Steam-pressure, lbs. per sq. in., 100. Inside diameter of shell 6' 0".
7/8" rivets, 15/16" holes. The Continental Iron Works.

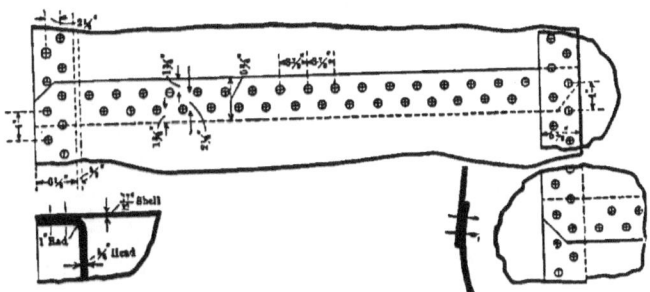

Fig. 165.

Steam-pressure, lbs. per sq. in., 180. Inside diameter of shell 6' 0".
1⅛" rivets, 1 3/16" holes. The Continental Iron Works.

312 FORM, STRENGTH, AND PROPORTIONS OF PARTS.

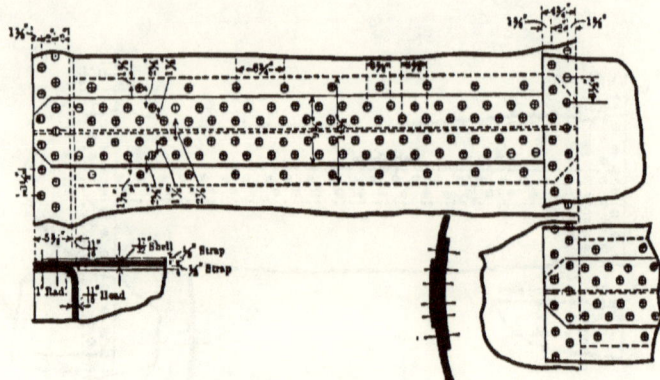

Fig. 166.

Steam-pressure, lbs. per sq. in., 160. Inside diameter of shell 6' 6'.
7/8" rivets, 15/16" holes. The Continental Iron Works.

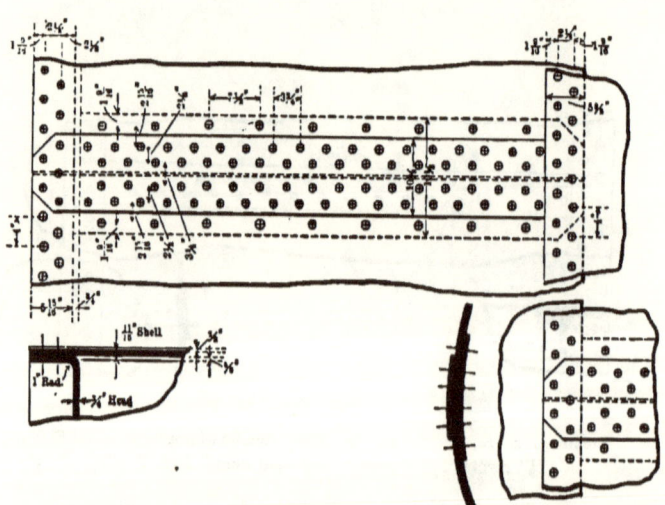

Fig. 167.

Steam-pressure, lbs. per sq. in., 200. Inside diameter of shell 6' 6".
1" rivets, $1\frac{1}{16}$" holes. The Continental Iron Works.

RIVETED JOINTS. 313

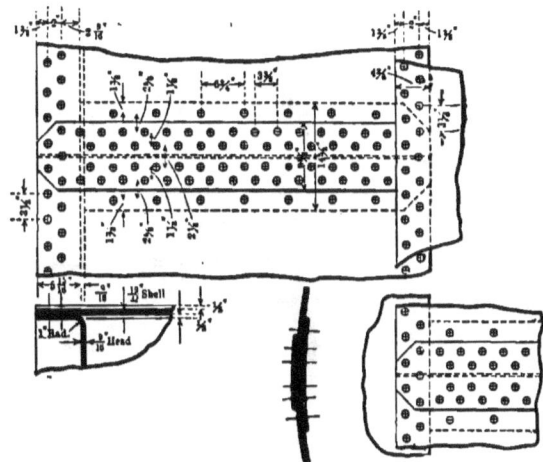

Fig. 168.

Steam-pressure, lbs. per sq. in., 100. Inside diameter of shell 11' 6".
7/8" rivets, 15/16" holes. The Continental Iron Works.

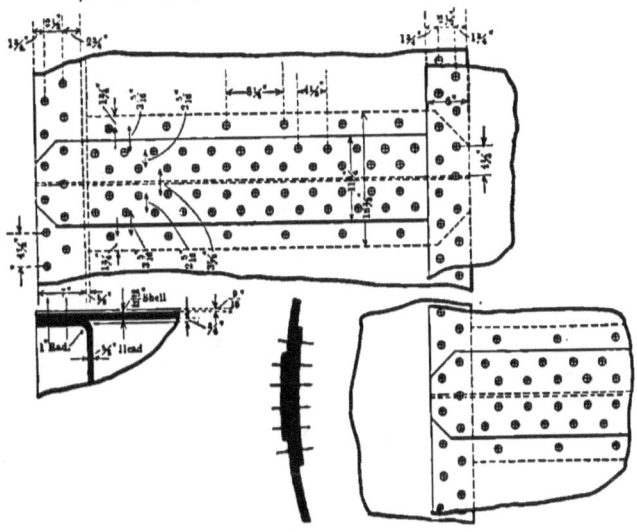

Fig. 169.

Steam-pressure, lbs. per sq. in., 130. Inside diameter of shell 11' 6".
1 1/8" rivets, 1 3/16" holes. The Continental Iron Works.

314 FORM, STRENGTH, AND PROPORTIONS OF PARTS.

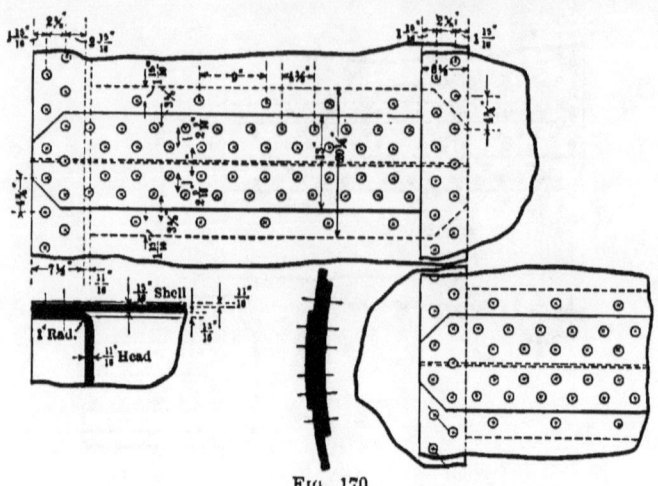

Fig. 170.

Steam-pressure, lbs. per sq. in., 160. Inside diameter of shell 11' 6".
1¼" rivets, 1 5/16" holes. The Continental Iron Works.

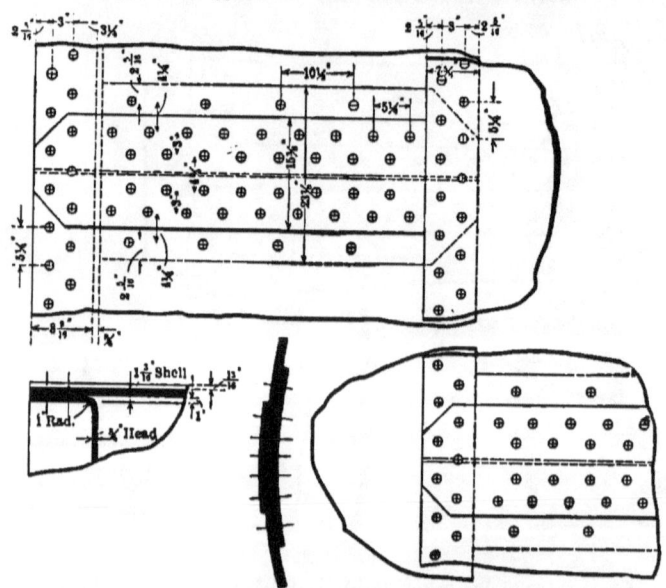

Fig. 171.

Steam-pressure, lbs. per sq. in., 200. Inside diameter of shell 11' 6".
1¼" rivets, 1 5/16" holes. The Continental Iron Works.

CHAPTER XIII.

FRAMES OF PUNCHING, SHEARING, AND RIVETING MACHINES.

PUNCHING AND SHEARING MACHINES.*

146. A C frame of the general form shown in Fig. 172 is very commonly used for punching and shearing machines. The punch

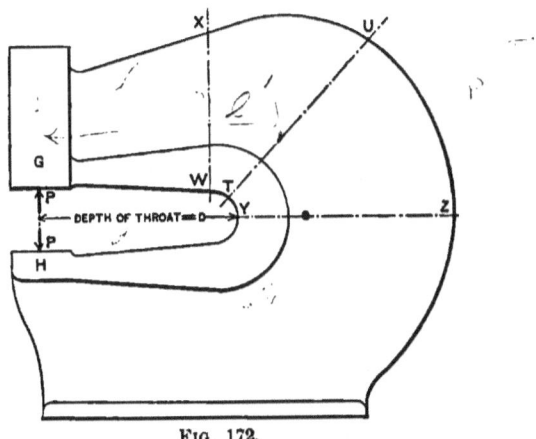

Fig. 172.

is attached to a slide which works in the guides G and operates against a die supported at H. The material to be punched or sheared is placed between the punch and die.

If the frame is for a stationary machine which is to set in the position shown in the figure, it is made with a flat base as shown. When the jaws are to stand vertically instead of horizontally, the base illustrated is not necessary, of course, but some suitable means

---

* It is believed that Professor Albert W. Smith was the first to present the method of dealing with the stresses in C frames that is given in this chapter.

must be supplied for supporting the frame. Portable punches are generally suspended by a chain or rope, hence no base is necessary.

The frames for heavy stationary punching and shearing machines are generally made of cast iron. Forgings and steel castings are generally used for portable machines.

In designing a punch-frame for strength it is necessary to consider the external forces acting upon it, together with the internal stresses due to the externally applied forces. Sections must be selected at different places along the frame, and each designed so as to give the required working stress in the material, due allowance being made for the localization of shrinkage stresses at re-entrant angles, and for the general appearance of the machine.

In general it will be found that for cast iron the sections near the punch will have to be made heavier than is necessary for the requisite strength in order to obtain walls thick enough to cast well in connection with the remainder of the frame.

The external forces acting on the frame when punching a piece of metal have practically the same effect on the material near the back part of the throat of the frame as if two opposite forces, both coincident with the axis of the punch, were applied to separate the jaws. In ordinary construction the centre line of the punch lies in a plane dividing the frame into two symmetrical halves. This median plane of the frame is parallel to the paper in Fig. 172. When the punch is so located, its pressure against the piece to be punched has no other tendency than to separate the jaws.

When used as a shearing-machine, the cut begins near one end of the shear-blades and travels across them to the other end as they are brought together. Assuming that the blades are set so that they are perpendicular to the median plane, it can be seen that at the beginning of the cut the pressure against the blades will be to one side of the median plane, although approximately parallel to it. On account of being to one side of the median plane, the pressure against the shears at the beginning of the cut has a tendency to open the frame more on one side than the other, and thus cause a torsional or twisting action upon it in the two parts extending forward to form the throat and jaws. When the cut has passed to the opposite end of the blades, a similar twisting action occurs, but in the opposite direction.

The frame should be of such a sectional form as will offer the greatest resistance to the twisting action just mentioned, as well as to the greatest (in ordinary cases) tendency to separate the jaws. This can be done, when the frame is cast, by making it of the hollow or box form.

If the shear-blades are placed parallel to the median plane with their cutting edges in it, and are sharp, there would probably be no great force acting on the frame when shearing other than one tending to separate the jaws in the same manner as when punching. As the blades become dull, however, there is a tendency for them to move sidewise relatively to each other, thus inducing a side bending action on the parts of the frame forming the jaws, and a twisting action on sections back of the throat. The box section is the best to resist this action, and is generally used for cast-iron frames.

The following notation may be used for punch-frames:

$A =$ total area of section of frame;
$I =$ moment of inertia of a section about its gravity axis which is normal to the median plane of the frame;
$I_g =$ moment of inertia, about its gravity axis normal to the median plane, of any part of the section;
$I_z =$ moment of inertia, about the gravity axis of the entire section, of any part of the section;
$P =$ force necessary to drive punch through the material to be perforated;
$S_c = I \div e_c =$ section modulus of the section with regard to compressive fibre-stress;
$S_t = I \div e_t =$ section modulus of the section with regard to tensile fibre-stress;
$a =$ area of any selected part of the section;
$b =$ greatest dimension of any partial section, measured parallel to the gravity axis about which the moment of inertia is required;
$d =$ depth of throat of frame;
$e =$ distance from line of action of P to gravity axis of section under consideration;
$e_c =$ distance from gravity axis of section to outermost fibre in compression;

318 FORM, STRENGTH, AND PROPORTIONS OF PARTS.

e_t = distance from gravity axis of section to outermost fibre in tension;
f_c = greatest compressive fibre-stress in section;
f_t = greatest tensile fibre-stress in section;
h = greatest dimension of any partial area measured perpendicular to the gravity axis about which the moment of inertia is required;
s = shearing-stress per unit area uniformly distributed over entire section;
t = tensile stress per unit area uniformly distributed over the entire section;
z = distance between gravity axis of any selected part of the area and the gravity axis of entire section.

147. Stresses in a section perpendicular to the motion of the punch.—The stress in the material on the section YZ, Fig. 172, taken normal to the pressure P against the punch and die, can be determined by considering the portion of the frame above YZ as a free body, as shown in Fig. 173.

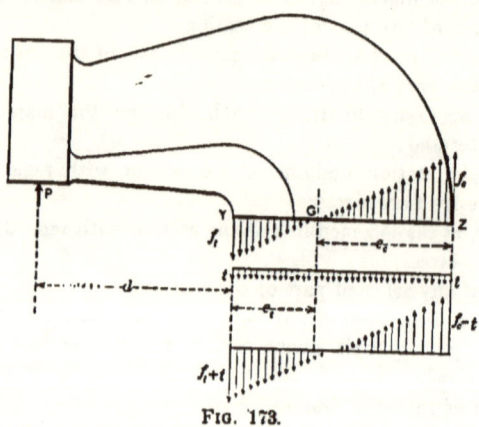

FIG. 173.

The forces acting on this part of the frame may be taken as those due to punching a plate. They are:

1st. The pressure P against the punch as it is forced through the piece to be perforated. The line of action of P coincides with the centre line of the punch.

2d. The fibre-stress in the material due to the bending action of P. P acts with a lever-arm whose length is the distance $d + e_t$ from its line of action to the centre of gravity of the section. The tensile fibre-stress f_t at Y, and the compressive fibre-stress f_c at Z, are:

$$f_t = \frac{P(d + e_t)}{S_t}; \quad \quad (187)$$

$$f_c = \frac{P(d + e_t)}{S_c}. \quad \quad (188)$$

3d. A uniformly distributed tensile stress of a value

$$t = P \div A$$

over the entire section.

By combining the stresses acting on the section YZ it can be seen that the maximum total tensile stress per unit area is at Y, and is equal to $f_t + t$.

The equation for obtaining the total stress is

$$\left. \begin{array}{c} \text{Maximum tensile stress} \\ \text{on section } YZ \end{array} \right\} = f_t + t = \frac{P(d + e_t)}{S_t} + \frac{P}{A}. \quad (189)$$

Similarly, the maximum total compressive stress per unit area on the section YZ is at Z, and is equal to the numerical difference of f_c and t. Therefore

$$\left. \begin{array}{c} \text{Maximum compressive} \\ \text{stress on section } YZ \end{array} \right\} = f_c - t = \frac{P(d + e_t)}{S_c} - \frac{P}{A}. \quad (190)$$

If the centre of gravity of the section lies midway between Y and Z, then $f_t = f_c$. Equations (189) and (190) show that if the centre of gravity is thus located the maximum tensile stress at Y will be greater than the maximum compressive stress at Z by an amount equal to $2t = 2P \div A$. This assumes that the modulus of elasticity of the material is the same for tension as for compression, which is practically true for the materials common to such construction.

If a material having equal strength in both tension and compression is used for the frame, the cross-section should be made heavier on the side next the punch or shear-blades in order that $(f_t + t)$ shall equal $(f_c - t)$, this being the condition for equal maximum tensile and compressive stresses.

For such a material as cast iron, which is much stronger in compression than tension, it becomes necessary for economy of material to make the section much heavier on the side next the punch than at the back. This is done to make $(f_t + t)$ smaller than $(f_c - t_1)$.

148. Numerical solution for a section perpendicular to the motion of the punch.—This corresponds to the section on YZ in Fig. 172.

Let the work required of the machine be to punch a 1-inch hole 19 inches from the edge of a plate $\frac{3}{4}$ of an inch thick, having a shearing-strength of 55000 pounds per square inch; also to be operated with shear-blades for shearing the same plate. The frame to be cast iron. The throat to be 20 inches deep, so as to allow the edge of the plate to clear it 1 inch, and to be stressed to practically 2000 pounds per square inch tension when punching the hole.

The force P necessary to drive the punch through the plate is found by multiplying the shearing-strength of the material by the area to be sheared by the punch. The sheared area is that of the wall of the hole made by the punch. Whence

$$P = 55000\pi \times 1 \times \tfrac{3}{4} = 129600 \text{ lbs.}$$

It can safely be assumed that the force exerted to separate the jaws of the frame will not be so great when shearing as when punching the size of hole specified. If the frame is made strong enough to operate as a punch, it is therefore only necessary to give it such a form of cross-section as will secure sufficient rigidity against twisting sidewise when shearing.

A drawing of any convenient size, and of a form considered suitable for its purpose, may be made for the required section. Fig. 174 may be taken as the sectional form selected for this problem. The moment of inertia I of this section, and thence its

FRAMES OF PUNCHING AND SIMILAR MACHINES. 321

section modulus, may now be found, the scale of the drawing being taken as unity (i.e., scale 1 inch = 1 inch). The section thus represented in Fig. 174 is not by any means large enough for the

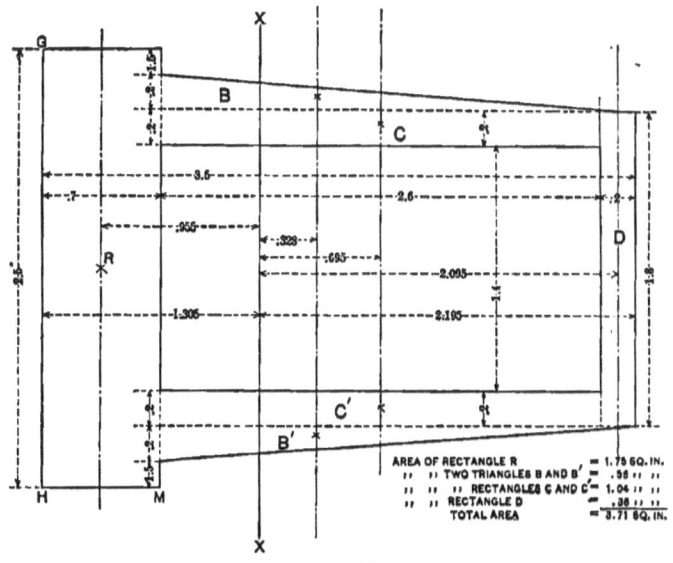

Fig. 174.

required strength but it is not necessary that it shall be, for the scale that must be applied to it to obtain the dimensions for the requisite strength can be determined by trial, as shown later.

In Fig. 174 the section has been taken of a simple form in order that the calculations for I might be readily made. The actual working form should have rounded corners and filleted re-entrant angles.

For convenience in determining I, the section is divided into six parts, namely:

1st. A rectangle R;
2d. Two similar and equal triangles B and B';
3d. Two similar and equal rectangles C and C';
4th. A rectangle D.

The divisions through the body of the section are indicated by broken lines. The centre of gravity of each partial area of the section is indicated by ×.

In order to determine I, the gravity axis XX of the section must first be located. This can be done conveniently by taking moments about the line GH, considering the areas as forces proportional to the areas, whence

$$e_t = 1.305 \text{ in.} \quad \text{and} \quad e_c = 2.195 \text{ in.}$$

For the rectangle R:

$$I_g = \tfrac{1}{12}bh^3 = \tfrac{1}{12} \times 2.5(.7)^3 = .0714;$$

$$az^2 = 1.75 \times (.955)^2 = 1.596;$$

$$I_x = I_g + az^2 = .0714 + 1.596 = 1.6674.$$

For the triangles B and B':

$$I_g = 2 \times \tfrac{1}{36}bh^3 = \tfrac{1}{18} \times .2(2.8)^3 = .2439;$$

$$az^2 = .56(.328)^2 = .0602;$$

$$I_x = .2439 + .0602 = .3041.$$

For the rectangles C and C':

$$I_g = 2 \times \frac{1}{12}bh^3 = \frac{.2(2.6)^3}{6} = .5856;$$

$$az^2 = 1.04(.695)^2 = .5023;$$

$$I_x = .5856 + .5023 = 1.0879.$$

For the rectangle D:

$$I_g = \tfrac{1}{12}bh^3 = \tfrac{1}{12} \times 1.8(.2)^3 = .0012;$$

$$I_h = az^2 = .36(2.095)^2 = 1.5800;$$

$$I_x = .0012 + 1.58 = 1.5812.$$

For the entire section:

$$I = 1.6674 + .3041 + 1.0879 + 1.5812 = 4.6406.$$

The section modulus for tension is

$$S_t = \frac{I}{e_t} = \frac{4.6406}{1.305} = 3.556.$$

This is the section modulus for the dimensions given in Fig. 174. It is clearly too small for the required service. A scale of drawing must therefore be assumed which will give a larger section.

Assume that the drawing Fig. 174 is one ninth size. The value of e_t for the enlarged section will therefore be 9×1.305, and S_t for the same will be $(9)^3 \times 3.556$, since the section moduli of similar sections are proportional to the cubes of their linear dimensions. The area will be $(9)^2 \times 3.71$.

The maximum tensile stress in the enlarged section will therefore be, by equation (189),

$$f_t + t = \frac{129600(20 + 9 \times 1.305)}{(9)^3 \times 3.556} + \frac{129600}{(9)^2 \times 3.71}$$

$$= 1590 + 430 = 2020 \text{ lbs. per sq. in.}$$

This value is practically satisfactory. Therefore by multiplying the linear dimensions of Fig. 174 by 9 a section of suitable size may be obtained. The dimensions obtained in this manner should be modified to agree with shop methods of measurement, of course. The corners should be rounded and re-entrant angles filleted. This need not be done to such an extent as to seriously affect the section modulus, however. It is often necessary to apply a fractional scale to the drawing, such, for instance, as 9.1, 8.7, etc.

If the section adopted is of such a form as to make it impossible to divide it into geometrical portions, the moment of inertia may be found graphically, the gravity axis being located during the process.* The section modulus can then be readily determined.

---

* For finding moment of inertia graphically see appendix, § B.

149. Section parallel to the motion of the punch.—This corresponds to a section on WX, Fig. 172. The portion of the frame to the left of WX may be considered a free body, as in Fig. 175. The

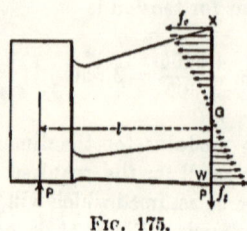

FIG. 175.

forces and stresses acting on it are indicated in the figure. They are:

1st. The force P, coincident with the centre line of the punch.

2d. The fibre-stress in the material due to the bending action of P. P acts on a lever-arm of length l, measured from the line of action of P to the plane of the section. The tensile fibre-stress is at W, and equals

$$f_t = \frac{Pl}{S_t}. \qquad (191)$$

The compressive fibre-stress is at X, and equals

$$f_c = \frac{Pl}{S_c}. \qquad (192)$$

3d. A shearing-stress uniformly distributed over the section. The value of this shear per unit area is

$$s = \frac{P}{A}. \qquad (193)$$

The maximum tension, compression, and shear are found by combining the stresses obtained by the last three equations. The formulas for maximum tension and compression, as given in works on the strength of materials, are:

Maximum tension $= \frac{f_t}{2} + \sqrt{s^2 + \left(\frac{f_t}{2}\right)^2}$;

Maximum compression $= \frac{f_c}{2} + \sqrt{s^2 + \left(\frac{f_c}{2}\right)^2}$.

Since the value of s is the same for all parts of the section, the maximum shear will occur where the fibre-stress due to bending is greatest. This will be at X if the gravity axis of the section is farther from X than from W. The equation is

$$\text{Maximum shear} = \sqrt{s^2 + \left(\frac{f_c}{2}\right)^2}. \quad . \quad . \quad (194)$$

The section must, of course, be designed strong enough to resist both tension and shear. In general, it is not possible to make it equally strong to resist both when such a material as cast iron is used. At a section near the punch the shearing-stress may be the one to fix the proportions, but when the section is taken near the back of the throat the tensile stress is more apt to be the one requiring consideration.

It should be noted that for the section WX the lever-arm l remains the same length whatever scale is applied to the drawing of the section.

The section WX must have such dimensions normal to the median plane of the frame as will allow it and the section on YZ to be joined together to form the frame.

In order to give a symmetrical appearance to the frame, and to obtain a form that will cast well, the sections near the guides at the working end of the frame are generally made much stronger than calculations show to be necessary.

150. Angular section of a punch-frame.—A section taken at an angle, as the one on TU, Fig. 172, has the following forces and stresses acting upon it. They may be seen by the aid of Fig. 176.

1st. The force P due to the pressure of the punch against the plate.

2d. The fibre-stress due to the bending action of P. The lever-arm of P is l, measured from the line of action of P to the gravity axis of the section which is projected as a point at G. The tensile fibre-stress at T, due to the bending action of P, is

$$f_t = \frac{Pl}{S_t}; \quad \ldots \quad (195)$$

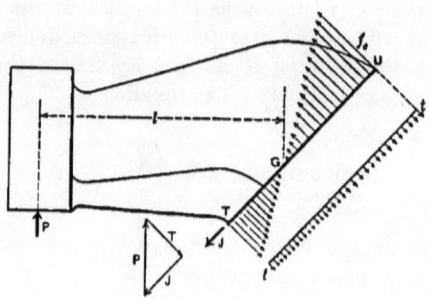

Fig. 176.

and the compressive fibre-stress at U, due to the same cause, is

$$f_c = \frac{Pl}{S_c}. \quad \ldots \quad (196)$$

3d. A shearing-stress J acting as shown in Fig. 176. The value of J is found by resolving P into two components, J and T, one parallel and the other perpendicular to the plane of the section. The shear per unit area due to J is

$$s = J \div A.$$

4th. A tensile stress T acting normal to the plane of the section. The tension per unit area due to T is

$$t = T \div A.$$

The formulas for maximum stresses are:

$$\text{Maximum tension} = \frac{f_t + t}{2} + \sqrt{s^2 + \left(\frac{f_t + t}{2}\right)^2};$$

Max. compression $= \dfrac{f_c - t}{2} + \sqrt{s^2 + \left(\dfrac{f_c - t}{2}\right)^2}$;

Maximum shear $= \sqrt{s^2 + \left(\dfrac{f_t + t}{2}\right)^2}$ or $= \sqrt{s^2 + \left(\dfrac{f_c - t}{2}\right)^2}$.

For the maximum shear the greater of the two quantities under the radicals is to be used. In many cases a numerical solution is necessary to determine which of these two quantities is the greater.

This section must have dimensions, normal to the median plane of the frame, intermediate between those of the sections on WX and YZ.

151. General form of a punch-frame.—In designing the frame enough sections should be tested to secure a nearly uniform strength throughout the frame, due allowance being made for the fact that the sections near the punch-guides may have to be heavier than is actually necessary for strength in order that the frame may have a symmetrical appearance. The walls of the frame should not be made so thin at any part that they will not cast well in connection with the heavier parts.

On account of the tendency of the shrinkage stresses in a casting to be greatest about re-entrant angles, it is generally advisable to make the sections bordering on the curve at the back of the throat somewhat stronger than those along the straight sides of the jaws. When the distance between the jaws is not greater than shown in Fig. 172, the back of the throat can be made an arc of a circle, or approximately so, as shown in the figure. When the jaws are far apart, however, this cannot generally be done well, but the re-entrant angles should be well filleted in order to prevent excessive shrinkage stresses in the material about them.

If the frame is made with an open flat bottom, suitable for resting on a foundation, the sections along the lower horizontal part, referring to Fig. 172, should be tested in the same way as those above, unless it is certain that each is at least as strong as the one immediately above it.

In most designs the frame tapers, somewhat in the manner of a pyramid, from the guides toward the back of the throat. The thickness HM, Fig. 174, of the rib generally decreases from the throat toward the punch, as indicated in Fig. 172; less frequently it is kept of the same thickness all the way around the throat and jaws.

Having obtained the general form of the frame with regard to strength and symmetry, the auxiliary parts, such as bearings, etc., can be added to complete the frame.

DIRECT-ACTING HYDRAULIC RIVETER.

152. A **hydraulic riveter** is required for driving rivets in boiler-plate up to 60 inches wide and $\frac{7}{8}$ of an inch thick, the largest rivet to be $\frac{5}{8}$ of an inch in diameter, corresponding to 0.307 of a square inch area.

Experiments show that a pressure of 160000 pounds per square inch on a hot rivet of this diameter will set the rivet and form the head satisfactorily. The pressure required for this case will therefore be

$$0.307 \times 160000 = 49120 = 50000 \text{ pounds, in round numbers.}$$

By calling this value 50000 there is no greater deviation from the correct value than is shown in the experiments on riveting.

Fig. 177 is a side elevation of the outlines of the most common form of such machines as they are found in practice. The cylinder Q contains a piston attached to a piston-rod, one end of which projects at R, where one of the dies for compressing the rivet and forming the head is attached. The piston-rod serves as a ram for driving the die. The stationary die is at S.

In order to rivet cylindrical shells of small diameters, such as flues, the stake should be made as slender as is possible for the required strength, so that the shell can be passed over it for driving the rivets in the longitudinal seams when riveting up a single section, and for the circular seams when riveting two sections together. It can be made of either a steel forging of rectangular cross-section or a steel casting having an I section.

The frame may be cast iron or a steel casting. Since the over-

FRAMES OF PUNCHING AND SIMILAR MACHINES. 329

reach of the riveter, which is the distance it will reach over a plate from its edge to drive a rivet, is small, the frame can be so made of cast iron as to have sufficient strength without excessive weight.

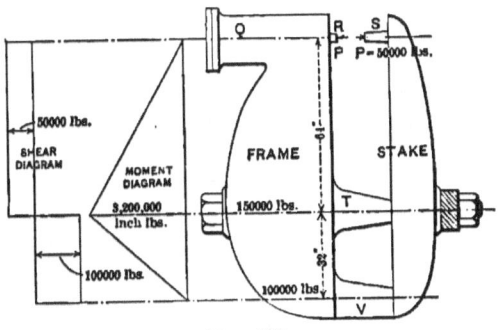

FIG. 177.

The stake and frame are held together by two bolts, one passing on each side of the stake and through a bar passing over the top. This bar, section-lined in the figure, clamps against the top or back of the stake.

The lugs T and V are a part of the frame, their office being to hold the stake in position.

In proportioning for strength the stake and frame can be dealt with separately, each as a beam supported at the ends and loaded eccentrically with a single load. The moment and shear diagrams, identical for both, are shown in the figure.

It will be noticed that the distance from the centre of the dies to the plane of the bolt centres is 64 inches, which is 4 inches in excess of the distance the centre of the rivet is to be from the edge of the widest plate. An examination of the figure shows that about this much must be allowed for the lug and bolts.

By taking moments about a point 64 inches from the centre line of the punch the pressure on the lug V is found. Its value is

$$V = (64 \div 32)50000 = 100000 \text{ pounds.}$$

That on T is

$$T = 50000 + 100000 = 150000 \text{ pounds.}$$

The bending moment is a maximum at the plane of the centre lines of the bolts, and equals

$$M = 50000 \times 64 = 3200000 \text{ inch-pounds.}$$

If the stake has a cross-section fairly thick, measured perpendicular to the plane of the paper, in proportion to its width, and is made of good steel, it will, on account of the length being large in proportion to its thickness and width, have fibre-stresses due to bending which are large in proportion to the shearing-stress uniformly distributed over the section. Under such conditions the shear can be neglected, and the stake designed to resist the stresses due to bending only. This is in accordance with the common method of dealing with beams whose height is small in proportion to their length. In general, the stake should taper from its maximum sectional dimensions at the bolts, growing smaller in both dimensions toward both ends. The thickness is generally reduced uniformly, but the height changes more rapidly as the distance from the largest part increases.

There should always be sufficient thickness to prevent the upper or stake die from springing sidewise to such an extent as to make the head of the rivet objectionably eccentric with the body.

The frame may be of either a solid I section or hollow. In either case the bending moment and shear can both be dealt with in the same manner as for the section WX of the punching and shearing machine.

The bolts must be tightened with a combined initial tension somewhat greater than that which is thrown upon them when a rivet is driven, in order to prevent the springing apart of the stake and lug. The lug T, therefore, must resist in this case a compression somewhat greater than 150000 pounds. The lug V must, of course, resist a compression of 100000 pounds.

The frame may either stand as in the figure, suitable flanges being provided for resting upon a foundation, or it may be placed with the axis of the cylinder vertical. The nature of the service required of the machine is the determining factor for the position in which it shall be placed.

CHAPTER XIV.

SELECTION OF MATERIALS.

153. In selecting materials for machine parts, the properties that must most commonly be considered with regard to their adaptability are as follows: strength, resilience, stiffness, coefficient of friction, durability, convenience and cost of working into the required form, and cost of the material as found on the market in the merchantable form. It is seldom that all of these properties need to be taken into account when deciding upon the material for any given machine-member.

A few examples may serve to show more clearly than any other means how the qualities have to be taken into account.

In an ordinary engine-lathe the bed is of a somewhat complicated form, having ribs, sometimes hollow, extending from side to side, bosses and other raised places for attaching parts, and the sides are ordinarily flanged or ribbed at both top and bottom in order to give rigidity. This complication at once sets forth the necessity of making the bed of some material that can be cast in moulds. There are only two materials, whose cost is such as to allow them to be used for such purposes, that can be formed by casting, namely, cast iron and steel. Cast iron is much the cheaper of the two, and can be planed and otherwise machined much more readily and, consequently, more cheaply than steel. It is a much weaker material than steel and is not so stiff. These two qualities apparently make it objectionable, but this is really not the case on account of the following reasons: In order for a lathe tool to take a broad cut when removing only a thin turning from the surface of the piece worked upon, it is necessary for the lathe-bed to be both rigid and heavy, so as to prevent the tool from springing away from its work and "chattering" against it; the mass of metal in the bed has much to do in the prevention of the small vibrations that cause the chattering. Therefore the mass of metal necessary to hold the tool steady for wide cuts will make the lathe strong and stiff enough for the heaviest cuts that the driving mechanism of the lathe can carry. Cast iron, therefore, in comparison with steel, has for the lathe-bed the advantage in cost and ease of machining, while its lower strength and moduli of elasticity are of small consideration, on account of the mass of metal to be used; clearly, cast iron is the best material for the purpose.

Even if the vibration of the tool had not been an item in the conditions above, the cast iron would have still been the best material on account of its softness making it easy to machine as well as the low cost of the raw material, both together making the finished product much less costly in cast iron than in steel. This is also true of numerous other machine-tool parts and machines.

The spindle or arbor of an ordinary engine-lathe rests in two bearings supported by the head-stock, and is driven by a gear attached to it near the bearing next to the face-plate or live-centre, as the case may be. A stepped pulley occupies all or most of the remaining length of the part of the spindle which lies between the bearings. When the lathe is working, all the power transmitted to overcome the resistance of the cutting edge of the tool must pass through the part of the spindle lying in the bearing just back of the face-plate, thus causing a torsional moment in the spindle. In order to withstand this twisting moment, and at the same time the weight of the face-plate and the work upon it or the centres, the spindle, unless large in diameter, must be made of some strong material, which should also be resilient in order to withstand any of the shocks that are apt to occur accidentally. By keeping the spindle small the frictional loss in the head-stock bearings is kept down, and there is probably less liability to cutting and seizing of the journal by the bearings when running at high speeds. There is also less frictional loss between the bearing surfaces of the stepped pulley and spindle when the latter is not larger than is necessary to give ample bearing surface. The best material for the spindle is, therefore, a moderately hard machine steel, used without hardening. In case great accuracy and very little wear of the bearing surfaces are desired, wrought iron or very low carbon steel can be used by case-hardening and then grinding to form in an emery grinding machine. This is the method frequently used in making milling-machine arbors.

In very heavy lathes, such as are used for turning locomotive drive-wheels, shafts for ocean vessels, or heavy ordnance, many designs have the driving-cone and back-gears placed on shafts serving to support them only, and the spindle is driven by a pinion meshing in a gear attached directly to the face-plate. By this means the spindle is relieved of all duty except supporting the faceplate and the load upon it or the live-centre. Since there is no torsional force to be resisted by the spindle, and furthermore because such lathes run at moderate speeds, the spindle can be enlarged and made of a weaker material than steel. Cast iron is found to give good service in such cases when made much larger than would be necessary for steel, lightness and rigidity being secured by making the spindle hollow. The large diameter gives a low pressure per unit area upon the bearing surfaces and, with proper care during the early life of the machine, they become glazed and wear well, even when the bearing as well as the spindle is of cast iron. The cost is much less for such a spindle of cast iron than for steel, even when the diameter is kept down with the latter.

In general, a steel lathe-spindle does not run satisfactorily upon cast iron, and since the head-stock of a lathe is almost invariably made of this material, it is necessary to provide some other material for the bearing surface upon which the journal of a steel spindle runs. In this case the material of the boxes is selected with regard to its coefficient of friction, durability in resisting frictional abrasion and wear, and possibly cost. The last item is generally not a very important one, however, since there is a comparatively small amount of material in them; brass, bronze, Babbitt metal, or some of the alloys resembling some one of them are the materials commonly chosen. Brass and bronze are more expensive, but they give a better and more finished appearance to the machine than Babbitt's or similar alloys which are used as linings for a cast-iron shell. The cost is far less for these metal linings with cast-iron casings than for brass or bronze.

In many cases the conditions to be fulfilled indicate clearly the necessary material; thus a spring that must have considerable strength and resilience must be made of steel containing enough carbon to cause it to harden and temper under proper heat treatment, or it must be made of brass or some similar alloy; in the general case, therefore, there are narrow limits to choose between. As soon as the purpose of the spring is defined, as for car- or wagon-springs, the cost immediately limits the selection to a single choice, steel being the only material admissible.

Hydraulic presses when made of cast iron allow the water to percolate through the walls of the cylinder when subjected to high pressures. This was formerly a source of great trouble, and various devices were invented to overcome it, one being to line the cylinder with copper, brass, or bronze. When steel castings came to be successfully made, it was found that they were water-tight even under pressures much higher than had been attempted with the cast iron. Naturally the material to be selected for a high-pressure hydraulic press is steel, the selection in this case being based mainly upon the density and consequent water-tightness of the metal, although the greater strength is a considerable factor, these two more than overbalancing the increased cost due to the higher price of steel castings and the more expensive work of machining.

TABLE XLIX.

GENERAL PROPERTIES AND METHODS OF WORKING THE MATERIALS MOST COMMON TO ENGINEERING CONSTRUCTION.

Material.	Tensile Strength.	Compressive Strength.	Shock Resisting Power.	Methods of Shaping for Use.
Cast iron..	Low.	Very high.	Low.	Casting.
Wrought iron, also called malleable iron.........	Medium.	Medium.	High.	Rolling, forging, and wire-drawing.
Bessemer and open-hearth steel, also called machinery steel, soft steel, mild steel, and ingot iron..	High.	High.	High.	Rolling, forging, and wire-drawing.
Crucible steel, also called cast steel and tool steel...	Very high.	Very high.	Medium.	Rolling, forging, and wire-drawing.
Steel castings..	High.	High.	High.	Casting.
Malleableized castings, also called malleable iron castings...	Medium.	High.	High.	Casting.
Brass..	Medium to low.	......	Medium.	Casting, rolling, forging, and wire-drawing.
Bronze..	Medium to low.	......	Medium.	Casting, rolling, forging, and wire-drawing.
Babbitt metal and other grades of white metal used for lining journal boxes......................	Very low.	......	......	Casting.

336 FORM, STRENGTH, AND PROPORTIONS OF PARTS.

TABLE L.

STRENGTHS AND MODULII OF ELASTICITY OF THE MATERIALS MOST COMMON TO ENGINEERING CONSTRUCTION.

Material.	Per Cent. Carbon.	Tensile Strength.		Compressive Strength.		Shearing Strength.	Flexure.	Modulus of Elasticity.	
		Elastic Limit.	Ultimate.	Elastic Limit.	Ultimate.	Ultimate.	Stress in Outer Fibre.	Tension.	Shearing.
Bessemer and open-hearth steel...	0.15	42,000	63,000	39,000		45,000			
	0.20	47,000	68,000	43,000		53,000			
	0.50	48,000	80,000	46,000		57,000		30,000,000	9,000,000
	0.70	53,000	89,000	53,000		60,000			
	0.80	57,000	103,000	63,000		68,000			
	0.96	69,000	118,000	71,000		83,000			
High-grade wrought iron		28,000	56,000	28,000		40,000	{ Elas. lim. 52,000	29,000,000	9,000,000
Common wrought iron		22,000	40,000	22,000		32,000		28,000,000	9,000,000
Crucible or tool steel		68,000	116,000	58,000				31,000,000	12,400,000
Malleable cast iron			35,000				{ Ultimate 70,000	19,000,000 to 31,000,000 Average 25,000,000	
Steel castings		29,000	47,000	29,000		18,000 to 20,000	{ Ultimate 30,000 to 54,000 Average 42,000		
Cast iron			10,000 to 35,000 Average 20,000		56,000 to 145,000 Average 90,000			13,000,000	7,000,000

APPENDIX.

A. Development of equations for an angular-thread screw.—
Fig. 178 shows a portion of an angular-thread screw of which HH'

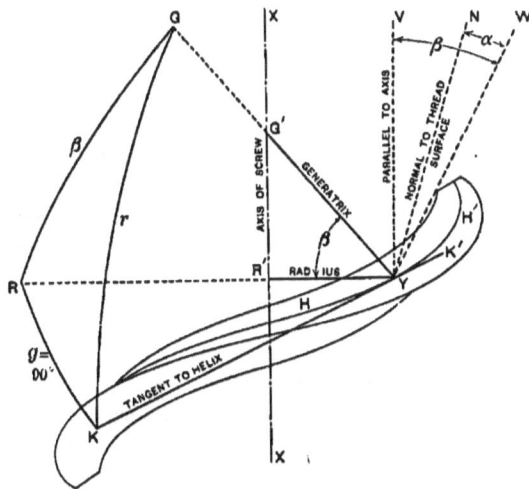

Fig. 178.

is the mean helix, and KK' a tangent to it at Y. Y is a point on HH', and VY is parallel to the axis of the screw. VY may be taken as representing an elementary force acting on the thread at Y. $G'Y$ is the position of the generating-line when passing through the point Y; the angle of $G'Y$ with the radial line $R'Y$ is β. $R'Y$ is, of course, normal to the axis XX of the screw.

The elementary force VY may be resolved into two forces, one radial along OR', and the other along the line WY, normal to $G'Y$ and lying in the axial plane containing $G'Y$ and $R'Y$. The radial component is annulled, so far as forces external to the screw are concerned, by the radial component of the elementary force corresponding to VY, and lying diametrically opposite it; therefore the radial component does not need further consideration.

The component acting along WY is held in equilibrium by two forces, one (NY) normal to the surface of the thread, and the other (not shown in the figure) perpendicular to the axial plane through Y (i.e., perpendicular to the plane containing $G'Y$ and $R'Y$). This latter force is the one which, acting with a lever-arm equal to the radius of the mean helix, tends to produce rotation of the screw.

In order to determine the relative values of this rotative force and that along WY it is necessary to know the value of the angle α = angle NYW in terms of the thread-angle β and the helix-angle θ. (θ is the angle between KK' and a normal to the axial plane through Y.)

To facilitate the determination of the value of α, the spherical triangle GRK may be formed by extending YG' and YR'. In this triangle the side $g = 90°$ and β are known, as well as the solid angle R along RY, which equals $90° + \theta$. The angles G and K are each less than $90°$.

The line NY is normal to the plane KYG, and since GY is the intersection of this plane with the axial plane RYG, α is the complement of the solid angle G, or $\alpha = (90° - G)$.

The determination of the value of α in terms of θ and β is as follows:

$$\cos r = \cos g \cos \beta + \sin g \sin \beta \cos R$$
$$= \cos 90° \cos \beta + \sin 90° \sin \beta \cos (90° + \theta)$$
$$= 0 - \sin \beta \sin \theta \quad \ldots \ldots \ldots \quad (197)$$

and

$$\sin G = \sin g \frac{\sin R}{\sin r} = \sin 90° \frac{\sin (90° + \theta)}{\sqrt{1 - \cos^2 r}}.$$

Substituting the value of $\cos r$ given in equation (197) gives

$$\sin G = \frac{\cos \theta}{\sqrt{1 - \sin^2 \beta \sin^2 \theta}}.$$

APPENDIX.

By trigonometrical relations

$$\cot G = \frac{\sqrt{1 - \sin^2 G}}{\sin G}$$

$$= \sqrt{1 - \frac{\cos^2 \theta}{1 - \sin^2 \beta \sin^2 \theta}} \times \frac{\sqrt{1 - \sin^2 \beta \sin^2 \theta}}{\cos \theta}$$

$$= \frac{\sqrt{1 - \sin^2 \beta \sin^2 \theta - \cos^2 \theta}}{\cos \theta}.$$

This reduces to

$$\cot G = \tan \theta \cos \beta = \tan \alpha. \quad \ldots \quad (198)$$

The relation between the force VY and the turning force through Y and normal to the axial plane RYG is as follows:

The force $WY = VY \sec \beta$, and the turning force $= WY \tan (\alpha + \phi)$, in which ϕ is the friction-angle. Therefore

$$\text{Turning force} = VY \sec \beta \tan (\alpha + \phi)$$

$$= VY \sec \beta \frac{\tan \alpha + \tan \phi}{1 - \tan \alpha \tan \phi}.$$

And by substituting the value of α given in equation (198)

$$\text{Turning force} = VY \sec \beta \frac{\tan \theta \cos \beta + \tan \phi}{1 - \tan \theta \cos \beta \tan \phi}$$

$$= VY \frac{\tan \theta + \sec \beta \tan \phi}{1 - \tan \theta \cos \beta \tan \phi}.$$

By adding together all the elementary forces VY and calling their sum $T =$ tension in bolt, the total turning force F becomes

$$F = T \frac{\tan \theta + \sec \beta \tan \phi}{1 - \tan \theta \cos \beta \tan \phi}.$$

B. **Graphical determination of the moment of inertia of a plane area. Approximate method.**—In the numerical solution given in § 148 it can be seen that the I_x of the rectangle R (referring to Fig. 174) is made up of the sum of two quantities, one of which is the moment of inertia I_g of the rectangle about its own gravity axis, and the other the product az^2 of the area of the rectangle R by the square of the distance of its centre of gravity from the gravity axis XX of the entire plane area. If the rectangle R were very narrow, as measured perpendicular to the gravity axis XX, its I_g would become so small as to be negligible for practical purposes. Its I_x would therefore practically be equal to az^2. The graphical solution which follows is a method of finding the sum of all the az^2's for the small areas into which a given plane area may be divided.

In Fig. 179 A is the plane section whose approximate moment of inertia about its gravity axis normal to the centre line of the area is required. This approximate moment of inertia will be given the same symbol I_x as the accurate moment of inertia about the gravity axis.

Divide the section A into a number of thin strips, 1, 2, 3, 4, . . . , 13, 14, parallel to the gravity axis about which the moment of inertia is to be determined. The dotted lines indicate the divisions. In Fig. 179 these strips are made somewhat wide in order to secure sufficient space for lettering and to make the drawing clear. For very accurate work they should be made narrower. These strips may be of the same or different widths. The position of the centre of gravity of each strip may now be estimated, and a line drawn through it parallel to the gravity axis about which I_x is to be found. The same can also be done for all the other strips. The area of strip 1 may now be represented by a line AB parallel to the given direction of the gravity axis. The line through the centre of gravity of strip 1, whose area is represented by the vector AB, is indicated by the letters ab written on opposite sides of the line. In the same manner bc is the line through the centre of gravity of strip 2, and BC the vector representing the area of 2. BC is a continuation of AB. By proceeding in this manner for all the strips, the line AO, which is the vector representing the entire area of the section A, is obtained.

Select a point P at a distance $\frac{1}{4}AO$ from the line AO, and from it draw rays to the points $A, B, C, \ldots, N, O$. One position of P may be readily located by drawing lines from A and O, each

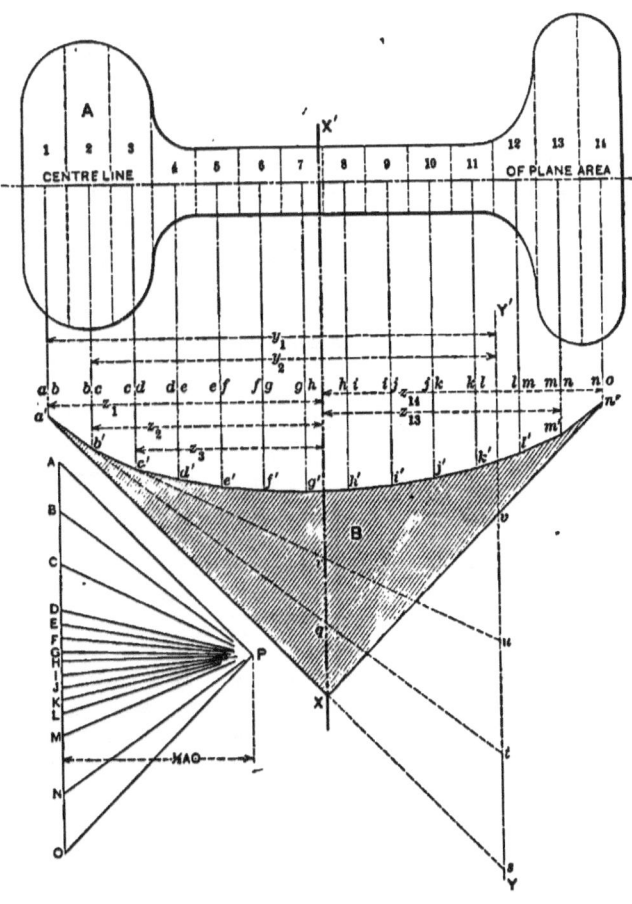

Fig. 179.

making an angle of 45° with AO; their intersection determines one position of P.

From any point a' on ab draw $a'X$ parallel to AP, and of indefinite length; from a' also draw $a'b'$ parallel to BP, intersecting bc at b'; through b' draw $b'c'$ parallel to CP, intersecting cd at c'; through c' draw $c'd'$ parallel to DP, intersecting de at d'. Continuing this operation, $n'X$ is finally drawn through n' parallel to OP, intersecting $a'X$ at X. The point X lies on the gravity axis of the entire section which is parallel to the strips into which the area was first divided. The gravity axis is therefore determined by drawing XX' perpendicular to the centre line of the area. (The proof that X determines the gravity axis is similar to that for finding a point through which the resultant of a number of parallel forces passes. It does not seem necessary to give it here.)

Extend $a'b'$ to intersect XX' at q. The triangles $a'Xq$ and PAB are similar, since their sides are parallel by construction. The altitude of $a'Xq$ is z_1; that of PAB is $\frac{1}{2}AO$.

Therefore

$$(AB) : (Xq) = (\tfrac{1}{2}AO) : z_1;$$

whence

$$(AB)z_1 = (\tfrac{1}{2}AO)(Xq).$$

By multiplying both sides of this equation by z_1 it becomes

$$(AB)z_1^2 = (AO)(\tfrac{1}{2}Xq \times z_1).$$

In this last equation the first member $(AB)z_1^2$ corresponds to the required product represented by the general expression az^2 for the strip under consideration; AO represents the entire area of the given section; and $\tfrac{1}{2}Xq \times z_1$ equals the area of the triangle $a'Xq$. Therefore for strip 1

Approximate I_x = (area of given section) × (area of triangle $a'Xq$).

In the same manner it may be shown for the strip 2 that its approximate I_x = (area of total section) × (area of triangle $b'qr$),

and so on for all the strips into which the given section is divided.

The sum of the areas of all the triangles of a nature similar to

$a'Xq$ and $b'qr$ equals the area of the funicular polygon B, shaded in the figure. Therefore for the entire section A,

Approx. $I =$ (area of given section) $\times$ (area of funicular polygon)
$= $ (area A) $\times$ (area B).

Both of these areas can be conveniently measured with a planimeter.

In a manner similar to the above it may be shown that the moment of inertia of the section A about any axis YY' parallel to XX' is

Approximate $I_y =$ (area A) $\times$ (area $a'b'c' \ldots m'n'vsa'$).

The points s, t, and u are obtained by prolonging $a'z$, $a'q$, and $a'r$ to intersect YY'.

INDEX.

Anti-friction curve, 58

Bearing alloys, tests of, 40, 41, 42
Bearings and lubrication, 1
Bearings, collar, 64
    conical, 74
    fibre-graphite, 37
    friction of, 43
    relative length of, 13
    roller-, 68
    roller thrust-, 72
    special forms of, 73
    thrust-, 64
    for engine cross-head, 12
        lathe head-stock, 22
            spindle, 333
            tool-rests, 10
        planer-saddles, 10
        rectilinear motion, 1
        rotary motion, 14
        shaper-ram, 8, 9
Belts, coefficient of friction for leather, 119
    cotton, 126
    diameters of pulleys for, 124
    effect of relative positions of pulleys on, 128
    efficiency of flat, 130
    elongation of leather, 120
    equations for power-transmission by flat, 102
    leather-link, 127
    pull of leather, 122
    rawhide, etc., 125
    rubber, 126
    slippage of leather, 119
    special system of, 129

Belts, strength of fastenings, 121
    tension in, 118
    thickness of leather, 125
    velocity of leather, 122
    wear of leather, 123
    weight of leather, 125
    working-strength of leather, 120
    for power-transmission, 102
        very high speed, 123
Bevel-gears, 98
        efficiency of, 104
        friction of, 109
        strength of teeth, 98
        wear of, 103
Blanton patent fastening, 205
Bolt-heads, proportions of, 183
Brake, friction-, 235
    Prony, 240
    strap, 239

Calking riveted joints, 286
Cap-screw, 185
Clutch-couplings, positive, 234
Coefficient of friction, 43
        for cotton rope, 141
            journals, 39
            Manila rope, 141
            non-metallic ropes, 141
            rawhide rope, 141
Collar-bearings, 64
        tables of working pressures, etc., 66, 67
Collar-friction of screws, 154
Coupling for shafts, Sellers, 233
            split-sleeve, 232
    multiple-ring friction-, 237
Couplings, coefficient of friction for, 239
        cone friction-, 235
        expansion-, 234
        materials for friction-, 239
        packing for pipe-, 282
        positive clutch-, 234
        for pipes, 282
            riveted, 284
        shafts, flexible, 233
            rigid, 231
Crank-pin, lubrication of, 31

Crown friction-gears, 110
Curve of constant tangent, 58
Cylinders, bursting tests of, 276
       gaskets for, 278
       tension in thick, 275
       tension in thin, 273
       tests of cast-iron, 278

Engine cross-head guides, 12

Factor of safety for tooth-gears, 96
Fibre gearing, 96
    graphite, 37
Flanges for pipes, 282
Fly-wheel, built-up plate, 262
       kinetic energy of, 244
       moment of inertia of, 244, 250
       numerical solution, 248, 249
       overhung, 270
       sectional-rim, 257
       stresses in, 252, 256
       wire-wound, 263
       with composite arms, 262
           tangent-arms, 262
Fly-wheels, 243
       bursting tests of, 258
       designs from practice, 265
       special forms, 263
Forced fits, 207
       allowance for, 211
       data from practice, 212–215
       stresses in parts, 207
       for heavy parts, 210
    lubrication, 55
Friction bevel-gears, 109
    of journal-bearings, 39
       screw-collars, 154
       square-thread screws, 157
Friction-brake, 235
       Prony, 240
       strap, 239
Friction-coupling, multiple-ring, 237
Friction-couplings, coefficient of friction, 239
           cone, 235
           materials for, 239
Friction-gears, 104
       coefficient of friction of, 104

Friction-gears, crown, 110
    cylindrical, 104
    efficiency of, 108
    grooved, 108
    for variable speed, 110, 111

Gaskets, 278, 282

Gears, bevel, 98
    coefficient of friction for friction-, 104
    efficiency of bevel-, 104
        friction-, 108
        spur-, 97
    factor of safety of, 96
    fibre, 96
    grooved friction-, 108
    material for mortise-, 96
    mortise-, 94
    rawhide, 96
    screw-, 166
    teeth of, 75
    wear of bevel-, 103

Gear-teeth, breaking loads for, 89
    buttressed, 92
    diagrams for relative strength, 80, 82, 83, 84, 85
    formulas for strength of, 87, 90
    method of strengthening, 91
    pressure on, 76
    short, 93
    shrouded, 92
    working loads for short, 94

Grooved friction-gears, 108

Hydraulic press, material for, 334
    riveter, 328

Journal-bearing, adjustable, 17
    cylindrical, 15
    for rod end, 20
    self-aligning, 21, 23, 24
    with collars, 21
    with conical box, 19
    with eccentric sleeves, 18

Journal-bearings, 14
    effect of changing proportions of, 45
    friction of, 39
    lubrication of, 24
    materials for, 34

INDEX. 349

Journal-bearings, pressure on, 28
   problem in design of, 48
   proportions of, 46
   rubbing surfaces of, 27
   submerged, 39
   working pressures of, 49
Journal-boxes, Babbitt metal, 35
   brass or bronze, 35
   cast-iron, 36
   fibre-graphite, 37
   hardened-steel, 37
   natural mineral, 38
Journals, cast-iron, 37
  hardened-steel, 37
  rolling or burnishing, 27
  seizure of, 15

Keys, 201
 Blanton patent fastening, 205
 eccentric, 205
 feather, 202
 flat, 201
 roller, 205
 square, 202, 203

Lathe-bed, material for, 331
Lathe-spindle, 332
Lathe-ways, angle of, 8
  or -V's, 1
Lubrication of crank-pin, 31
   journal-bearings, 24
   non-metallic ropes, 143
   reciprocating journal, 33
   rotating pulley, 30
   step-bearings, 52, 54, 55
   with oil-bath, 26
    oil-pad, 26
    oil-ring or -chain, 29
    wicking, 30
   special devices for, 29

Machine bolt, 185
Materials, general properties of, tables, 335, 336
  selection of, 331
Moment of inertia by graphical method, 340
Mortise-gearing, 94

Nut-locks, 193

Oil-bath lubrication, 26
Oil-pad lubrication, 26

Packing, 278, 282
Pins, cylindrical and taper, 204
Pipe-couplings, 282
    expansion, 284
    packing for, 282
    riveted, 284
Pipe-flanges, 282
Pipes, bursting tests of, 276
  special forms of, 281
  spiral riveted, 282
Pivot-bearings, conical, 56
Planer-saddle bearings, 10
Planer-ways, flat, 4
    lubrication of, 5
    or -V's, 1
    special form, 5
Prony brake, 240
Pulley, sectional-rim, 257
  Walker differential, 148
  with composite arms, 262
Pulleys, 243
  designs from practice, 265
  diameters of, 124
  dimensions of, tables, 264, 265
  effect of, upon ropes, 145
  for ropes, 144
    wire ropes, 148
  grooves for ropes, 136
  idler-, for ropes, 140
  lubrication of, 30
  special forms, 263
  with tangent-arms, 262
Punching, effect of, 299
Punching-machine, 315
    form of frame, 327
    stresses in frame, 318

Rawhide gearing, 96
Riveted joints, 286
    bending of plates, 301
    efficiency of, 298
    examples from practice, 301-314
    faulty construction of, 300

INDEX.

Riveted joints, grooving of, 300
    margin, 291
    overlap, 291
    pitch of rivets, 291
    preparation of plates, 299
    single-riveted, 292
    stress in, 301
Rivets, pitch of, 295
    size of, 295
    strength of, 293
Roller-bearings, 68
    angle of conical rollers, 73
    coefficient of friction of, 71, 73
    roller for, 70
    tests of, 71
    with conical rollers, 72
Roller thrust-bearing, 72
Rope belting, efficiency of, 145
Ropes, coefficient of friction, 141
    diameter of, 144
        pulleys for, 144
    effect of position of pulleys upon, 145
    equations for power-transmission, 134
    for power-transmission, 131
    grooves for, 140
        non-metallic, 136
    idler-pulleys for, 140
    lubrication of non-metallic, 143
    speed of, 142
    strength of non-metallic, 142
    tension with varying loads, 132
    unequal driving of, 138
    water-proof coating for, 143
    wear of non-metallic, 143
    weight of, 144
    wire, 145
        distance between pulleys, 147
        distance between supports, 146
        pulleys for, 148
        sheaves for, 146
        size, speed, and power transmitted, 146

Schiele's anti-friction curve, 58
Screw-bolt, effect of fine threads, 198
Screw-bolts, endurance of, 199
    strength of, 195

Screw-bolts, strengthening by reducing diameter of body, 200
Screw-fastenings, 181
Screw-gears, 166
    coefficient of friction of, 180
    efficiency of, 178, 179
    strength of, 178
    turning force, 178
Screw-thread, angular, 337
    buttress, 184
    cold-rolled, 199
    self-locking, 194
    square, 185
    undercut, 195
    Whitworth, 184
Screw-threads, forms of, 181
    pitch of, 182
    proportions of U.S. Standard, 182
Screws, angular-thread, 163
  application of formula for maximum stress in, 162
  coefficient of friction of, 157
  collar-friction of, 154
  design of, 159
  efficiency of angular-thread, 165
    square-thread, 156
  for power-transmission, 150
  maximum stress in, 161
  pitch of, 151
  square-thread, 152
  turning moment and axial force, 152, 154
        in angular-thread, 163
Self-aligning bearing, 21
Set-screws, 188
    for pulleys, 192
    holding power of, 189
Shaft-coupling, Sellers, 232
Shaft couplings, flexible, 233
    rigid, 231
    split sleeve, 232
Shafts, bending-strength of, 220
  combined stresses in, 223
  deflection of, 223
  effect of hub on, 222
  experiments for endurance of round, 229
  hollow round, 228
    strength of, 218
  other than round, 231

INDEX.

Shafts, overhanging, single force, 226
    strength of round, 217
    subjected to more than one force, 224
        shocks, 223
    twist of, 219
Shaper-ram bearing, 8
Shearing, effect of, 299
Shearing-machine, 315
        form of frame, 327·
        stresses in frame, 318
Shrinkage-fit, numerical example, 210
Shrinkage-fits, 207
        allowance for, 211
        for heavy parts, 210
        stresses in parts, 207
Speed of ropes, 142
Spur-gears, 75
Steel and iron, properties of, 335, 336
Step-bearings, 51
        lubrication of, 52
        water-lubrication of, 54
Strap brake, 239
Strength of non-metallic ropes, 142
Stud-bolts, 185

Thrust-bearings, 64
Thrust-bearing with conical rollers, 72
Toothed gears, strength of teeth, 75
Tractrix, 58
        co-ordinates for, 63
Tractrix-bearing, wear of, 59
Tubing, bursting tests of, 276, 277

V bearings, 1
    angle of, 3, 4
    pressure on, 3

Wire ropes, 145
Worm and wheel, 166
        double, 167
        efficiency of, 172, 174, 175, **176**
            Hindley, 177
        pitch of, 171
        speeds and pressures, 173
        strength of, 166
        tests of, 171
        thrust-bearings for, 174
        turning force and efficiency of, **167**

SHORT-TITLE CATALOGUE

OF THE

PUBLICATIONS

OF

JOHN WILEY & SONS,

NEW YORK.

LONDON: CHAPMAN & HALL, LIMITED.

ARRANGED UNDER SUBJECTS.

Descriptive circulars sent on application.
Books marked with an asterisk are sold at *net* prices only.
All books are bound in cloth unless otherwise stated.

AGRICULTURE.

CATTLE FEEDING—DAIRY PRACTICE—DISEASES OF ANIMALS—
GARDENING, ETC.

Armsby's Manual of Cattle Feeding................12mo,	$1	75
Downing's Fruit and Fruit Trees......................8vo,	5	00
Grotenfelt's The Principles of Modern Dairy Practice. (Woll.) 12mo,	2	00
Kemp's Landscape Gardening..........................12mo,	2	50
Loudon's Gardening for Ladies. (Downing.)...........12mo,	1	50
Maynard's Landscape Gardening......................12mo,	1	50
Steel's Treatise on the Diseases of the Dog............8vo,	3	50
" Treatise on the Diseases of the Ox................8vo,	6	00
Stockbridge's Rocks and Soils........................8vo,	2	50
Woll's Handbook for Farmers and Dairymen............12mo,	1	50

ARCHITECTURE.

BUILDING—CARPENTRY—STAIRS—VENTILATION—LAW, ETC.

Berg's Buildings and Structures of American Railroads.....4to,	7	50
Birkmire's American Theatres—Planning and Construction.8vo,	3	00
" Architectural Iron and Steel....................8vo,	3	50
" Compound Riveted Girders......................8vo,	2	00
" Skeleton Construction in Buildings.............8vo,	3	00

Birkmire's Planning and Construction of High Office Buildings.
8vo, $3 50
Carpenter's Heating and Ventilating of Buildings..........8vo, 3 00
Freitag's Architectural Engineering.... ....8vo, 2 50
Gerhard's Sanitary House Inspection....................16mo, 1 00
" Theatre Fires and Panics....................12mo, 1 50
Hatfield's American House Carpenter.....................8vo, 5 00
Holly's Carpenter and Joiner.. ....18mo, 75
Kidder's Architect and Builder's Pocket-book...16mo, morocco, 4 00
Merrill's Stones for Building and Decoration.............8vo, 5 00
Monckton's Stair Building—Wood, Iron, and Stone........4to, 4 00
Wait's Engineering and Architectural Jurisprudence.......8vo, 6 00
Sheep, 6 50
Worcester's Small Hospitals—Establishment and Maintenance, including Atkinson's Suggestions for Hospital Architecture... ....12mo, 1 25
World's Columbian Exposition of 1893.............Large 4to, 2 50

ARMY, NAVY, Etc.
MILITARY ENGINEERING—ORDNANCE—LAW, ETC.

Bourne's Screw Propellers...................................4to, 5 00
* Bruff's Ordnance and Gunnery..........................8vo, 6 00
Chase's Screw Propellers..................................8vo, 3 00
Cooke's Naval Ordnance 8vo, 12 50
Cronkhite's Gunnery for Non-com. Officers.....32mo, morocco, 2 00
* Davis's Treatise on Military Law.......................8vo, 7 00
Sheep, 7 50
* " Elements of Law..............................8vo, 2 50
De Brack's Cavalry Outpost Duties. (Carr.)....32mo, morocco, 2 00
Dietz's Soldier's First Aid....................16mo, morocco, 1 25
* Dredge's Modern French Artillery....Large 4to, half morocco, 15 00
" Record of the Transportation Exhibits Building, World's Columbian Exposition of 1893..4to, half morocco, 10 00
Durand's Resistance and Propulsion of Ships..............8vo, 5 00
Dyer's Light Artillery..................................12mo, 3 00
Hoff's Naval Tactics......................................8vo, 1 50
* Ingalls's Ballistic Tables..............................8vo, 1 50
" Handbook of Problems in Direct Fire............8vo, 4 00

Mahan's Permanent Fortifications. (Mercur.).8vo, half morocco,	$7 50
Mercur's Attack of Fortified Places....................12mo,	2 00
" Elements of the Art of War....................8vo,	4 00
Metcalfe's Ordnance and Gunnery..........12mo, with Atlas,	5 00
Murray's A Manual for Courts-Martial........16mo, morocco,	1 50
" Infantry Drill Regulations adapted to the Springfield Rifle, Caliber .45....................32mo, paper,	10
*,Phelps's Practical Marine Surveying...................8vo,	2 50
Powell's Army Officer's Examiner.....................12mo,	4 00
Sharpe's Subsisting Armies..................32mo, morocco,	1 50
Very's Navies of the World...............8vo, half morocco,	3 50
Wheeler's Siege Operations..............................8vo,	2 00
Winthrop's Abridgment of Military Law...............12mo,	2 50
Woodhull's Notes on Military Hygiene..................16mo,	1 50
Young's Simple Elements of Navigation.......16mo, morocco,	2 00
" " " " " first edition........	1 00

ASSAYING.

SMELTING—ORE DRESSING—ALLOYS, ETC.

Fletcher's Quant. Assaying with the Blowpipe..16mo, morocco,	1 50
Furman's Practical Assaying............................8vo,	3 00
Kunhardt's Ore Dressing................................8vo,	1 50
O'Driscoll's Treatment of Gold Ores....................8vo,	2 00
Ricketts and Miller's Notes on Assaying.................8vo,	3 00
Thurston's Alloys, Brasses, and Bronzes...............8vo,	2 50
Wilson's Cyanide Processes..........................12mo,	1 50
" The Chlorination Process....................12mo,	1 50

ASTRONOMY.

PRACTICAL, THEORETICAL, AND DESCRIPTIVE.

Craig's Azimuth...4to,	3 50
Doolittle's Practical Astronomy.........................8vo,	4 00
Gore's Elements of Geodesy.............................8vo,	2 50
Hayford's Text-book of Geodetic Astronomy............8vo,	3 00
* Michie and Harlow's Practical Astronomy...............8vo,	3 00
* White's Theoretical and Descriptive Astronomy........12mo,	2 00

BOTANY.

GARDENING FOR LADIES, ETC.

Baldwin's Orchids of New England..............Small 8vo,	$1 50
Loudon's Gardening for Ladies. (Downing.)............12mo,	1 50
Thomé's Structural Botany............................16mo,	2 25
Westermaier's General Botany. (Schneider.)............8vo,	2 00

BRIDGES, ROOFS, Etc.

CANTILEVER—DRAW—HIGHWAY—SUSPENSION.
(See also ENGINEERING, p. 7.)

Boller's Highway Bridges................................8vo,	2 00
* " The Thames River Bridge..................4to, paper,	5 00
Burr's Stresses in Bridges..... ....8vo,	3 50
Crehore's Mechanics of the Girder.......................8vo,	5 00
Dredge's Thames Bridges...... ....7 parts, per part,	1 25
Du Bois's Stresses in Framed Structures............Small 4to,	10 00
Foster's Wooden Trestle Bridges..........................4to,	5 00
Greene's Arches in Wood, etc............................8vo,	2 50
" Bridge Trusses.................................8vo,	2 50
" Roof Trusses...................................8vo,	1 25
Howe's Treatise on Arches 8vo,	4 00
Johnson's Modern Framed Structures..............Small 4to,	10 00
Merriman & Jacoby's Text-book of Roofs and Bridges. Part I., Stresses..8vo,	2 50
Merriman & Jacoby's Text-book of Roofs and Bridges. Part II., Graphic Statics 8vo,	2 50
Merriman & Jacoby's Text-book of Roofs and Bridges. Part III., Bridge Design............................8vo,	2 50
Merriman & Jacoby's Text-book of Roofs and Bridges. Part IV., Continuous, Draw, Cantilever, Suspension, and Arched Bridges..................................8vo,	2 50
*Morison's The Memphis Bridge..................Oblong 4to,	10 00
Waddell's Iron Highway Bridges.... ....8vo,	4 00
" De Pontibus (a Pocket-book for Bridge Engineers). 16mo, morocco,	3 00
Wood's Construction of Bridges and Roofs...............8vo,	2 00
Wright's Designing of Draw Spans. Parts I. and II..8vo, each	2 50
" " " " " ' Complete...........8vo,	3 50

CHEMISTRY.

QUALITATIVE—QUANTITATIVE—ORGANIC—INORGANIC, ETC.

Adriance's Laboratory Calculations..................12mo,	$1 25
Allen's Tables for Iron Analysis.......................8vo,	3 00
Austen's Notes for Chemical Students..................12mo,	1 50
Bolton's Student's Guide in Quantitative Analysis........8vo,	1 50
Classen's Analysis by Electrolysis. (Herrick and Boltwood.).8vo,	3 00
Crafts's Qualitative Analysis. (Schaeffer.).............12mo,	1 50
Drechsel's Chemical Reactions. (Merrill.).............12mo,	1 25
Fresenius's Quantitative Chemical Analysis. (Allen.).......8vo,	6 00
" Qualitative " " (Johnson.).....8vo,	3 00
" " " " (Wells.) Trans. 16th German Edition............................8vo,	5 00
Fuertes's Water and Public Health....................12mo,	1 50
Gill's Gas and Fuel Analysis...........................12mo,	1 25
Hammarsten's Physiological Chemistry. (Mandel.)........8vo,	4 00
Helm's Principles of Mathematical Chemistry. (Morgan).12mo,	1 50
Kolbe's Inorganic Chemistry...........................12mo,	1 50
Ladd's Quantitative Chemical Analysis.................12mo,	1 00
Landauer's Spectrum Analysis. (Tingle.)................8vo,	3 00
Löb's Electrolysis and Electrosynthesis of Organic Compounds. (Lorenz.)...12mo,	1 00
Mandel's Bio-chemical Laboratory....................12mo,	1 50
Mason's Water-supply................................8vo,	5 00
" Examination of Water......................12mo,	1 25
Meyer's Organic Analysis. (Tingle.) (*In the press.*)	
Miller's Chemical Physics.............................8vo,	2 00
Mixter's Elementary Text-book of Chemistry............12mo,	1 50
Morgan's The Theory of Solutions and Its Results.......12mo,	1 00
" Elements of Physical Chemistry..............12mo,	2 00
Nichols's Water-supply (Chemical and Sanitary)..........8vo,	2 50
O'Brine's Laboratory Guide to Chemical Analysis.........8vo,	2 00
Perkins's Qualitative Analysis.........................12mo,	1 00
Pinner's Organic Chemistry. (Austen.)................12mo,	1 50
Poole's Calorific Power of Fuels........................8vo,	3 00
Ricketts and Russell's Notes on Inorganic Chemistry (Non-metallic)................................Oblong 8vo, morocco,	75
Ruddiman's Incompatibilities in Prescriptions............8vo,	2 00

Schimpf's Volumetric Analysis............12mo,	$2 50
Spencer's Sugar Manufacturer's Handbook.....16mo, morocco,	2 00
" Handbook for Chemists of Beet Sugar Houses. 16mo, morocco,	3 00
Stockbridge's Rocks and Soils............8vo,	2 50
Tillman's Descriptive General Chemistry. (*In the press.*)	
Van Deventer's Physical Chemistry for Beginners. (Boltwood.) 12mo,	1 50
Wells's Inorganic Qualitative Analysis............12mo,	1 50
" Laboratory Guide in Qualitative Chemical Analysis. 8vo,	1 50
Whipple's Microscopy of Drinking-water............8vo,	3 50
Wiechmann's Chemical Lecture Notes............12mo,	3 00
" Sugar Analysis............Small 8vo,	2 50
Wulling's Inorganic Phar. and Med. Chemistry............12mo,	2 00

DRAWING.

ELEMENTARY—GEOMETRICAL—MECHANICAL—TOPOGRAPHICAL.

Hill's Shades and Shadows and Perspective............8vo,	2 00
MacCord's Descriptive Geometry............8vo,	3 00
" Kinematics............8vo,	5 00
" Mechanical Drawing............8vo,	4 00
Mahan's Industrial Drawing. (Thompson.)........2 vols., 8vo,	3 50
Reed's Topographical Drawing. (H. A.)............4to,	5 00
Reid's A Course in Mechanical Drawing............8vo,	2 00
" Mechanical Drawing and Elementary Machine Design. 8vo. (*In the press.*)	
Smith's Topographical Drawing. (Macmillan.)............8vo,	2 50
Warren's Descriptive Geometry............2 vols., 8vo,	3 50
" Drafting Instruments............12mo,	1 25
" Free-hand Drawing............12mo,	1 00
" Linear Perspective............12mo,	1 00
" Machine Construction............2 vols., 8vo,	7 50
" Plane Problems............12mo,	1 25
" Primary Geometry............12mo,	75
" Problems and Theorems............8vo,	2 50
" Projection Drawing............12mo,	1 50

Warren's Shades and Shadows............................8vo,	$3	00
" Stereotomy—Stone-cutting......................8vo,	2	50
Whelpley's Letter Engraving.........................12mo,	2	00

ELECTRICITY AND MAGNETISM.

ILLUMINATION—BATTERIES—PHYSICS—RAILWAYS.

Anthony and Brackett's Text-book of Physics. (Magie.) Small 8vo,	3	00
Anthony's Theory of Electrical Measurements...........12mo,	1	00
Barker's Deep-sea Soundings...........................8vo,	2	00
Benjamin's Voltaic Cell................................8vo,	3	00
" History of Electricity........................8vo,	3	00
Classen's Analysis by Electrolysis. (Herrick and Boltwood.) 8vo,	3	00
Cosmic Law of Thermal Repulsion......................12mo,		75
Crehore and Squier's Experiments with a New Polarizing Photo-Chronograph......................................8vo,	3	00
Dawson's Electric Railways and Tramways. Small, 4to, half morocco,	12	50
* Dredge's Electric Illuminations....2 vols., 4to, half morocco,	25	00
" " " Vol. II..................4to,	7	50
Gilbert's De maguete. (Mottelay.)......................8vo,	2	50
Holman's Precision of Measurements....................8vo,	2	00
" Telescope-mirror-scale Method...........Large 8vo,		75
Löb's Electrolysis and Electrosynthesis of Organic Compounds. (Lorenz.)..12mo,	1	00
*Michie's Wave Motion Relating to Sound and Light,......8vo,	4	00
Morgan's The Theory of Solutions and Its Results........12mo,	1	00
Niaudet's Electric Batteries. (Fishback.)..............12mo,	2	50
Pratt and Alden's Street-railway Road-beds..............8vo,	2	00
Reagan's Steam and Electric Locomotives...............12mo,	2	00
Thurston's Stationary Steam Engines for Electric Lighting Purposes..8vo,	2	50
*Tillman's Heat......................................8vo,	1	50

ENGINEERING.

CIVIL—MECHANICAL—SANITARY, ETC.

(*See also* BRIDGES, p. 4; HYDRAULICS, p. 9; MATERIALS OF ENGINEERING, p. 10; MECHANICS AND MACHINERY, p. 12; STEAM ENGINES AND BOILERS, p. 14.)

Baker's Masonry Construction................................8vo,	$5	00
" Surveying Instruments.........................12mo,	3	00
Black's U. S. Public Works......................Oblong 4to,	5	00
Brooks's Street-railway Location...............16mo, morocco,	1	50
Butts's Civil Engineers' Field Book............16mo, morocco,	2	50
Byrne's Highway Construction............................8vo,	5	00
" Inspection of Materials and Workmanship........16mo,	3	00
Carpenter's Experimental Engineering 8vo,	6	00
Church's Mechanics of Engineering—Solids and Fluids....8vo,	6	00
" Notes and Examples in Mechanics...............8vo,	2	00
Crandall's Earthwork Tables................................8vo,	1	50
" The Transition Curve...............16mo, morocco,	1	50
*Dredge's Penn. Railroad Construction, etc. Large 4to, half morocco,	20	00
*Drinker's Tunnelling......................4to, half morocco,	25	00
Eissler's Explosives—Nitroglycerine and Dynamite........8vo,	4	00
Folwell's Sewerage...8vo,	3	00
Fowler's Coffer-dam Process for Piers......................8vo,	2	50
Gerhard's Sanitary House Inspection......................12mo,	1	00
Godwin's Railroad Engineer's Field-book......16mo, morocco,	2	50
Gore's Elements of Geodesy...............................8vo,	2	50
Howard's Transition Curve Field-book.........16mo, morocco,	1	50
Howe's Retaining Walls (New Edition.).................12mo,	1	25
Hudson's Excavation Tables. Vol. II.....................8vo,	1	00
Hutton's Mechanical Engineering of Power Plants........8vo,	5	00
Johnson's Materials of Construction...................Large 8vo,	6	00
" Stadia Reduction Diagram..Sheet, 22½ × 28½ inches,		50
" Theory and Practice of Surveying........Small 8vo,	4	00
Kent's Mechanical Engineer's Pocket-book.....16mo, morocco,	5	00
Kiersted's Sewage Disposal......... ....12mo,	1	25
Mahan's Civil Engineering. (Wood.)......................8vo,	5	00
Merriman and Brook's Handbook for Surveyors....16mo, mor.,	2	00
Merriman's Geodetic Surveying............................8vo,	2	00
" Retaining Walls and Masonry Dams..........8vo,	2	00
" Sanitary Engineering.........................8vo,	2	00
Nagle's Manual for Railroad Engineers........16mo, morocco,	3	00
Ogden's Sewer Design. (*In the press.*)		
Patton's Civil Engineering..................8vo, half morocco,	7	50

Patton's Foundations............................8vo,	$5	00
Pratt and Alden's Street-railway Road-beds..............8vo,	2	00
Rockwell's Roads and Pavements in France........ ....12mo,	1	25
Searles's Field Engineering ....16mo, morocco,	3	00
" Railroad Spiral................. .....16mo, morocco.	1	50
Siebert and Biggin's, Modern Stone Cutting and Masonry...8vo,	1	50
Smart's Engineering Laboratory Practice...............12mo,	2	50
Smith's Wire Manufacture and Uses................Small 4to,	3	00
Spalding's Roads and Pavements......................,12mo,	2	00
" . Hydraulic Cement..........................12mo,	2	00
Taylor's Prismoidal Formulas and Earthwork............8vo,	1	50
Thurston's Materials of Construction...................8vo,	5	00
* Trautwine's Civil Engineer's Pocket-book....16mo, morocco,	5	00
* " Cross-section.........................Sheet,		25
* " Excavations and Embankments............8vo,	2	00
* " Laying Out Curves............12mo, morocco,	2	50
Waddell's De Pontibus (A Pocket-book for Bridge Engineers).		
16mo, morocco,	3	00
Wait's Engineering and Architectural Jurisprudence.......8vo,	6	00
Sheep,	6	50
" Law of Field Operation in Engineering, etc........8vo.		
Warren's Stereotomy—Stone-cutting.....................8vo,	2	50
*Webb's Engineering Instruments............16mo, morocco,		50
" " " New Edition.............	1	25
Wegmann's Construction of Masonry Dams..............4to,	5	00
Wellington's Location of Railways.... ....Small 8vo,	5	00
Wheeler's Civil Engineering.............................8vo,	4	00
Wolff's Windmill as a Prime Mover....................8vo,	3	00

HYDRAULICS.

WATER-WHEELS—WINDMILLS—SERVICE PIPE—DRAINAGE, ETC.

(*See also* ENGINEERING, p. 7.)

Bazin's Experiments upon the Contraction of the Liquid Vein.		
(Trautwine.)................................8vo,	2	00
Bovey's Treatise on Hydraulics........................8vo,	4	00
Coffin's Graphical Solution of Hydraulic Problems.......12mo,	2	50
Ferrel's Treatise on the Winds, Cyclones, and Tornadoes...8vo,	4	00
Fuertes's Water and Public Health....................12mo,	1	50
Ganguillet & Kutter's Flow of Water. (Hering & Trautwine.)		
8vo,	4	00
Hazen's Filtration of Public Water Supply.............8vo,	2	00
Herschel's 115 Experiments..........................8vo,	2	00

Kiersted's Sewage Disposal...........................12mo, $1 25
Mason's Water Supply.................................8vo, 5 00
" Examination of Water. 12mo, 1 25
Merriman's Treatise on Hydraulics....................8vo, 4 00
Nichols's Water Supply (Chemical and Sanitary)........8vo, 2 50
Wegmann's Water Supply of the City of New York......4to, 10 00
Weisbach's Hydraulics. (Du Bois.)....................8vo, 5 00
Whipple's Microscopy of Drinking Water..............8vo, 3 50
Wilson's Irrigation Engineering......................8vo, 4 00
" Hydraulic and Placer Mining...............12mo, 2 00
Wolff's Windmill as a Prime Mover.....................8vo, 3 00
Wood's Theory of Turbines............................8vo, 2 50

MANUFACTURES.

BOILERS—EXPLOSIVES—IRON—STEEL—SUGAR—WOOLLENS, ETC.

Allen's Tables for Iron Analysis......................8vo, 3 00
Beaumont's Woollen and Worsted Manufacture........12mo, 1 50
Bolland's Encyclopædia of Founding Terms..........12mo, 3 00
" The Iron Founder.......................12mo, 2 50
" " " " Supplement..................12mo, 2 50
Bouvier's Handbook on Oil Painting..................12mo, 2 00
Eissler's Explosives, Nitroglycerine and Dynamite........8vo, 4 00
Fodr's Boiler Making for Boiler Makers................18mo, 1 00
Metcalfe's Cost of Manufactures......................8vo, 5 00
Metcalf's Steel—A Manual for Steel Users.............12mo, 2 00
* Reisig's Guide to Piece Dyeing......................8vo, 25 00
Spencer's Sugar Manufacturer's Handbook....16mo, morocco, 2 00
" Handbook for Chemists of Beet Sugar Houses.
16mo, morocco, 3 00
Thurston's Manual of Steam Boilers..................... 8vo, 5 00
Walke's Lectures on Explosives.........................8vo, 4 00
West's American Foundry Practice....................12mo, 2 50
" Moulder's Text-book 12mo, 2 50
Wiechmann's Sugar Analysis.................. Small 8vo, 2 50
Woodbury's Fire Protection of Mills....................8vo, 2 50

MATERIALS OF ENGINEERING.

STRENGTH—ELASTICITY—RESISTANCE, ETC.

(*See also* ENGINEERING, p. 7.)

Baker's Masonry Construction..........................8vo, 5 00
Beardslee and Kent's Strength of Wrought Iron..........8vo, 1 50
Bovey's Strength of Materials.........................8vo, 7 50
Burr's Elasticity and Resistance of Materials............8vo, 5 00
Byrne's Highway Construction.........................8vo, 5 00

Church's Mechanics of Engineering—Solids and Fluids.....8vo,	$6 00
Du Bois's Stresses in Framed Structures............Small 4to,	10 00
Johnson's Materials of Construction.....................8vo,	6 00
Lanza's Applied Mechanics..............................8vo,	7 50
Martens's Materials. (Henning.).........8vo. (*In the press.*)	
Merrill's Stones for Building and Decoration.............8vo,	5 00
Merriman's Mechanics of Materials......................8vo,	4 00
" Strength of Materials.....................12mo,	1 00
Patton's Treatise on Foundations........................8vo,	5 00
Rockwell's Roads and Pavements in France............12mo,	1 25
Spalding's Roads and Pavements.......................12mo,	2 00
Thurston's Materials of Construction....................8vo,	5 00
" Materials of Engineering.............3 vols., 8vo,	8 00
Vol. I., Non-metallic 8vo,	2 00
Vol. II., Iron and Steel.........................8vo,	3 50
Vol. III., Alloys, Brasses, and Bronzes............8vo,	2 50
Wood's Resistance of Materials.........................8vo,	2 00

MATHEMATICS.
CALCULUS—GEOMETRY—TRIGONOMETRY, ETC.

Baker's Elliptic Functions.............................8vo,	1 50
Ballard's Pyramid Problem............................8vo,	1 50
Barnard's Pyramid Problem...........................8vo,	1 50
*Bass's Differential Calculus...........................12mo,	4 00
Briggs's Plane Analytical Geometry...................12mo,	1 00
Chapman's Theory of Equations......................12mo,	1 50
Compton's Logarithmic Computations.................12mo,	1 50
Davis's Introduction to the Logic of Algebra.............8vo,	1 50
Halsted's Elements of Geometry........................8vo,	1 75
" Synthetic Geometry........................8vo,	1 50
Johnson's Curve Tracing..............................12mo,	1 00
" Differential Equations—Ordinary and Partial. Small 8vo,	3 50
" Integral Calculus.........................12mo,	1 50
" " " Unabridged. Small 8vo. (*In the press.*)	
" Least Squares............................12mo,	1 50
*Ludlow's Logarithmic and Other Tables. (Bass.).......8vo,	2 00
* " Trigonometry with Tables. (Bass.)............8vo,	3 00
*Mahan's Descriptive Geometry (Stone Cutting) 8vo,	1 50
Merriman and Woodward's Higher Mathematics........8vo,	5 00
Merriman's Method of Least Squares....................8vo,	2 00
Parker's Quadrature of the Circle.......................8vo,	2 50
Rice and Johnson's Differential and Integral Calculus, 2 vols. in 1, small 8vo,	2 50

Rice and Johnson's Differential Calculus............Small 8vo,	$3 00
" Abridgment of Differential Calculus.	
Small 8vo,	1 50
Totten's Metrology..8vo,	2 50
Warren's Descriptive Geometry..................2 vols., 8vo,	3 50
" Drafting Instruments..........................12mo,	1 25
" Free-hand Drawing............................12mo,	1 00
" Higher Linear Perspective......................8vo,	3 50
" Linear Perspective............................12mo,	1 00
" Primary Geometry............................12mo,	75
" Plane Problems...............................12mo,	1 25
" Problems and Theorems........................8vo,	2 50
" Projection Drawing...........................12mo,	1 50
Wood's Co-ordinate Geometry............................8vo,	2 00
" Trigonometry...................................12mo,	1 00
Woolf's Descriptive Geometry......................Large 8vo,	3 00

MECHANICS—MACHINERY.

TEXT-BOOKS AND PRACTICAL WORKS.

(See also ENGINEERING, p. 7.)

Baldwin's Steam Heating for Buildings..................12mo,	2 50
Benjamin's Wrinkles and Recipes......................12mo,	2 00
Chordal's Letters to Mechanics........................12mo,	2 00
Church's Mechanics of Engineering....................8vo,	6 00
" Notes and Examples in Mechanics.............8vo,	2 00
Crehore's Mechanics of the Girder.....................8vo,	5 00
Cromwell's Belts and Pulleys.........................12mo,	1 50
" Toothed Gearing...........................12mo,	1 50
Compton's First Lessons in Metal Working.............12mo,	1 50
Compton and De Groodt's Speed Lathe................12mo,	1 50
Dana's Elementary Mechanics.........................12mo,	1 50
Dingey's Machinery Pattern Making...................12mo,	2 00
Dredge's Trans. Exhibits Building, World Exposition.	
Large 4to, half morocco,	10 00
Du Bois's Mechanics. Vol. I., Kinematics 8vo,	3 50
" " Vol. II., Statics.................8vo,	4 00
" " Vol. III., Kinetics...............8vo,	3 50
Fitzgerald's Boston Machinist.........................18mo,	1 00
Flather's Dynamometers..............................12mo,	2 00
" Rope Driving...............................12mo,	2 00
Hall's Car Lubrication................................12mo,	1 00
Holly's Saw Filing....................................18mo,	75
Johnson's Theoretical Mechanics. An Elementary Treatise	
(In the press.)	
Jones's Machine Design. Part I., Kinematics............8vo,	1 50

Jones's Machine Design. Part II., Strength and Proportion of Machine Parts	8vo,	$3 00
Lanza's Applied Mechanics	8vo,	7 50
MacCord's Kinematics	8vo,	5 00
Merriman's Mechanics of Materials	8vo,	4 00
Metcalfe's Cost of Manufactures	8vo,	5 00
*Michie's Analytical Mechanics	8vo,	4 00
Richards's Compressed Air	12mo,	1 50
Robinson's Principles of Mechanism	8vo,	3 00
Smith's Press-working of Metals	8vo,	3 00
Thurston's Friction and Lost Work	8vo,	3 00
" The Animal as a Machine	12mo,	1 00
Warren's Machine Construction	2 vols., 8vo,	7 50
Weisbach's Hydraulics and Hydraulic Motors. (Du Bois.)	8vo,	5 00
" Mechanics of Engineering. Vol. III., Part I., Sec. I. (Klein.)	8vo,	5 00
Weisbach's Mechanics of Engineering. Vol. III., Part I., Sec. II. (Klein.)	8vo,	5 00
Weisbach's Steam Engines. (Du Bois.)	8vo,	5 00
Wood's Analytical Mechanics	8vo,	3 00
" Elementary Mechanics	12mo,	1 25
" " " Supplement and Key	12mo,	1 25

METALLURGY.

IRON—GOLD—SILVER—ALLOYS, ETC.

Allen's Tables for Iron Analysis	8vo,	3 00
Egleston's Gold and Mercury	Large 8vo,	7 50
" Metallurgy of Silver	Large 8vo,	7 50
* Kerl's Metallurgy—Copper and Iron	8vo,	15 00
* " " Steel, Fuel, etc	8vo,	15 00
Kunhardt's Ore Dressing in Europe	8vo,	1 50
Metcalf's Steel—A Manual for Steel Users	12mo,	2 00
O'Driscoll's Treatment of Gold Ores	8vo,	2 00
Thurston's Iron and Steel	8vo,	3 50
" Alloys	8vo,	2 50
Wilson's Cyanide Processes	12mo,	1 50

MINERALOGY AND MINING.

MINE ACCIDENTS—VENTILATION—ORE DRESSING, ETC.

Barringer's Minerals of Commercial Value	Oblong morocco,	2 50
Beard's Ventilation of Mines	12mo,	2 50
Boyd's Resources of South Western Virginia	8vo,	3 00
" Map of South Western Virginia	Pocket-book form,	2 00

Brush and Penfield's Determinative Mineralogy. New Ed. 8vo,	$4 00
Chester's Catalogue of Minerals................8vo,	1 25
" " " " ....Paper,	50
" Dictionary of the Names of Minerals........8vo,	3 00
Dana's American Localities of Minerals.........Large 8vo,	1 00
" Descriptive Mineralogy. (E. S.)...Large half morocco,	12 50
" Mineralogy and Petrography. (J. D.).......12mo,	2 00
" Minerals and How to Study Them. (E. S.)....12mo,	1 50
" Text-book of Mineralogy. (E. S.)...New Edition. 8vo,	4 00
* Drinker's Tunnelling, Explosives, Compounds, and Rock Drills. 4to, half morocco,	25 00
Egleston's Catalogue of Minerals and Synonyms........8vo,	2 50
Eissler's Explosives—Nitroglycerine and Dynamite......8vo,	4 00
Hussak's Rock-forming Minerals. (Smith.).......Small 8vo,	2 00
Ihlseng's Manual of Mining..................8vo,	4 00
Kunhardt's Ore Dressing in Europe..............8vo,	1 50
O'Driscoll's Treatment of Gold Ores..............8vo,	2 00
* Penfield's Record of Mineral Tests........Paper, 8vo,	50
Rosenbusch's Microscopical Physiography of Minerals and Rocks. (Iddings.)....................8vo,	5 00
Sawyer's Accidents in Mines...............Large 8vo,	7 00
Stockbridge's Rocks and Soils..................8vo,	2 50
Walke's Lectures on Explosives.................8vo,	4 00
Williams's Lithology........................8vo,	3 00
Wilson's Mine Ventilation....................12mo,	1 25
" Hydraulic and Placer Mining.............12mo,	2 50

STEAM AND ELECTRICAL ENGINES, BOILERS, Etc.

STATIONARY—MARINE—LOCOMOTIVE—GAS ENGINES, ETC.

(*See also* ENGINEERING, p. 7.)

Baldwin's Steam Heating for Buildings..........12mo,	2 50
Clerk's Gas Engine....................Small 8vo,	4 00
Ford's Boiler Making for Boiler Makers..........18mo,	1 00
Hemenway's Indicator Practice..............12mo,	2 00
Hoadley's Warm-blast Furnace..............8vo,	1 50
Kneass's Practice and Theory of the Injector........8vo,	1 50
MacCord's Slide Valve....................8vo,	2 00
Meyer's Modern Locomotive Construction..........4to,	10 00
Peabody and Miller's Steam-boilers.............8vo,	4 00
Peabody's Tables of Saturated Steam............8vo,	1 00
" Thermodynamics of the Steam Engine.......8vo,	5 00
" Valve Gears for the Steam Engine.........8vo,	2 50
Pray's Twenty Years with the Indicator........Large 8vo,	2 50
Pupin and Osterberg's Thermodynamics..........12mo,	1 25

Reagan's Steam and Electric Locomotives............ ....12mo,	$2	00
Röntgen's Thermodynamics. (Du Bois.)................8vo.	5	00
Sinclair's Locomotive Running.........................12mo,	2	00
Snow's Steam-boiler Practice..... ....8vo. (*In the press.*)		
Thurston's Boiler Explosions..... ....12mo,	1	50
" Engine and Boiler Trials. 8vo,	5	00
" Manual of the Steam Engine. Part I., Structure and Theory............................8vo,	6	00
" Manual of the Steam Engine. Part II., Design, Construction, and Operation..............8vo,	6	00
2 parts,	10	00
Thurston's Philosophy of the Steam Engine.............12mo,		75
" Reflection on the Motive Power of Heat. (Carnot.) 12mo,	. 1	50
" Stationary Steam Engines...................8vo,	2	50
" Steam-boiler Construction and Operation.......8vo,	5	00
Spangler's Valve Gears..8vo,	2	50
Weisbach's Steam Engine. (Du Bois.)...................8vo,	5	00
Whitham's Constructive Steam Engineering..............8vo,	6	00
" Steam-engine Design........................8vo,	5	00
Wilson's Steam Boilers. (Flather.)....................12mo,	2	50
Wood's Thermodynamics, Heat Motors, etc..............8vo,	4	00

TABLES, WEIGHTS, AND MEASURES.

FOR ACTUARIES, CHEMISTS, ENGINEERS, MECHANICS—METRIC TABLES, ETC.

Adriance's Laboratory Calculations.....................12mo,	1	25
Allen's Tables for Iron Analysis.........................8vo,	3	00
Bixby's Graphical Computing Tables......... ....Sheet,		25
Compton's Logarithms......................................12mo,	1	50
Crandall's Railway and Earthwork Tables..... ....8vo,	1	50
Egleston's Weights and Measures................. ....18mo,		75
Fisher's Table of Cubic Yards.....................Cardboard,		25
Hudson's Excavation Tables. Vol. II..... ....8vo,	1	00
Johnson's Stadia and Earthwork Tables................8vo,	1	25
Ludlow's Logarithmic and Other Tables. (Bass.).......12mo,	2	00
Totten's Metrology......................................8vo,	2	50

VENTILATION.

STEAM HEATING—HOUSE INSPECTION—MINE VENTILATION.

Baldwin's Steam Heating..............................12mo,	2	50
Beard's Ventilation of Mines...........................12mo,	2	50
Carpenter's Heating and Ventilating of Buildings..........8vo,	3	00
Gerhard's Sanitary House Inspection....................12mo,	1	00
Reid's Ventilation of American Dwellings 12mo,	1	50
Wilson's Mine Ventilation..............................12mo,	1	25

MISCELLANEOUS PUBLICATIONS.

Alcott's Gems, Sentiment, Language.............Gilt edges,	$5 00
Bailey's The New Tale of a Tub........................8vo,	75
Ballard's Solution of the Pyramid Problem............8vo,	1 50
Barnard's The Metrological System of the Great Pyramid..8vo,	1 50
Davis's Elements of Law..............................8vo,	2 00
Emmon's Geological Guide-book of the Rocky Mountains..8vo,	1 50
Ferrel's Treatise on the Winds........................8vo,	4 00
Haines's Addresses Delivered before the Am. Ry. Assn...12mo.	2 50
Mott's The Fallacy of the Present Theory of Sound..Sq. 16mo,	1 00
Perkins's Cornell University....................Oblong 4to,	1 50
Ricketts's History of Rensselaer Polytechnic Institute.....8vo,	3 00
Rotherham's The New Testament Critically Emphasized. 12mo,	1 50
" The Emphasized New Test. A new translation. Large 8vo,	2 00
Totten's An Important Question in Metrology............8vo,	2 50
Whitehouse's Lake Mœris.............................Paper,	25
* Wiley's Yosemite, Alaska, and Yellowstone............4to,	3 00

HEBREW AND CHALDEE TEXT-BOOKS.
FOR SCHOOLS AND THEOLOGICAL SEMINARIES.

Gesenius's. Hebrew and Chaldee Lexicon to Old Testament. (Tregelles.)...................Small 4to, half morocco,	5 00
Green's Elementary Hebrew Grammar................12mo,	1 25
" Grammar of the Hebrew Language (New Edition).8vo,	3 00
" Hebrew Chrestomathy..........................8vo,	2 00
Letteris's Hebrew Bible (Massoretic Notes in English). 8vo, arabesque,	2 25

MEDICAL.

Bull's Maternal Management in Health and Disease......12mo,	1 00
Hammarsten's Physiological Chemistry. (Mandel.).......8vo,	4 00
Mott's Composition, Digestibility, and Nutritive Value of Food. Large mounted chart,	1 25
Ruddiman's Incompatibilities in Prescriptions............8vo,	2 00
Steel's Treatise on the Diseases of the Ox................8vo,	6 00
" Treatise on the Diseases of the Dog...............8vo,	3 50
Woodhull's Military Hygiene..........................16mo,	1 50
Worcester's Small Hospitals—Establishment and Maintenance, including Atkinson's Suggestions for Hospital Architecture..................................12mo,	1 25

www.ingramcontent.com/pod-product-compliance
Lightning Source LLC
Chambersburg PA
CBHW051248300426
44114CB00011B/940